IRON—
Binary Phase Diagrams

Ortrud Kubaschewski

IRON–
Binary Phase Diagrams

With a Preface by
Oswald Kubaschewski

With 103 Figures and 3 Tables

1982
Springer-Verlag Berlin Heidelberg GmbH

Dr. rer.-nat. Ortrud Kubaschewski-von Goldbeck

Lehrstuhl für Metallurgie der Kernbrennstoffe
und Theoretische Hüttenkunde
RWTH Aachen
Kopernikusstrasse 16
D — 5100 Aachen
Federal Republic of Germany

ISBN 978-3-662-08026-9   ISBN 978-3-662-08024-5 (eBook)
DOI 10.1007/978-3-662-08024-5

Library of Congress Cataloging in Publication Data

Kubaschewski, Ortrud, 1922—. Iron — binary phase diagrams.
Bibliography: p. 1. Iron. 2. Binary systems
(Metallurgy). 3. Phase diagrams. I. Title.
TN693.17K84   1982   669'.961   82-10263

Originally published by Springer-Verlag Berlin/Heidelberg, and Verlag Stahleisen mbH, Düsseldorf in 1982
Softcover reprint of the hardcover 1st edition 1982

Offsetprinting and bookbinding: Konrad Triltsch, Würzburg
2060/3020—543210

Preface

At the official dinner of a meeting in May 1939, I was seated next to Max Hansen. When I congratulated him on the well deserved success of his "Aufbau der Zweistoff-Legierungen", he smiled: "yes, it was a struggle with the hydra, and so it has taken me seven years", meaning that whenever he had thought to have finished the phase diagram of a particular system, new evidence would turn up like the new heads of the Greek monster.

There is no need to point out the importance of assessed phase diagrams to metallurgists or even anyone concerned with the technology and application of metals and alloys. The information contained therein is fundamental to considerations concerning the chemical, physical and mechanical properties of alloys. Hansen's German monograph was followed by a revised English edition in 1958 with K. Anderko and the supplements by R.P. Elliott (1965) and F.A. Shunk (1969). All those who have made use of these volumes will admit that much diligent labour has gone into this work, necessary to cope with the ever increasing number of publications and the consequent improvements.

In view of the large amount of experimental evidence it has become advisable to subdivide the subject matter and to deal with the binary phase diagrams of individual metals in turn. Mrs Kubaschewski had already elaborated the phase diagrams of a number of metals of interest to reactor technology for the International Atomic Energy Agency and then turned her attention to iron. I have been in the position to bear witness to the painstaking work that she has put into the construction of such diagrams shown on the following pages. A vast amount of experimental, often conflicting, evidence had to be sifted. Thoughts had to be exerted on the position and the limits of nearly every phase, the thermodynamic conditions being always observed. And all this is presented in an improved form on so-called 'raster paper'. The phase boundaries are based mainly on the results of conventional experimental methods, but there are certain boundaries which are better ascertained by thermochemical calculation. The author has taken this into account.

It may well be asked whether in these days of thermochemical calculations of phase diagrams (e.g. CALPHAD), one might not store the relevant heat and entropy values of each phase on a computer and let this do the working out. This is certainly true for the evaluation of ternary and higher phase diagrams. However, the binary systems should be treated separately, namely in the way chosen here. In the earlier days when we developed the thermochemical method for describing equilibrium diagrams (in which, incidentally, Miss von Goldbeck was also involved), William Hume-Rothery would protest by saying: "you may be able to correct certain phase boundaries by calculation but you cannot *predict* the existence of intermetallic phases". This is correct, at least at the present state of knowledge concerning the causes of chemical stability of phases. One may add that thermochemical calculations cannot be based on the results of calorimetric and Gibbs energy measurements alone, in particular when the diagrams are more complicated. So I used to reply to Hum-Rothery that the real value of the thermochemical method lay in the possibility of extrapolating from the binary into the multicomponent ranges, whereas in the binaries it is merely used for adjustments. The problem of predicting phases that do not appear in the binary border systems becomes less and less prominent the greater the number of component elements there are per system. The reason for this is the increasing influence of the entropy of mixing which favours the stability of the disordered solutions and thus depresses ($T\Delta S$ term!) the appearance of any ordered phase to lower, in practice less important, temperatures.

I usually distinguish between 'phase diagrams' and 'equilibrium diagrams', the former being derived wholly from conventional measurements, the latter from thermochemical calculations. One may safely say that the present series of binaries approaches the status of *equilibrium diagrams*. Hence, I submit that the author has coped with the hydra and expect that many of these diagrams will remain valid in detail for quite a number of years to come.

Aachen, May 1982   Oswald Kubaschewski

Acknowledgements

The author is most grateful to Prof. Dr. O. Knacke who offered her the hospitality of his laboratory and his assistance in many ways, to Prof. Dr. Drs. h.c. O. Kubaschewski for continued interest, advice and, in particular, for meticulously checking the English, to Dipl. Ing. R. Steffen for critically reading the manuscript, to Mrs. G. Stüsser and Mrs. E. Klein who made such an excellent job in drawing the raster diagrams and to the Commission of the European Communities, Division Steel, without whose financial assistance this work could not have been accomplished.

Ortrud Kubaschewski

Contents

Contents

Notation

For the convenience of the reader, the following drawing indicates the various lines that are being used for the description of the phase boundaries in the diagrams of this monograph:

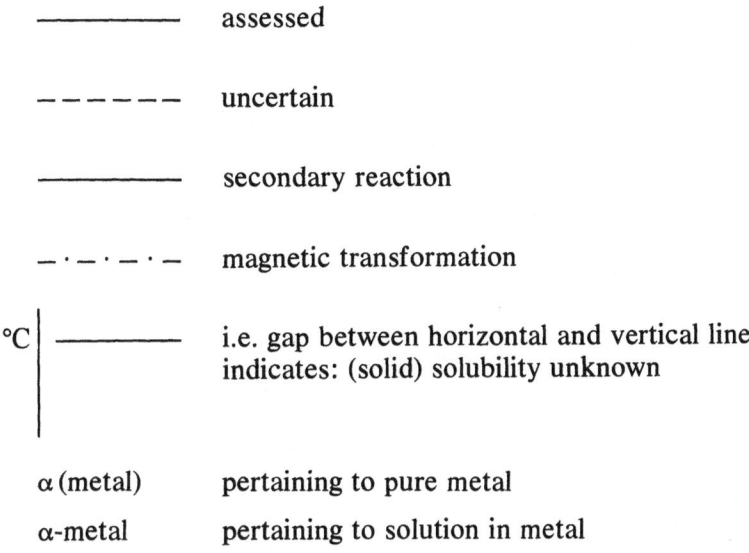

———————— assessed

— — — — — — uncertain

———————— secondary reaction

— · — · — · — magnetic transformation

i.e. gap between horizontal and vertical line indicates: (solid) solubility unknown

α (metal) pertaining to pure metal

α-metal pertaining to solution in metal

Introduction

This monograph is concerned with the binary equilibrium diagrams of iron with all the elements of the Periodic Chart except the halogens. Innumerable publications concerning these diagrams are available, and here is an attempt to correlate and assess the literature data. The monograph may be to some extent incomplete, but it is hoped to be a help to all who have to deal with the metallurgy of iron and steel and even metallurgists in general.

The procedure of compilation has been as follows. The starting point has naturally been the classic work of Hansen [A] with its successors published by Elliott [B] and Shunk [C]. The thermochemical tables of Hultgren et al. [D] have also frequently been consulted. From there on (say, 1967) the literature has been searched and recorded well into 1981. A series of reviews edited by Ageev, 1968 to 1977 [E], has been a great help, in particular with regard to those Russian publications on which to lay hands had been difficult. Since the number of publications concerning the binary phase diagrams of iron is astronomic, a strict selection of references had to be made so as not to overload the text. Earlier publications (if not quoted) may be traced *via* the compilations just mentioned. Some of the recent publications may have been overlooked but others, if they are just repetitive, have been disregarded deliberately.

The melting point of iron accepted here is 1,536°C, and the transformation temperatures are 911°C and 1,392°C, and these temperatures apply to all the phase diagrams to follow, original results being consequently adjusted where necessary. The ferromagnetic transformation has its maximum at 767–771°C although the corresponding heat capacity *versus* temperature hump is spread over a much wider range. These temperatures as well as those for the other elements (see Table 1, Appendix) are based on the International Practical Temperature Scale 1968, revised 1975 and denoted IPTS-68. Secondary reference points have been assessed and listed by Crovini et al. [F]. Other melting and transformation points are from Hultgren's compilation [D] (updated to IPTS-68, where necessary) and from more recent re-determinations quoted in the original literature. Unfortunately, many investigators do not state the temperature scale they have used.

Table 1 (Appendix), in addition to the transition temperatures, also lists the relative atomic mass, crystal structure and density of the elements.

The arrangement of the systems is the alphabetical order in terms of chemical symbols of the component elements except for the alkali metals (appearing under 'Alk') and the rare earth metals (Fe-R), and here Eu and Yb are exceptions in that they are listed in the normal alphabetical sequence. The Fe-R systems are discussed in the following sequence: Ce, Pr, Nd, Pm, Sm, Gd, Tb, Dy, Ho, Er, Tm, Lu (see 'Contents').

It should be noted that the pertinent references are listed immediately at the end of the text for every system.

Care has been taken to mention in the texts the modes of alloy preparation and the purity of the component metals as quoted by the investigators. Unless stated otherwise, % always indicate weight percent. –SI units are used where applicable.

Compared with the earlier compilations, the following advances are claimed. Phase boundaries could be pursued to lower temperatures thanks to improved experimental methods and longer times of annealing. Partly, new data for the magnetic properties are included in the text and in the diagrams. The emphasis is still on results obtained by conventional methods. However, results of thermochemical calculations are taken into consideration. (For a discussion of the thermochemical approach, see for instance [G].) This is particularly useful, no, indispensable when it comes to the extrapolation of solid solubilities to lower temperatures or to the assessment of solidus curves which are prone to slow equilibration (e.g. Cr-Fe).

It is generally more difficult to assess accurately the temperature and composition of solid-solid transformations because the differences in Gibbs energy, representing 'driving force', may be very small and the rates of equilibration consequently very slow. This is particularly true when a compound exhibits two or three modifications of the Laves structures. These are so similar in co-ordination that little energy is needed for the re-arrangement of the atoms. (See also [H].)

Only stable phases are included in the diagrams—with a few exceptions (e.g. Fe-C, Fe-N). Martensitic transformations, metallic glasses and the so-called ω phases are not included but are occasionally mentioned in the text. In future works of this nature, it may be desirable to say more about metastable phases. However, since their number is infinite, it is difficult to see where to draw the line.

Concerning the notation of phases, the one accepted by the majority of authors is also used here, but it should be borne in mind that it is sometimes controversial. Agreement will eventually be reached by general consent but is not a matter for the present compiler. Greek letters are used to denote solid solutions. Intermetallic phases are designated by Greek letters or stoichiometric compositions or both.

Since the present author has been asked to continue this work by updating the binary phase diagrams of iron whenever desirable, she would be grateful for any reprints of new investigations pertaining to the subject matter of this monograph.

The following references pertain to the above but will also appear in the following texts with their Latin capitals.

References

A   Hansen, M.; Anderko, K.: Constitution of Binary Alloys. New York: McGraw-Hill 1958

B   Elliott, R.P.: Constitution of Binary Alloys, First Supplement New York: McGraw-Hill 1965

C   Shunk, F.A.: Constitution of Binary Alloys, Second Supplement New York: McGraw-Hill 1969

D   Hultgren, R.; Orr, R.L.; Anderson, P.D.; Kelley, K.K.: Selected Values of Thermodynamic Properties of Metals and Alloys. New York: Wiley 1963 and Supplements

E   Ageev, N.V. (Ed.): Phase Diagrams of Metallic Systems 1968–1977, Acad. Sci. USSR, Moscow 1979

F   Crovini, L.; Bedford, R.E.; Moser, A.: Extended List of Secondary Reference Points. Metrologia 13 (1977) 197

G   Kubaschewski, O.; Barin, I.: Phase Equilibria in Condensed Systems. Pure Appl. Chem. 38 (1974) 469

H   Elliott, R.P.: Trans. ASM 53 (1960) 321

Information on the solubility of silver in liquid and solid iron is fragmentary and conflicting. From earlier studies it may be concluded that silver and iron are virtually immiscible in the liquid state. These results have been confirmed by Vogel and von Mässenhausen [1] and Gibson and Hume-Rothery [2].

The only apparently reliable data pertaining to the solubility of silver in solid iron have been presented by Wriedt et al. [3]. They measured the solubility of Ag in γ(Fe) between 1,366 and 1,561 K by an isopiestic technique in which purified iron and silver alloy were equilibrated in a sealed isothermal capsule. Two grades of purified iron and 99.9999% silver were used. According to this the silver content of γ(Fe) in equilibrium with iron-saturated liquid silver may be expressed in the form of the following formula:

$$\lg(\text{wt.\% Ag}) = -6{,}027\,T^{-1} + 2.289 \quad (1{,}234\text{--}1{,}665\ \text{K})$$

Expressed in the form of actual solubility values:

Temp. °C	wt.% Ag	at.% Ag		Temp. °C	wt.% Ag	at.% Ag
961	0.0025	0.0013		1,200	0.0157	0.0081
1,000	0.0036	0.0019		1,300	0.0287	0.015
1,100	0.0079	0.0041		1,392	0.0470	0.024

Data for the solid solubility of Fe in *monocrystalline* Ag have been supplied by Bernardini et al. [4]. According to their results the solubility increases from 10^{-6} at 650 °C to 40×10^{-6} at 920 °C. It was noticed that even small traces of oxygen reduce the solubility considerably.

The solubility of Ag in *liquid* Fe is likely to be larger than that given above for solid Fe; even so, "virtual immiscibility" was a fair statement of the earlier observers.

References

1 Vogel, R.; von Mässenhausen, W.: Arch. Eisenhüttenwes. 27 (1956) 143
2 Gibson, W.S.; Hume-Rothery, W.: J. Iron Steel Inst. 189 (1958) 243
3 Wriedt, H.A.; Morrison, W.B.; Cole, W.E.: Metall. Trans. 4 (1973) 1453
4 Bernardini, J.; Combe-Brun, A.; Cabane, J.: C.R. Acad. Sci. Paris 269 (1969) Ser. C: 287

Fe–Li
Na
K

Iron–Alkaline Metals

The solubilities of lithium, sodium and potassium in iron are obviously so low that their determination at normal pressure is beyond experimental facilities, as reported by Wever [1], for instance. The same may safely be assumed to be the case for rubidium and cesium.

However, the solubilities of iron in the *liquid* alkaline metals are within reach of experimental determination. Results for α- and γ(Fe) in *lithium* have been summarized by Elliott [B: Refs. 1–5] and Shunk [C: Refs. 1, 2]. Of these, the data of Bychkov et al. [2] for γ(Fe) and of Beskorovainyi and Yakovlev [3] as well as Weeks [4] for α(Fe) have been selected on the grounds that they used the purer materials (e.g. 99.9% Li [2]). For the solubility of α(Fe) in *sodium*, the results of Baus [5] who employed a radiochemical technique have been accepted. A comparative study of the solubility of Fe, Cr and Ni in Na at $T < 600°C$ by Singer et al. [6] may also be mentioned. Swisher [7] has measured the solubility of γ(Fe) in *potassium* by analysis of melts equilibrated in Mo containers. The results must be regarded as approximations owing to the influence of oxygen present.

All these results may be expressed in the form of the following formulae:

Solubility in at.% Fe

a) α(Fe) in Li: lg at.% = $- 700\,T^{-1} - 2.46$ (673–1,184 K)
b) γ(Fe) in Li: lg at.% = $-9,000\,T^{-1} + 4.75$ (1,273–1,473 K)
c) α(Fe) in Na: lg at.% = $-1,137\,T^{-1} - 5.0$ (503– 817 K)
d) α(Fe) in K: lg at.% = $-6,600\,T^{-1} + 4.43$ (943–1,198 K)

Expressed in the form of actual solubility values:

	Temp. °C	400	500	600	700	800	900
a)	at.% α(Fe)	3.16×10^{-4}	4.3×10^{-4}	5.5×10^{-4}	6.6×10^{-4}	7.7×10^{-4}	8.8×10^{-4}
c)	at.% α(Fe)	2.05×10^{-7}	3.4×10^{-7}				
d)	at.% α(Fe)				7.4×10^{-4}	4.4×10^{-3}	1.9×10^{-2}	6.3×10^{-2}

	Temp. °C	1,000	1,100	1,200
b)	at.% γ(Fe)	0.005	0.016	0.043

References

1 Wever, F.: Arch. Eisenhüttenwes. 2 (1928/29) 739
2 Bychkov, Yu.V.; Rozanov, A.N.; Yakovleva, V.B.: Sov. At. Energy, USSR 7 (1961) 987
3 Beskorovainyi, N.M., Yakovlev, E.I.: Met. i. Metalloved. Chistykh Metal., Sb. Nauchn., Rabot 2 (1960) 189
4 Weeks, J.R.: NASA Spec. Publ. NASA-SP-41 (1963) 21
5 Baus, R.A.: Proc. U.N. Int. Conf. Peaceful Uses At. Energy, Geneva 9 (1955) 356
6 Singer, R.M.; Fleitmann, A.H.; Weeks, J.R.; Isaaks, H.S.: Corrosion by Liquid Metals (1970) 561
7 Swisher, J.H.: NASA: Tech. Note, NASA-TN-D-2734 (1965) 18

Work on phase relationships of the system Fe-Al has been reviewed by Hansen [A], Elliott [B] and Shunk [C]. Fig. 1 shows the modified phase diagram which includes a re-investigation of the liquidus/solidus curves by Lee [1] and Schürmann and Kaiser [2], the α-Fe solubility range suggested by Köster and Gödecke [3] and new determinations of the γ-loop by Rocquet et al. [4] and Rocquet and Petit [5].

The system is characterized by a wide α-Fe solid solution range. The ordered α_2(FeAl) region is devided into different fields. The system entails five stable phases, namely Fe$_3$Al, ε, FeAl$_2$, Fe$_2$Al$_5$ and FeAl$_3$, each of them with a homogeneity range.

The melting point of Al has been accepted as 660.46°C (IPTS-68).

The shape of the liquidus/solidus curve shown by [A] (Fig. 54) has, on the whole, been confirmed by [1] and [2]. However, there still exists disagreements about the mode of formation of FeAl$_3$. Measurements based on thermal analyses of alloys made of "H"-iron and 99.99% Al carried out by [1] indicate a eutectic formation of the congruently melting FeAl$_3$ phase, while recent results (thermal analysis and isothermal holding tests supplemented by electron micro-analysis (EMA)) by [2] confirmed previous findings, namely the peritectic formation of this phase. The experimentally established data reported by [2] which were also used in a thermodynamic assessment of the Fe-Al phase diagram [6] have been adopted in Fig. 1.

The addition of Al to Fe hardly changes the course of the liquidus/solidus between 0 and 10 at.% Al. They then fall smoothly to 1,310°C, where a peritectic reaction occurs [1, 2] corresponding to:

$$L(50.7 \text{ at.\% Al}) + \alpha\text{-Fe}(45.2 \text{ at.\% Al}) \leftrightarrows FeAl(\alpha_2).$$

In Fig. 1, the formation of α_2(FeAl) is not shown as a peritectic reaction, but as a second order type reaction concordant with the Fe-Si and Fe-Ga system and a suggestion by [3].

γ-loop

Rocquet et al. [4] undertook dilatometric and micrographic examinations of about 100 quenched samples after isothermal treatment, to clarify the uncertainties pertaining to the position of the γ-loop. Values for the limits of the α- and γ-region were found to be 0.625 and 0.95 wt.% Al (1.285 and 1.95 at.% Al): Fig. 2. These results are in agreement with those calculated by Oelsen [7], who used 21.36 J/g Fe as the heat of transformation of pure Fe at the point A$_4$[1], and of [A] and were confirmed by magnetic measurements carried out by [5]. Results published by Hirano and Hishinuma [8] are inconsistent with these values. Their findings are based on electron probe microanalyses of diffusion couples and are regarded by [5] as questionable.

---

1 However, this corresponds to 1,194 J/g-atom Fe whereas the accepted value is 838 J/g-atom

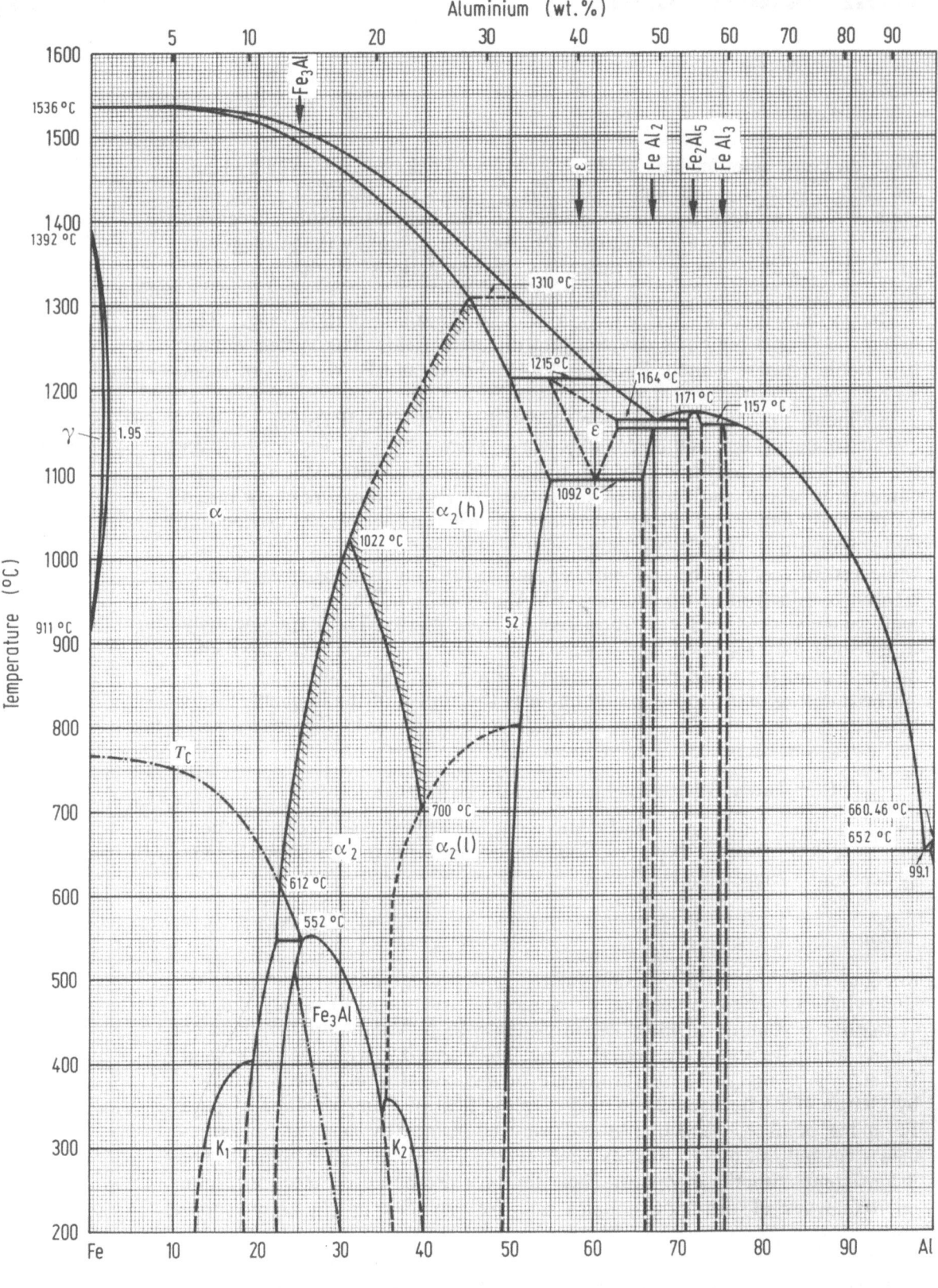

Fig. 1. Fe-Al

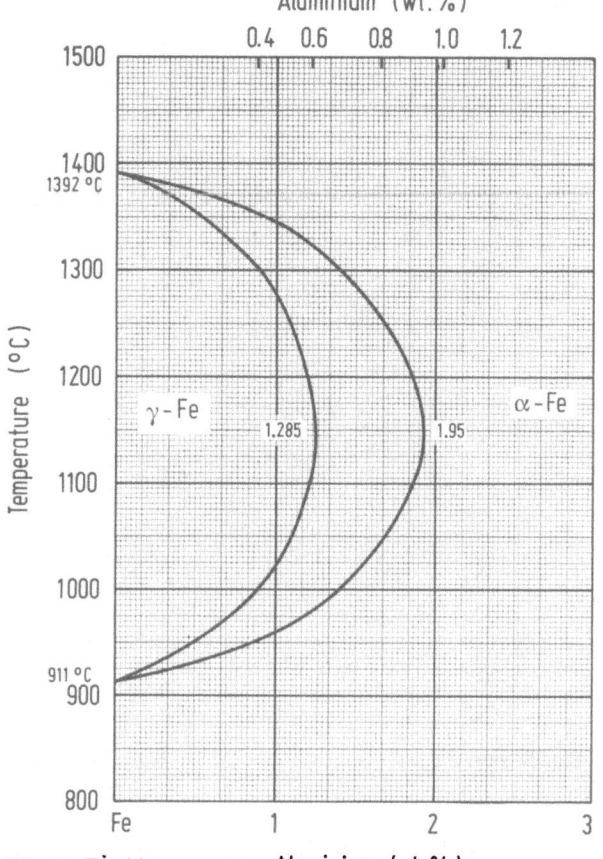

Fig. 2. Fe-Al

Aluminium (at.%)

Order/disorder transformation

A transformation in the α-Fe solution region, an ordered bcc structure (B2), was first mentioned in 1930 (see [A: Ref. 8]). Since then, numerous studies on the α-Fe solution range have been published. Recently (1980), [3] summarized and investigated the 0–50 at.% Al region using thermal analysis, dilatometric and *E*-modul measurements of samples prepared from pure iron (0.02% impurities) and 99.999% aluminium which were homogenized and annealed up to 6 months. They confirmed the phase diagram accepted here in the range 22–35 at.% Al below 600°C. However, the α_2(B2) region, which up to now was considered to be homogenous, is subdivided; α_2(FeAl) occurring reversibly in a high and low temperature modification. The transformation takes place at 803°C and is temperature-dependent. On the other hand, α_2 (high) transforms into α_2' which was demonstrated by thermal analysis and a change in volume; and α_2' again changes into the ordered structure $D0_3$ (Fe$_3$Al). The transformation of the latter takes place at 552°C and 26.8 at.% Al. The phase boundaries, α in α_2'; α_2' in Fe$_3$Al and the Curie temperatures of α and Fe$_3$Al confirm previous work. (See also Oki et al. [9].) The so-called "double Curie point" phenomenon is primarily caused by the coexistence of the ferro and paramagnetic phases.

The relationship between α, α_2' and Fe$_3$Al have been investigated by Warlimont and Lütjering [10] and Warlimont [11] (electron microscopy) and by Morgand [12] and Swann et al. [13] (transmission electron micro-

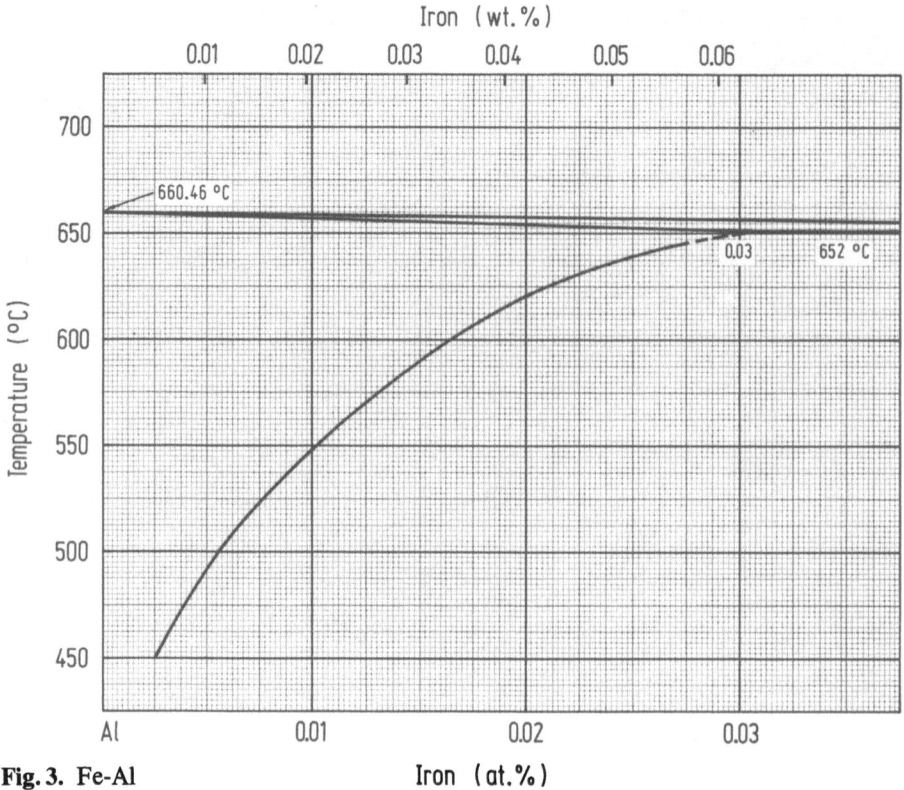

Fig. 3. Fe-Al

scopy). A study on the mechanisms of phase transformations within the miscibility gap of Fe-rich Fe-Al alloys has been undertaken by Allen and Cahn [14] who developed rules of general applicability. Electron microscopy has been used to determine the reactions experimentally, which are in agreement with the theoretical predictions (see also [15, 16]).

The transition α/α_2 (B2) is a second order type reaction (see FeGa, FeSi). The transition α/Fe_3Al was found to be a classical, first order reaction [10, 12, 13]. The transition $Fe_3Al/FeAl$ according to [12] is a homogenous one.

This is in accordance with the experimental work of [13] and theoretical predictions of Guttman et al. [17] and Rudman [18].

The two phase fields, indicated as K_1 and K_2, appear to exist below 400°C. The boundaries are easy to identify by dilatometric measurements. The structure corresponds to the "phenomena" discovered by Thomas [19, 20]: Fig. 1. [19] explained this as a first step to an ordering structure.

The solubility of Fe in Al has been determined by Edgar [21] and Nishio et al. [22]. They used the Mössbauer effect with Fe^{57} [22] and resistivity measurements and microscopy [20] and their results are in excellent agreement. Data obtained by [22] are represented in Fig. 3. The maximum solid solubility of Fe in Al is about 0.03 at.% at 652°C.

Compounds

Fe_3Al (DO_3) is a low temperature phase. It is formed at 552°C and 26.5 at.% Al by a first order reaction from FeAl (α_2') [3]. The phase boundaries of ϵ are still somewhat uncertain. It is a high temperature phase, formed by peritectic reaction (ε(54.33 at.% Al) $\rightleftharpoons$ α_2(49.9 at.% Al) + L(60.18 at.% Al)) at 1,215°C [2] and has a complex bcc structure ([A: Refs. 26, 45]) or, according to Taylor and Jones [23], a nearly hexagonal structure with parameters very similar to $FeAl_2$. $FeAl_2$ is formed by a peritectoidal reaction at about 1,153°C and crystallizes in a complex rhombohedral structure. For experiments to clarify this structure, see Mayer and Morandini [24]. There is general agreement about the congruent formation of Fe_2Al_5. A melting point of 1,171°C [2], and 1,173°C [A] respectively was found for the stoichiometric composition, while [1] reported 1,156°C at 69.6 at.% Al the highest melting point for the 69–74 at.% Al concentration range. The homogeneity range is still uncertain. As a compromise, 70–72.5 at.% was accepted by [A]. The Al-richest phase $FeAl_3$ is formed peritectically at 1,157°C [2] (1,160°C [A]) and crystallizes in a monoclinic structure. Precipitates may form a star-shaped cluster which can be explained in terms of the twinning mechanism of the 100 plane, see Lous et al. [25].

A set of interdependent thermodynamic values, consistent with the Fe-Al phase diagram, based on a critical assessment of all the published phase diagram and thermodynamical data has recently been submitted by Ansara [26].

References

1 Lee, R.J.: J. Iron Steel Inst. 194 (1960) 222
2 Schürmann, E.; Kaiser, H.P.: Arch. Eisenhüttenwes. 51 (1980) 325
3 Köster, W.; Gödecke, T.: Z. Metallkd. 71 (1980) 765
4 Rocquet, P.; Jegaden, G.; Petit, J.C.: J. Iron Steel Inst. 205 (1967) 437
5 Rocquet, P.; Petit, J.C.: J. Iron Steel Inst. 209/1 (1971) 69
6 Schürmann, E.; Kaiser, H.P.: Arch. Eisenhüttenwes. 52 (1981) 127
7 Oelsen, W.: Stahl Eisen 69 (1949) 468
8 Hirano, K.I.; Hishinuma, A.: J. Jpn. Inst. Met. 32 (1968) 516
9 Oki, K.; Yamamura, A.; Hasaka, M.; Eguchi, T.: J. Jpn. Inst. Met. 41.1 (1977) 3
10 Warlimont, H.; Lütjering, G.: Acta Metall. 12 (1964) 1460
11 Warlimont, H.: Z. Metallkd. 60 (1969) 195
12 Morgand, P.: Metall. Trans. 1 (1970) 2331
13 Swann, P.R.; Duff, W.R.; Fischer, R.M.: Trans. AIME 245 (1969) 851
14 Allen, S.M.; Cahn, J.W.: Acta Metall. 24 (1976) 425
15 Sagane, H.; Oki, K.: J. Jpn. Inst. Met. 43 (1979) 569
16 Oki, K.; Yamamura, A.; Kudo, K.; Eguchi, T.: Trans Jpn. Inst. Met. 20 (1979) 451
17 Guttmann, L.: Solid State Phys., 3 (1956) 145
18 Rudman, P.S.: Acta Metall. 8 (1960) 321
19 Thomas, H.: Z. Metallkd. 41 (1950) 185
20 Thomas, H.: Z. Phys. 129 (1951) 219
21 Edgar, I.K.: Trans. AIME 180 (1949) 225
22 Nishio, M.; Nasu, S.; Murakami, Y.: J. Jpn. Inst. Met. 34 (1970) 1173
23 Taylor, A.; Jones, R.M.: J. Phys. Chem. Solids 6 (1958) 21
24 Mayer, A.; Morandini, U.: Z. Metallkd. 62 (1971) 633
25 Lous, R.; Mora, R.; Pastor, J.: Met. Sci. J. 14 (1980) 557
26 Ansara, I.: Commission des Communautés Européennes. CECA No. 7210-CA/3/303, Nov. 1981

Fe–Am   Iron–Americium

Phase relationships have not been established. Only the following intermediate Laves phase of the Fe-Am system has been prepared and analysed: Fe_2Am of the ($MgCu_2$) type with the lattice parameter $a = 0.730$ nm.

Reference

1   Aldred, A.T.; Dunlap, B.D.; Lam, D.J.; Shenoy, G.K.: Crystal Structures and Magnetic Properties of Americium Laves Phases. Paper presented at 4th International Transplutonium Element Symposium Baden-Baden, Germany; Sept. 1975, available NTIS: Conf-750913-15 (NTIS: National Technical Information Service)

Fe–As   Iron–Arsenic
Fig. 4

The tentative phase diagram shown in Fig. 4 is based on Hansen [A] in the range Fe-FeAs, but the α(Fe) solvus is due to Svechnikov et al. [1], Svechnikov and Shurin [2] and Predel and Frebel [16]. The critical points in the range Fe_2As-FeAs are taken from Sawamura et al. [3]; the range FeAs-As from results of Clark [4]. Arsenic sublimes without melting at atmospheric pressure. The hypothetical melting point at 1 bar = 808 °C (Hultgren [D]) has been adopted in Fig. 4.

The component elements form four compounds, namely Fe_2As, tetragonal [5, 8, 9, 11]; Fe_3As_2 [A], FeAs, orthorhombic, MnP-type structure [7, 11, 12] and $FeAs_2$, orthorhombic [5, 6].

[4], investigating the FeAs-As region studied alloys by thermal analysis. $FeAs_2$ appears to be the only intermediate phase in this region. It occurs according to [4] at the stoichiometric composition, while Heyding and Calvert [5] observed a $FeAs_2$ phase with a slight As deficiency. FeAs does not show a significant homogeneity range. It forms eutectics with As at 800 °C at > 93.4 at.% As and with FeAs at 1,008 °C and about 58 at.% As.

Lattice parameters for $FeAs_2$ reported by Swanson et al. [6] are in excellent agreement with those established by [5].

Magnetic structure and properties of FeAs were recently studied by Selte et al. [7]. They employed X-ray, neutron diffraction, magnetic susceptibility, thermal and diffuse reflectance measurements. They confirmed the earlier reported lattice parameters of the orthorhombic unit cell, which remains essentially of the MnP-type between 12 and 1,300 K. Below a Néel temperature of

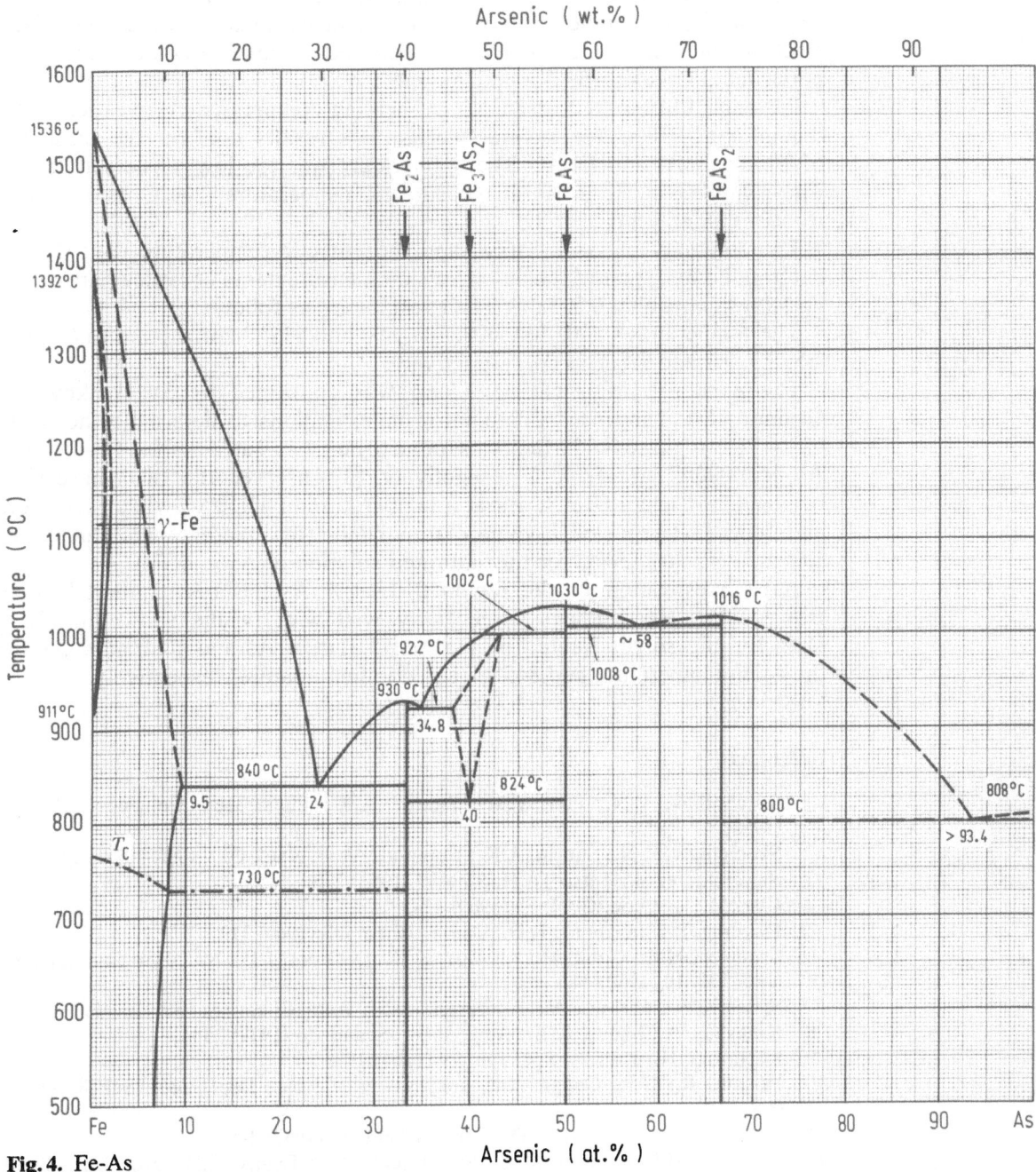

Fig. 4. Fe-As

77 K a helimagnetic structure is adopted. The magnetic susceptibility of FeAs, determined from 4.2 to ~1,000 K, satisfies the Curie-Weiss Law between ~300 and 650 K.

The composition range and magnetic properties of Fe_2As was studied by Kjekshus and Skaug [8]. They found no appreciable range of homogeneity and suggest a stoichiometric composition of Fe 2.00 ± 0.05 As. The unit cell dimensions are reasonably consistent with those reported in the literature. The magnetic properties—Fe_2As is antiferromagnetic—are well established by susceptibility measurements and neutron diffraction studies [8, 9].

Fe–As   The nucleation, morphology and kinetics of the precipitation reaction of Fe_2As has been studied by means of metallographic methods (light and electron microscopy) of long time annealed samples prepared from high purity metals (Fe = > 99.7 % and As = 99.9995 %).

Pastant [10] prepared arsenides by heating compressed As powder in Fe crucibles at 470°C for several weeks. His results confirm the existence of $FeAs_2$, FeAs and Fe_2As. The temperature was too low, to detect Fe_3As_2. The solubility of Fe in As at the eutectic temperature (800°C) is estimated to be ~ 0.07 at.% [4].

The solubility of As in α(Fe) and the eutectic temperature (840°C) was determined by [1] and [2] using precision measurements of lattice parameters on alloys containing 0.02 % C. Their results are as follows:

The temperatures 400, 500, 600, 700, 800 and 835°C correspond to 6.1, 6.8, 7.5, 8.3, 9.1 and 9.5 at.% As respectively.

The existence of a closed γ-loop was confirmed by Svechnikov and Gridnev [13]. They used dilatometric investigations to determine the width of the (α + γ) phase field which was found to be 1.1–1.9 at.% As and 1.8–2.5 at.% As at 1,000°C and 1,100°C respectively. [2] observed the two phase field between 1.31 and 1.80 at.% As at a temperature of 1,150°C, while Sawamura and Mori [14] reported a value extrapolated from micrographic studies, of 1.65 at.% As at 1,150°C for the saturated solution of As in γ(Fe).

A re-investigation of the region 0–15 at.% As has been undertaken by Bovzic [15]. His findings agree fairly well with results shown in Fig. 4. Data for the solvus quoted by [15] are 5.31 at.% As and 10.43 at.% As for the room temperature and 845°C (eutectic temperature) respectively.

References

1 Svechnikov, V.N.; Pan, V.M.; Shurin, A.K.: Phys. Met. Metallogr. USSR 6 (1958) 80
2 Svechnikov, V.N.; Shurin, A.K.: Dopovidi Akad. Nauk Ukr. RSR (1957) 27
3 Sawamura et al.: Tetsu To Hagane 39 (1953) 776
4 Clark, L.A.: Econ. Geol. 55 (1960) 1345
5 Heyding, R.D.; Calvert, L.D.: Can. J. Chem. 38 (1960) 313
6 Swanson, H.E.; Cook, M.I.; Evans, E.H.; DeGroot, J.H.: Nat. Bur. Stand. (U.S.) Circ. 539, 10 (1960) 34
7 Selte, K.: Kjekshus, A.; Andresen, A.F.: Acta Chem. Scand. 26 (1972) 3101
8 Kjekshus, A.; Skaug, K.E.: Acta Chem. Scand. 26 (1972) 2554
9 Katsuraki, H.: J. Phys. Soc. Jpn. 19 (1964) 1988
10 Pastant, R.: C.R. Congr. Nat. Soc. Savantes Paris, Dept. Sect. Sci. 87 (1963) 605
11 Heyding, R.D.; Calvert, L.D.: Can J. Chem. 35 (1957) 449
12 Swanson, H.E.; Morris, M.C.; Stinchfield, R.P.; Evans, E.H.: Nat. Bur. Stand. (U.S.) Monogr. 25, Sect. 1 (1962) 19
13 Svechnikov, V.N.; Gridnev, V.N.: Metallurgia 1 (1938) 13
14 Sawamura, H.; Mori, T.: Mem. Fac. Eng. Kyoto Univ. 14 (1952) 129
15 Bovzic, B.: Bull Acad. Serbe Sci. Arts, Cl. Sci. Tech., 14 (1979) (in German)
16 Predel, B.: Frebel, M.: Arch. Eisenhüttenwes. 42 (1971) 365

Figure 5 represents the phase relationships in the system Fe-Au, constructed from data reported in the literature [1–7].

The diagram is based mainly on experimental results obtained by Raub and Walter [3] except for the Fe-rich region which was adopted from data quoted by Buckley and Hume-Rothery [6]. The solid solubilities of Au in δ(Fe), γ(Fe) and α(Fe) were taken from observations by [6, 3], Jette et al. [4] and Royen and Reinhardt [7].

The melting point of gold (1,064.43°C) is a primary fixed point on the International Practical Temperature Scale (IPTS) of 1968.

There are no compounds in this system. The intermediary phase Fe_3Au reported in earlier investigations has not been confirmed ([A: Refs. 4–7]).

Thermal analysis carried out by [3] confirmed the shape of the liquidus curve already established by Isaac and Tammann [1]. Addition of Au decreases the liquidus temperature to a minimum at about 80 at.% Au at 1,035°C.

Fe-rich region

Addition of Au increases the γ/δ(Fe) transformation temperature [6] but decreases that of the α/γ(Fe) transformation. The former was determined by thermal analysis, the latter by dilatometric measurement [3]. [6] determined the part phase diagram above 1,160°C in the concentration range 0–18 at.% Au by means of thermal and metallographic analyses of alloys prepared from 99.95% Fe (melting point, 1,535°C) and 99.99% Au. The maximum solid solubility of Au in δ(Fe) and γ(Fe) was found to be 2.3 at.% at 1,431°C and 4.1 at.% at 1,171°C, respectively. The work of [6] indicates that a redetermination of the γ(Fe) solvus in the range 903–1,171°C is advisable. In Fig. 5 the solid solubility data of Au in α(Fe) for the temperature range 500–903°C are the ones taken from X-ray measurements of [3]. The results are in good agreement with values reported by [7] (H_2O/H_2 equilibrium pressures). At 500°C the solid solubility of Au in Fe was found to be 0.14 [3] and 0.1 [7] at.% respectively. The maximum solubility of Au in α(Fe) is placed at 2.3 at.% Au at 903°C [3].

The phase boundary of the Au-rich (Fe-Au) solid solution was established by means of X-ray measurements in the temperature range 400–1,100°C and agrees very well with X-ray data recorded by [4].

The magnetic properties of Au-rich Fe-Au alloys have interested a number of investigators. The influence of ordering, spinorientation, impurities etc. have been studied during the last years by Wenger and Keesom [8], Dawes and Voles [9], Murani [10], Smith and Liu [11] and de Mayo [12]. Resistivity measurements and electron microscopy [13] and metallographic methods [14] applied by Higgins and Wilkes [13] and Frebel and Predel [14] have achieved much to clarify the morphology and kinetics of the precipitation in Fe-Au alloys. For further information the reader may be referred to the publications of [8–14].

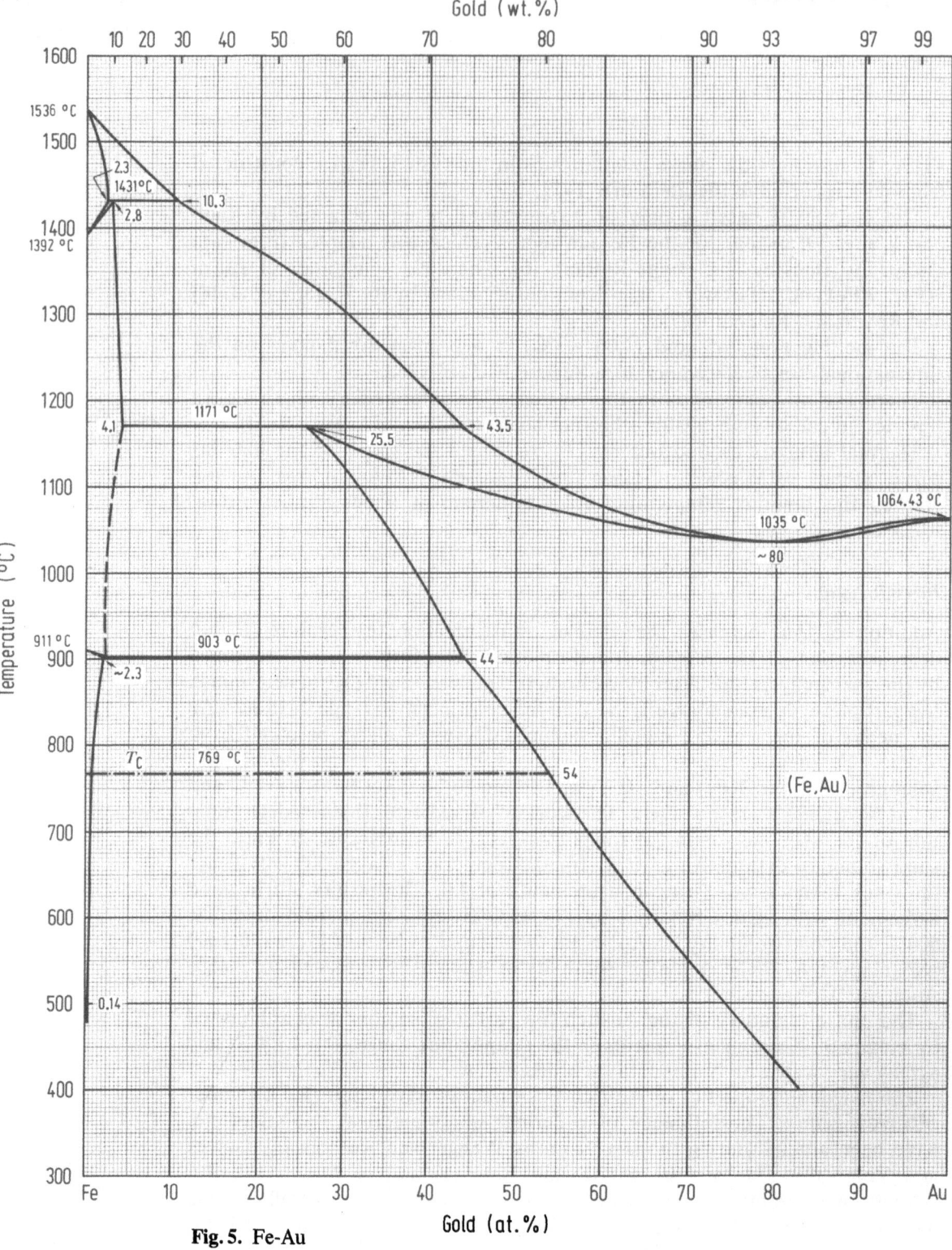

Fig. 5. Fe-Au

1   Isaac, E.; Tammann, W.: Z. Anorg. Chem. 53 (1907) 291
2   Fallot, M.: Ann. Phys. 6 (1936) 376
3   Raub, E.; Walter, P.: Z. Metallkd. 41 (1950) 234
4   Jette, E.R.; Brunner, W.L.; Foote, F.: Trans. AIME 111 (1934) 354
5   Seigle, L.L.: Trans. AIME 206 (1956) 91
6   Buckley, R.A.; Hume-Rothery, W.: J. Iron Steel Inst. 201 (1963) 121
7   Royen, P.; Reinhardt, H.: Z. Anorg. Allg. Chem. 281 (1955) 18
8   Wenger, L.E.; Keesom, P.H.: Phys. Rev. B11 (1975) 3497
9   Dawes, D.G.; Coles, B.R.: J. Phys. F: Met. Phys. 9 (1979) L 215
10 Murani, A.P.: J. Phys. F: Met. Phys. 4 (1974) 757
11 Smith, F.W.; Liu, J.C.: Solid State Commun. 26 (1978) 91
12 de Mayo, B.: J. Phys. Chem. Solids 35 (1974) 1525
13 Higgins, J.; Wilkes, P.: Philos. Mag., Ser. 8 25.1 (1972) 599
14 Frebel, M.; Predel, B.: Z. Metallkd. 64 (1973) 913

Iron–Boron

Fe–B
Fig. 6

A critical assessment of the phase diagram and thermodynamical data has been carried out by Chart [1]. The diagram has been based primarily on the diagrams suggested by Portnoi et al. [2, 3], Voroshnin et al. [4], Hansen and Anderko [A] supplemented by findings obtained by authors [B, C] and [5–11]. The melting point of boron (2,350 K) selected by Chart is 50 K higher than the value chosen by Hultgren [D] and 100 K lower than JANAF [12].

The system is characterised by small mutual solid solubility and two compounds, Fe_2B and FeB respectively. Fe_2B is formed peritectically at 1,407°C and shows a bc tetragonal, C 16, $CuAl_2$-type structure with $a = 0.5109$ nm, $c = 0.4249$ nm, $c/a = 0.832$.

The eutectic and boride decomposition temperatures are uncertain to the extent of ca. 30–50 K.

FeB melts congruently at 1,590°C. It has a homogeneity range of about 1 at.% B. The crystal structure according to Pearson [13] is orthorhombic, B 27, with $a = 0.551$, $b = 0.295$ and $c = 0.406$ nm. Evidence for the existence of a high-temperature modification of FeB is inconclusive ([2, 3] and [C]). The higher boride FeB_x where x = 20–40 reported by Portnoi et al. [2, 3] has been taken to be a solid solution of Fe in rhombohedral boron [14, 15].

There is contradictory evidence for the existence of a diboride FeB_2 [2–4, 11] most of which indicates, that this compound is not stable. It is not included in Fig. 6.

Information relating to the crystal structures and nature of bonding of iron-borides has been given and discussed by authors (see references [16–21]).

By the liquid quenching technique amorphous phases in the Fe-B system may be obtained. For references pertaining to the formation of those phases the papers quoted in [26–31] may be consulted.

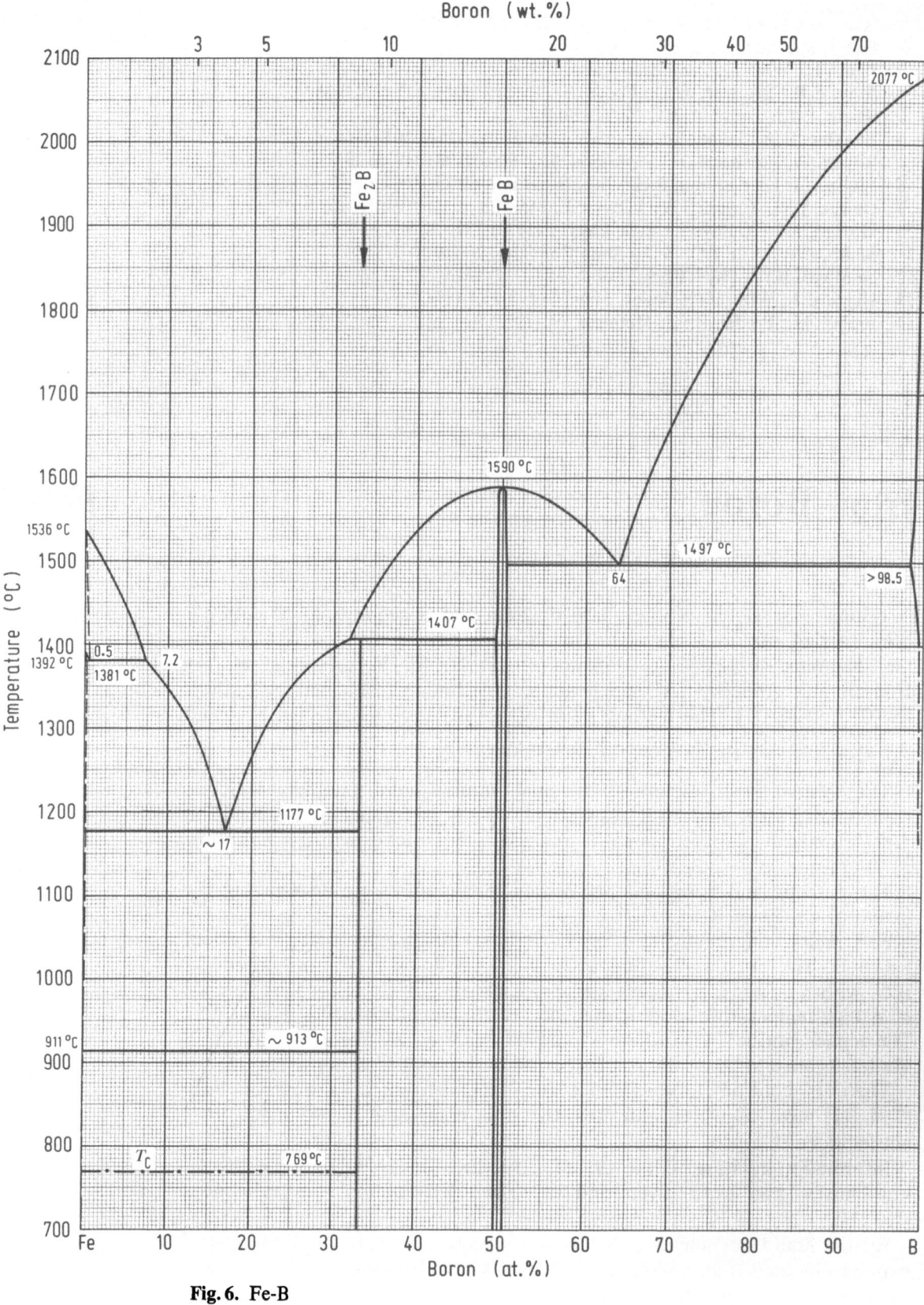

Fig. 6. Fe-B

The maximum solid solubility of iron in boron, according to [2] is < 1.5 at.%.

More studies have been undertaken to determine the solubility boundaries of boron in iron. Hansen and Anderko [A] report a maximum value for B in δ(Fe) of ~ 0.5 at.%. Brown et al. [22] in a recent publication, reviewed earlier investigations pertaining to the solubility of B in α- and γ(Fe) and for clarification carried out new measurements in the 500–1,300 °C temperature range. They employed the technique of boron autoradiography, which can distinguish between boron in solution, boron-rich precipitates and boron segregated to grain boundaries. However, the solubility limits determined by this method should be regarded as upper limits, as the technique cannot differentiate between soluble boron and very fine boride particles (< 60.0 nm). According to [22], the boron segregates in γ(Fe) uniformly on the grain boundaries, a thermodynamically normal lattice solution of boron occurs, however on the evidence available, no conclusion can be reached as to whether this is an interstitial or substitutional solid solution. In α-Fe, bcc, boron is *not* an interstitial solute and it is possible to argue that the total solubility measured could be due to the stabilisation of B by interstitial impurities. It seems likely that lower solubility limits would be measured in Fe of even greater purity. The maximum solubility of B in γ(Fe) and α(Fe) are about 0.025 and 0.01 at.% respectively [22]. Considering the experimental difficulties and very low absolute solubilities, results of all teams agree fairly well [22–25].

References

1 Chart, G.T.: Commission des Communautés Européennes, CECA No. 7210-CA/3/303, Nov. 1981

2 Portnoi, K.I.; Romashov, V.M.: Sov. Powder Metall. Met. Ceram. 11 (1972) 378

3 Portnoi, K.I.; Levinskaya, M.Kh.; Romashov, V.M.: Sov. Powder Metall. Met. Ceram. 8 (1969) 657

4 Voroshnin, L.G.; Lyakhovich, L.S.; Panich, G.G.: Protasevich, G.F.: Met. Sci. Heat Treat. Met. USSR 1970) (9) 732

5 Kiessling, R.; Liu, Y.H.: J. Met. 3 (1951) 639

6 Krishtal, M.A.; Turkel'taub, G.M.: Met. Sci. Heat Treat. Met. (USSR) (1967) (8) 620

7 Krishtal, M.A.; Svobadov, A.N.: 9 (1968) 678

8 Kunitskii, Yu.A.; Marek, E.V.: Sov. Powder Metall. Met. Ceram. 10 (1971) 216

9 Kostetskii, I.I.; L'vov, S.N.; Kunitskii, Yu.A.: Inorg. Mater, 7 USSR (1971) 839

10 Plotnikova, A.F.; Ilyushchenko, N.G.; Anfinogenov, A.I.; Belyaeva, C.I.; Finkel'shteyn, S.D.: Tr. Inst. Elektrochim. Ural. Nauchn. Tsentr. Akad. Nauk. SSSR 18 (1972) 112

11 Sidorenko, F.A.; Serebrennikov, N.N.; Budozhanov, V.D.; Putintsev, Yu.V.; Trushevskii, S.N.; Korabanova, V.D.; Geld, P.V.: High Temp. 15 (1977) 36

12 Stull, D.R.; Prophet, H.: JANAF Thermochemical Tables, 2nd ed. Washington: Nat. Bur. Stand. 1971

13 Pearson, W.B.: A Handbook of Lattice Spacings and Structures of Metals and Alloys, Vol. 2. Oxford: Pergamon 1967

14 Carlsson, J.O.; Lundström, T.: J. Less Common Met. 22 (1970) 317

15 Spear, K.E.: In: Applications of Phase Diagrams in Metallurgy and Ceramics, Vol. 2, ed. G.C. Carter NBS Spec. Publ. 496 (1978) 744

16 Kiessling, R.: Acta Chem. Scand. 4 (1950) 209

Fe—B
17 Kiessling, R.: J. Electrochem. Soc. 98 (1951) 166
18 Cooper, J.D.; Gibb, T.C.; Greenwood, N.N.; Parish, R.V.: Trans. Farad. Soc. 60 (1964) 2097
19 Zhurakovs'ky, E.A.; Kotlyar, V.I.; Shashkina, T.B.: Dopov. Akad. Nauk Ukr. RSR (1969) (7) 654
20 Shevelev, A.K.: Phys. Met. Metallogr. 22 (1966) (2) 50
21 Greenwood, N.N.; Parish, R.V.; Thornton, P.: Q. Rev. Chem. Soc. 20 (1966) 441
22 Brown, A.; Garnish, J.D.; Honeycombe, R.W.K.; Met. Sci. J. 8 (1974) 317
23 Hasegawa, M.; Okamoto, M.: Nippon Kinzoku Gakkaishi 29 (1965) 328
24 Lucci, A.; Della Gatta, G.; Venturello, G.: Met. Sci. J. 3 (1969) 14
25 Lucci, A.; Venturello, G.: Scr. Metall. 5 (1971) 17
26 Fukamichi, K.; Kikuchi, M.; Arakawa, S.; Masumoto, T.: Solid State Commun. 23 (1977) 955
27 Takahasi, M.; Koshimura, M.: Jpn. J. Appl. Phys. 16 (1977) 1711
28 Hasegawa, R.; Ray, R.: J. Appl. Phys. 49 (1978) 4174
29 Walter, J.L.; Bartram, S.F.; Mella, I.: Mater. Sci. Eng. 36 (1978) 193
30 Herold, U.; Köster, U.: Z. Metallkd. 69 (1978) 326
31 Miroshnichenko, I.S.; Bashev, V.F.: Russ. Metall. (1977/3) 178

Fe–Ba Iron–Barium

One may safely assume that iron and barium do not alloy significantly, neither in the solid nor in the liquid state. Any statements to the contrary (see Hansen [A]) are likely to be false. This conclusion is based on an argument of Kubaschewski [1].

Reference

1 Kubaschewski, O.: Phase Stability in Metals and Alloys. Proc. Battelle Colloquium, Geneva and Villars 1967, p. 63.

Fe–Be Iron–Beryllium

Figs. 7, 8

The Fe-Be diagram in Fig. 7 is the one published by Aldinger and Petzow [1] except for the solid solubility of Fe in Be. It is a revised version of a diagram proposed by Hansen [A]. The solubility of Fe in α(Be) has been calculated using data reported by Myers and Smugeresky [2], Hammond et al. [3], Donze et al. [4], Jacobson and Hammond [5] and Gelles et al. [6] and is represented in Fig. 8.

The system is characterized by a large solid solution α, δ-Fe and four intermediary phases, Fe_3Be, $FeBe_2$, $FeBe_5$ and $FeBe_7$, each with a range of

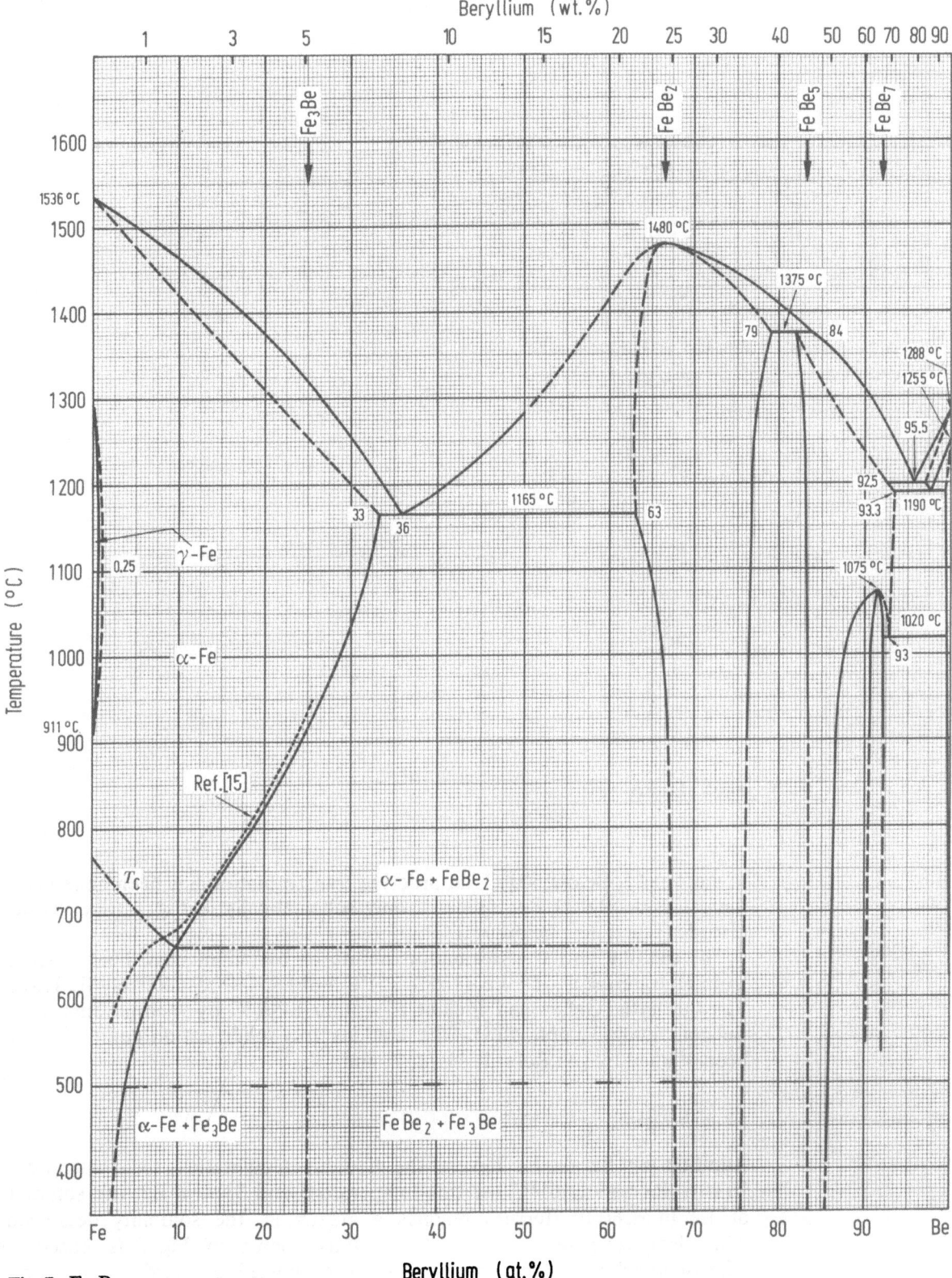

Fig. 7. Fe-Be

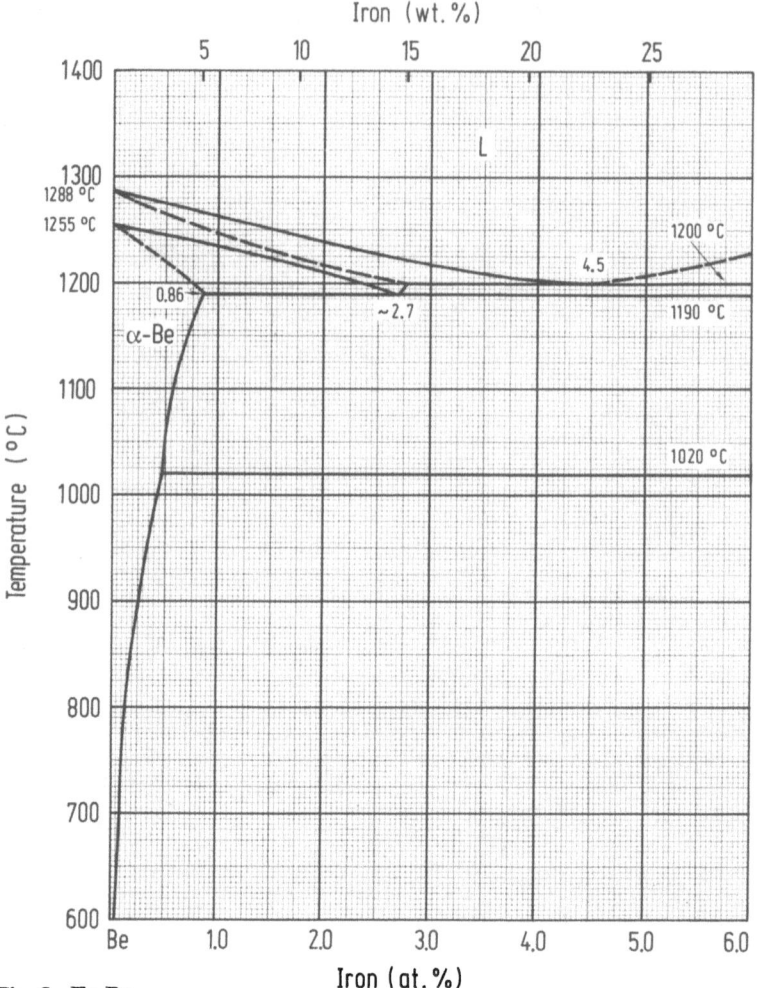

Fig. 8. Fe-Be

homogeneity. Below 600°C, an ordering transformation has been observed in the range 0–50 at.% Be, indicated only schematically in Fig. 7. (See, 'Order/disorder reactions', below.)

The melting and transformation points of beryllium have been selected from Hultgren et al. [D], i.e. (1,288 ± 5)°C and 1,255°C respectively (IPTS-68). For the latter Gelles and Pickett [7] found 1,262°C by thermal analysis of Be containing 3.1 ppm BeO. The liquidus and solidification equilibria have been studied by [A: Refs. 1–3], [1, 7] and [5].

Fe-rich region

Addition of Be to Fe lowers the liquidus temperature to a eutectic at 36 at.% Be and 1,165°C. At this temperature and 33 at.% Be, the solubility of Be in α, δ(Fe) (ferrite) reaches a maximum, the solubility decreasing with decreasing temperature. The solvus shown in Fig. 7 is based on microscopical and X-ray measurements supported by hardness tests and electrical resistivity studies carried out by Heubner [9] as well as on data reported by [1].

Recent X-ray diffraction measurements and electron-probe micro-analyses by Takayama et al. [15], indicated by a dashed line in Fig. 7, illustrate the influence of the magnetic transition on the solubility limits of Be in α(Fe). They used samples prepared from 99.95 % Fe and >99.5 % Be and annealed for 150–7,000 hours in the temperature range 600–950 °C. The addition of Be lowers the Curie temperature remarkably, and according to [15] the solubility range of the ferromagnetic region is much lower than expected from the extent of the paramagnetic regions.

Be-rich region

The addition of Fe to Be lowers its liquidus temperature to a eutectic at 95.5 at.% Be and 1,200 °C [7], while the β(Be) transformation temperature is lowered to a eutectoid (α-Be/FeBe$_5$) at 1,190 °C and 97.3 at.% Be (Fig. 8).

The solubility of Fe in Be is very low and the determination complicated by the role played by aluminium in Be of commercial purity [2]. [5] redetermined the solubility in the temperature range 800–1,000 °C on high purity alloys (zone-refined Be) using diffusion "sandwiches" of Be with two-phase alloys in the Fe-Be system. It was analysed by an electron beam microprobe technique. Their results agree very well with those reported by [6] and [4], who obtained data on diffusion couples by X-ray absorption and microprobe analyses respectively. [3] measured the solubility of Fe in Be at temperatures as low as 773 K by applying the high energy ion beams technique. A comparison with data obtained at higher temperatures [4–6] shows exellent agreement. The solvus calculated from the experimentally established values is shown in Fig. 8. The maximum solubility is 0.86 at.% Fe at 1,190 °C.

Fe$_3$Be is a low temperature phase formed by an ordering reaction.

FeBe$_2$ is formed by congruent melting at 1,480 °C, has a wide range of homogeneity at higher temperatures and crystallizes in the hexagonal MgZn$_2$ structure. FeBe$_5$ formed by peritectic reaction at 1,375 °C has a fcc MgCu$_2$ type structure.

The existence of a hexagonal phase in the concentration range 91.67–92.31 at.% Be was first mentioned by [A: Ref. 5] and would correspond to an index FeBe$_{11}$ or FeBe$_{12}$. According to [1] this phase richest in Be is quoted as being FeBe$_7$. For further information, see [1], Levine and Lütjering [8] and "Beryllium" [9] 'Crystal Structures'.

Order/disorder reactions

Precipitation reactions in the Fe-Be system have been studied by a number of investigators. In spite of this, some confusion remains concerning the mechanism of the various reactions, the temperatures and concentration regions within which they occur. Ino [10] calculated the decomposition and ordering processes in bcc binary alloys by using a pairwise interaction model. Experimental results (electron microscope and Mössbauer effect studies) of Fe-Be alloys carried out by Sumitomo et al. [11] and Yagisawa and Yoshida (12) are well explained by applying this model. In his report, [10] summarized work undertaken previously on Fe-Be alloys (e.g. v. Heimendahl and Heubner [13] and Higgins et al. [14]), explained the ageing processes and proposed a part phase diagram. According to him, both phase decomposition and B2-type ordering occur simultaneously. The D0$_3$-type superlattice is not stable at any composition or temperature.

References

1 Aldinger, F.; Petzow, G.: Constitution of Beryllium and its Alloys, Chap. 7, p. 235 in: Beryllium Science and Technology, Vol. 1, New York. Plenum Press 1979

2 Myers, S.M.; Smugeresky, J.E.: Metall. Trans. 7A (1976) 795; NBS Spec. Publ. 496 (1977)

3 Hammond, M.L.: Davinroy, A.T.: Jacobson, M.I.: Beryllium rich End of Five Binary Systems. Air Force Materials, Lab. Rep. AFML-Tr-65-223 (1965)

4 Donze, G.; LeHazif, R.; Dutilloy, D.; Adda, Y.: Compt. Rendus 254 (1962) 2328

5 Jacobson, M.I.; Hammond, M.L.: Trans. AIME 242 (1968) 1385

6 Gelles, S.H.; Pickett, J.J.: Livine, E.D.; Nowak, W.B.: The Metallurgy of Beryllium. Inst. Metals, London (1963) 588

7 Gelles, S.H.; Pickett, J.J.: U.S.At. Energy Comm. NMI-1218 (1960) 44

8 Levine, E.D.; Lütgering, G.: J. Less Common Met. 23 (1971) 343

9 'Beryllium', At. Energy Rev. Spec. Issue No. 4, Part III, International Atomic Energy Agency, Vienna 1973

10 Ino, H.: Acta Metall. 26 (1978) 827

11 Sumitomo, Y.; Matsuhiro, K.; Ura, M.; Fujita, F.E.: Report on the Annual Meeting of Japan Inst. Metals (1971)

12 Yagisawa, K.: Yoshida, H.: Jpn. J. Appl. Phys. 8 (1969) 179

13 von Heimendahl, M.; Heubner, U.: Acta Metall. 11 (1963) 1115

14 Higgins, J.; Nicholson, R.B.; Wilkes, P.: Acta Metall. 22 (1974) 201

15 Takayama, T.; Wey, M.Y.; Nishizawa, T.: Trans. Jpn. Inst. Met. 22 (1981) 315

Fe–Bi Iron–Bismuth

Stable intermetallic phases in the system Fe-Bi are not formed. Mutual solubilities in the liquid and solid states are very small. Apparently reliable values for the solubility of Fe in *liquid* Bi have been supplied by Weeks [1]. He applied a sampling and chemical analysis technique of high purity Fe and 99.999% Bi. His results may be expressed by the equation:

$$\lg (\text{at.}\% \text{ Fe}) = -3{,}490\, T^{-1} - 2.26.$$

From this the following values are tabulated:

Temp. °C	at.% Fe	Temp. °C	at.% Fe
400	0.0012	600	0.0183
450	0.0028	650	0.030
500	0.0056	700	0.047
550	0.0105		

Reference

1 Weeks, J.R.: Trans. ASM 58 (1965) 302

Because of the technical importance of the phase relationships in the iron-carbon system, the number of its investigations is astronomical. Even so, few changes to the phase diagrams of the stable (Fe-graphite) and the metastable (Fe-cementite) systems had to be made in the past (Hansen [A], Elliott and Gleiser [1]). However, a novel feature of the phase boundary evaluation has been introduced by the application of thermochemical calculations, leading to the evaluation of true equilibrium diagrams. In this way, Chipman [2] dealt with the ranges up to 12 wt.%, that is 39 at.% C at 0–3,000 °C. Schürmann and Schmid [3] went further by calculating the equilibria of graphite and cementite with ferrite and austenite separately. Because of the small differences in Gibbs energy, the metastable equilibria are of particular significance.

The "classical" phase diagram is shown in Fig. 9, superimposed by the various metastable equilibria due to metastable ferrite [3, 4]. Schürmann and Schmid [4] have dismantled this complex assembly of phase boundaries based on their thermochemical evaluations. Figure 12 shows the austenite-graphite (resp. cementite) diagram with the melting point of γ (iron) (1,526 °C) 10 K lower than the normal melting point. The austenite-graphite, that is, the stable boundaries are indicated by full lines, those of the austenite-cementite equilibrium by dashed lines. The melting point of Fe_3C is being accepted as 1,252 °C in contrast to the value of 1,227 °C suggested by [2]. Figures 10 and 11 represent the results of the calculations introduced by [4] pertaining to ferrite, that is the α, δ (iron) phase, assuming that the austenite may be disregarded. It may be repeated that only the small differences in Gibbs energy, implying that all the metastable equilibria indicated in Fig. 9 may occur in practice, justify the presentation of the additional diagrams.

Figure 10 indicates the equilibria between metastable α, δ (ferritic) iron and, respectively, cementite and graphite. It may be seen that the "cementite and graphite phase boundaries" cross at 1,138 °C. Thus a uniform stable or metastable system cannot exist in equilibrium with α, δ. Instead cementite appears in this metastable system in the temperature range 855–1,138 °C, whereas α, δ is in equilibrium with graphite above and below these temperatures.

From this, the phase relationships in the ferritic Fe-C system (after elimination of the austenitic phase) may be deduced. They are represented in Fig. 11.

Readers who are simply faced with the question which phase may be observed under given conditions of temperature and concentration may be referred back to Fig. 9. Although most of these phase mixtures are metastable, in view of the small energy differences, they may occur nevertheless. And here, the advantage of the thermochemical approach over the conventional methods becomes apparent. The actual phenomena are of course determined by the kinetics, formulated by "Ostwald's Stufenregel", which thermodynamics alone cannot predict.

As for the earlier references pertaining to the Fe-C phase relationships, the papers quoted in [A] and [1—4] may be consulted.

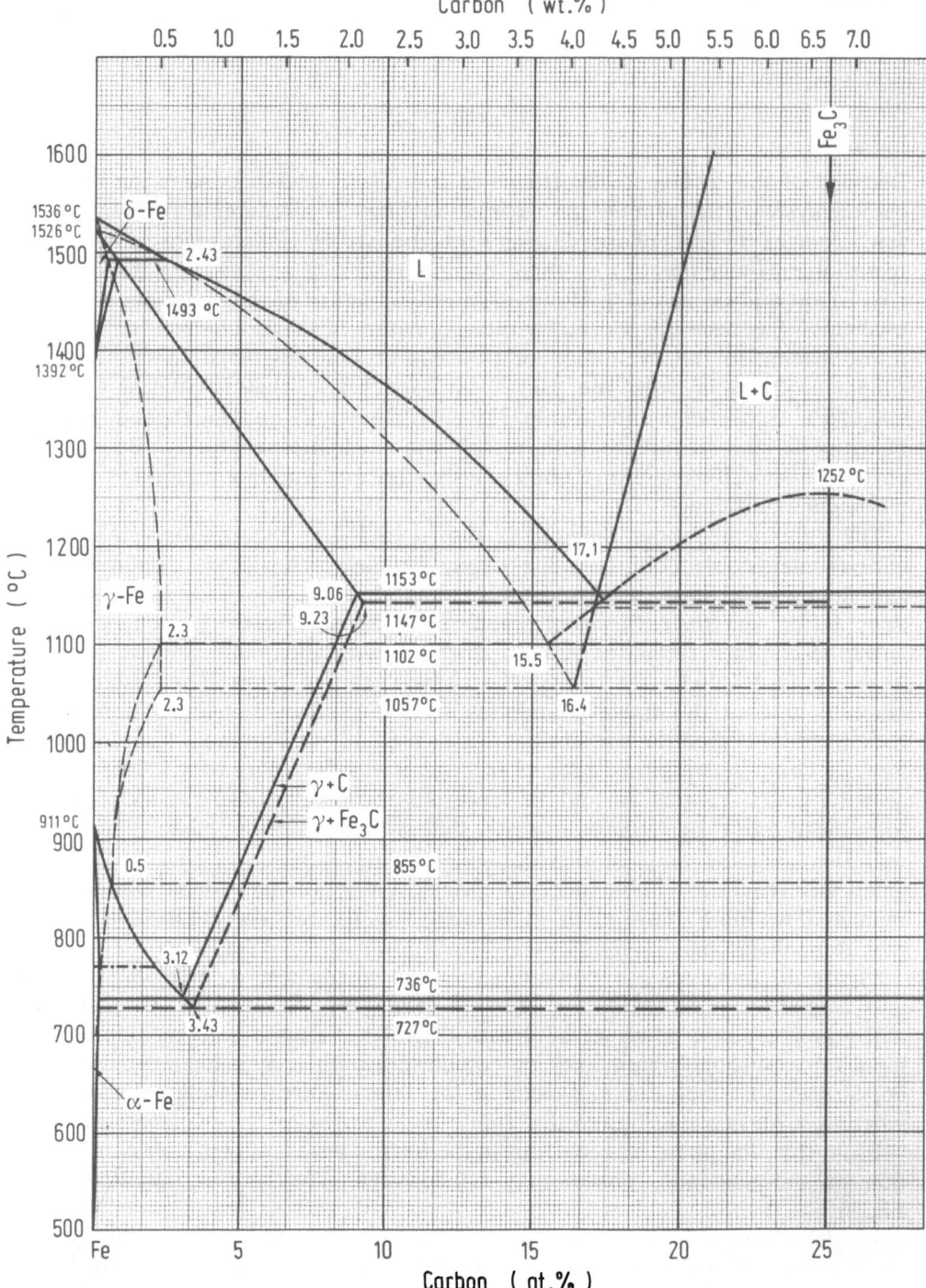

Fig. 9. Fe-C. A summary of stable and metastable equilibria

Numerous carbides are reported in the literature but only two of them have been investigated under normal pressure and metastable equilibrium conditions: firstly cementite, which is usually assigned the formula Fe_3C and crystallizes in an orthorhombic structure and secondly between room temperature and 230°C $Fe_{2.2}C$, generally called the Hägg carbide [5]. Fe_7C_3

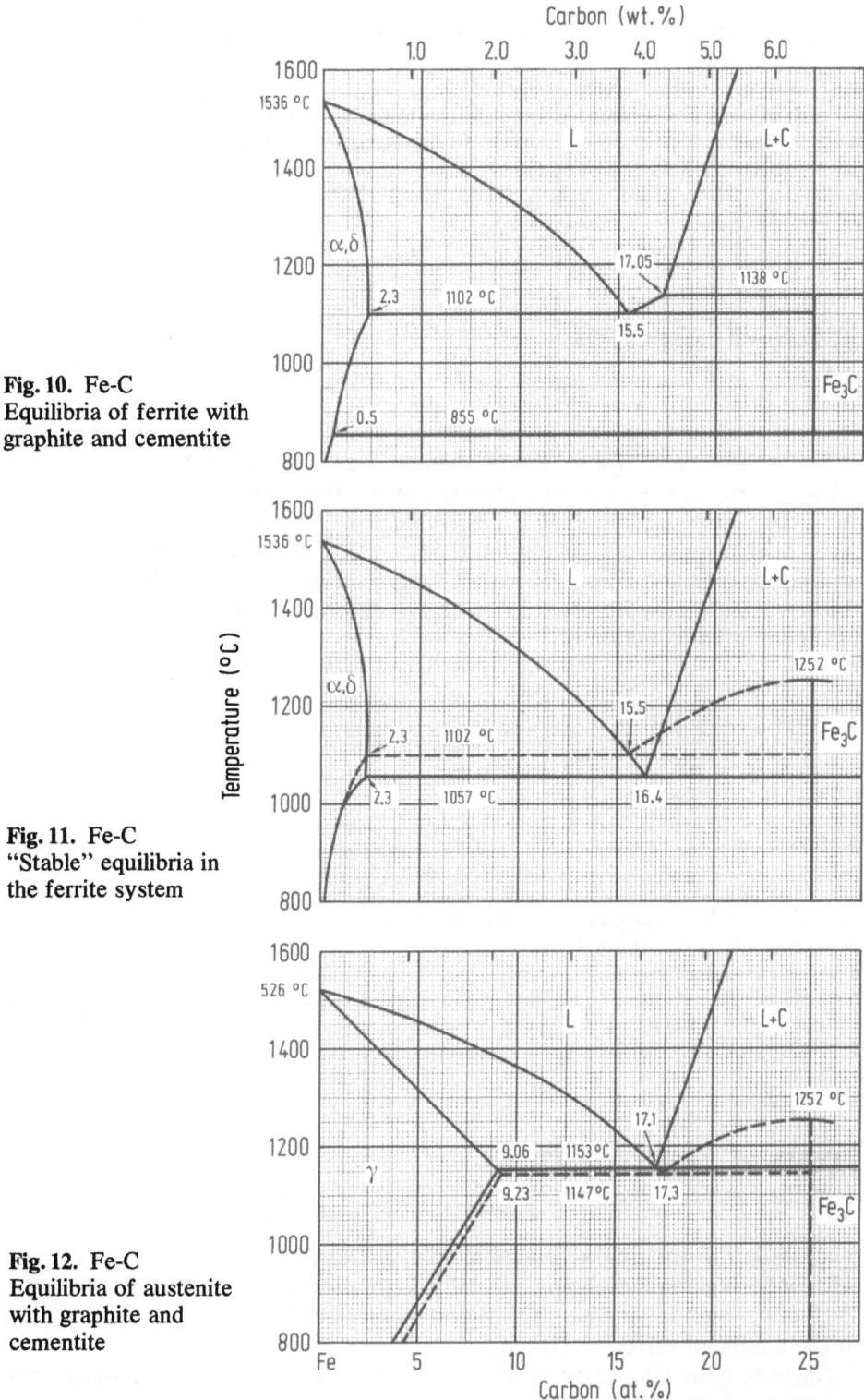

Fig. 10. Fe-C
Equilibria of ferrite with
graphite and cementite

Fig. 11. Fe-C
"Stable" equilibria in
the ferrite system

Fig. 12. Fe-C
Equilibria of austenite
with graphite and
cementite

and carbon in the form of diamond appear only at high pressures and carbon
concentrations exceeding 30–35 at.% [6]. Zhukov and Snezhnoy suggest
a hypothetical metastable Fe-C (diamond) phase diagram for normal pressure [6], to whose paper the reader may be referred.

References

1 Elliott, J.F.; Gleiser, M.: Thermochemistry for Steelmaking, Vol. 1. Reading: Addison-Wesley 1960
2 Chipman, J.: Met. Trans. 3 (1972) 55
3 Schürmann, E.; Schmid, R.: Arch. Eisenhüttenwes. 50 (1979) 101
4 Schürmann, E.; Schmid, R.: Arch. Eisenhüttenwes. 50 (1979) 185
5 Hägg, G.: Z. Kristallogr. 89 (1934) 92
6 Zhukov, A.A.; Snezhnoy, R.L.: Doklady Akad. Nauk SSSR 211, 1 (1973) 145

Fe—Ca Iron—Calcium

Iron and Calcium are virtually immiscible [1–3].

The solubility of Fe in *liquid* calcium was quoted by Schürmann et al. [1] as <0.01 wt.% in the temperature range 900–1,200°C (thermal analysis). At 1,600°C the solubility of Ca in *liquid* iron was found to be <10^{-4} wt.% and under a pressure of 14 bar 0.032 wt.% (0.045 at.%) by Philbrock et. al. [2] and Sponseller and Flinn [3] respectively (see also Joyant [5]). If required, this last figure may be provided with a temperature coefficient with the aid of an empirical equation [4]

$$\overline{H}_{Ca} = 3,400 \, \overline{S}_{Ca}^{E}$$

to yield a solubility equation for liquid calcium in liquid iron at 14 bar, i.e.

$$\lg (\text{at.\% Ca}) = -14,000 \, T^{-1} + 2.12.$$

References

1 Schürmann, E.; Fünders, P.; Litterscheidt, H.: Arch. Eisenhüttenwes. 46 (1975) 427
2 Philbrock, W.O.; Goldman, K.M.; Helzel, M.M.; J. Met., Trans. 188 (1950) 361
3 Sponseller, D.L.; Flinn, R.A.: Trans. AIME 230 (1964) 876
4 Kubaschewski, O.: High Temp. High Press. 13 (1981) 435
5 Joyant, M.: Commission des Communautés Européennes CECA No. 7210-CA/3/303, Nov. 1981

Fe—Cd Iron—Cadmium

Very little information is available relevant to the Fe-Cd system. Cadmium was once regarded as being virtually insoluble in solid iron [1]. However, Parfenova and Izgaryshev [2] found some evidence of mutual solubility in the course of diffusion studies.

According to Chasanov et al. [3] stable compounds do not exist, thus confirming Scheil [4].

Solubility data for Fe in *liquid* Cd have been reported by Tammann and Oelsen [5] and by [3] who carried out magnetic and radiochemical measurements. Values like 2 to 3×10^{-4} wt.% Fe in the temperature range 400–700°C [5] and 2.4×10^{-4} to 4.4×10^{-3} at.% Fe at 647°C [3] have been quoted.

References

1 Wever, F.: Naturwiss. 17 (1929) 304
2 Parfenova, M.I.; Izgaryshev, N.A.: J. Appl. Chem. USSR 25 (1952) 831
3 Chasanov, M.G.; Hunt, P.D.; Johnson, I.; Feder, H.M.: Trans. Metall. Soc. AIME 224 (1962) 935
4 Scheil, E.: Z. Metallkd. 38 (1947) 320
5 Tammann, G.; Oelsen, W.: Z. Anorg. Allg. Chem. 186 (1930) 277

Iron-Cerium, see Fe-R: Iron-Rare Earth Metals, Figs. 56, 57

Fe–Ce

Iron–Cobalt

Fe–Co

Figs. 13, 14

The system is characterized by an extensive solid solution range between fcc γ(Fe) and fcc γ(Co) and a wide α-Fe solid solution region which transforms via a second order reaction into the ordered CsCl type phase FeCo. Two additional ordered phases near 28 and 74 at.% Co, found by electrical resistivity measurements only (Viting [1]), could not be detected by any other method. (He may have measured the second order A2/B2 transition which appears possible considering the phase diagram in Fig. 13). Since the publication of the Fe-Co system by Hansen [A], the more recent investigations concentrated mainly on the clarification of second order reaction (A2/B2 and magnetic transformations) as well as the low temperatures Co-rich region.

The phase diagram in Fig. 13 confirms in its essential features the one published by [A]. However, more recent information has been used, mainly data pertaining to the order/disorder reactions.

Liquidus/solidus curves

Both curves fall, via a peritectic reaction at 1,494°C, smoothly to a shallow minimum at 1,476°C and about 65 at.% Co. They then rise to the melting point of Co (1,495°C). (The melting point of Co, 1,495°C is a secondary reference point on the International Practical Temperature Scale (IPTS-68).) The values are based on the calorimetric measurements by Predel and Mohs [2] which confirm excellently the course of the liquidus and solidus curve selected by [A].

The transformation point of cobalt has not been fixed so far. The allotropic transformation γ/ε occurs over a wide range of temperature (around 400°C), obviously owing to the small differences in Gibbs energy of the two modifications and the relatively low temperature. Thus, the scatter of the phase boundary data shown in Fig. 14 does not contradict the phase rule but simply reflects various experimental results of which those obtained by Masumoto and Watanabe [3] based on X-ray diffraction and thermal expansion measurements supplemented by microscopy, have been adopted in Figs. 13 and 14. Masumoto and Watanabe [3] confirmed earlier results by Masumoto [4] which in turn are in fair agreement with observations by Hashimoto [5]. A thermodynamic analysis of the fcc/hcp transformation in ternary Fe-alloys leading to the relative stability of ε(Fe) was undertaken by Ishida and Nishizawa [6]. By using this it is possible to evaluate

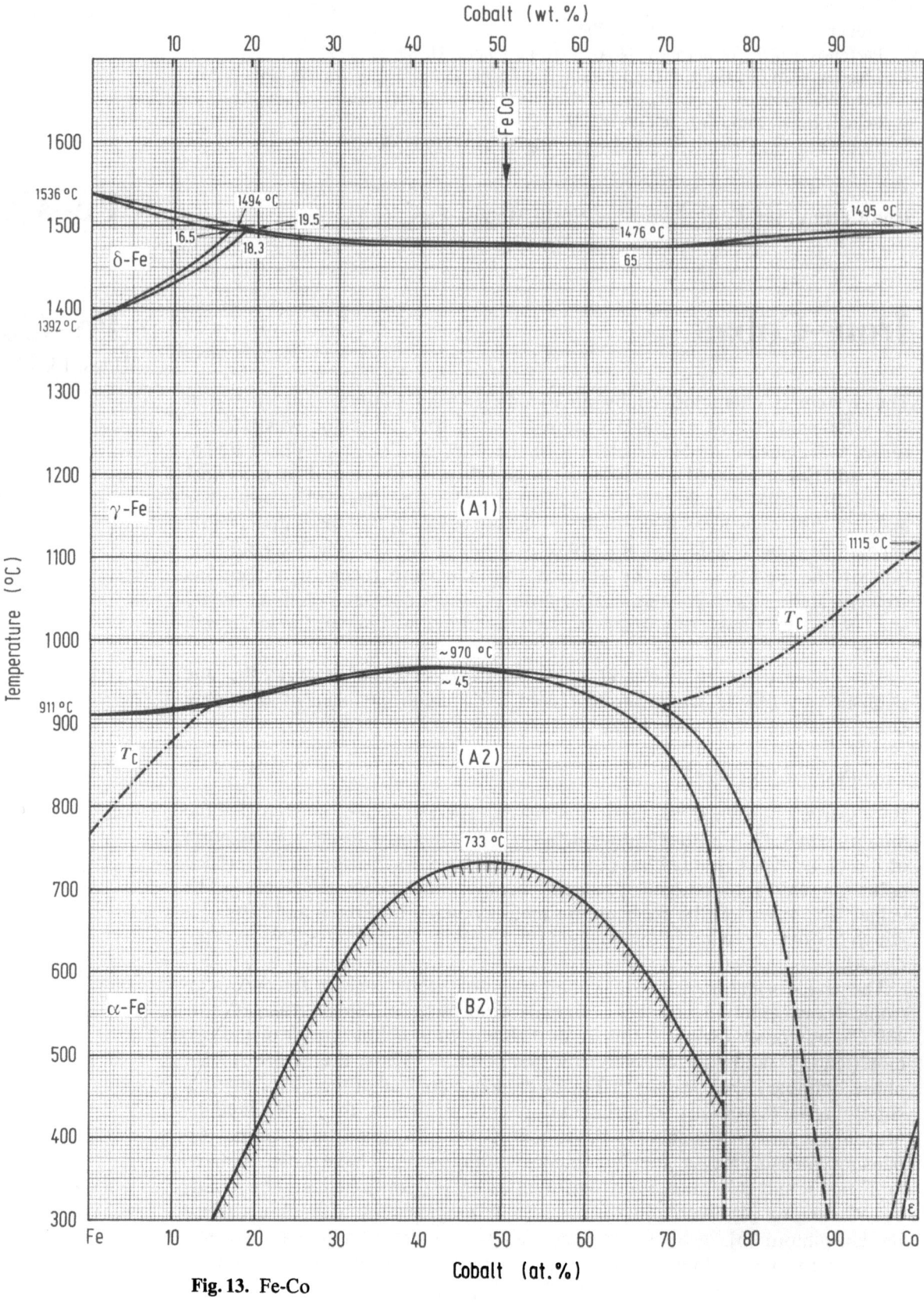

Fig. 13. Fe-Co

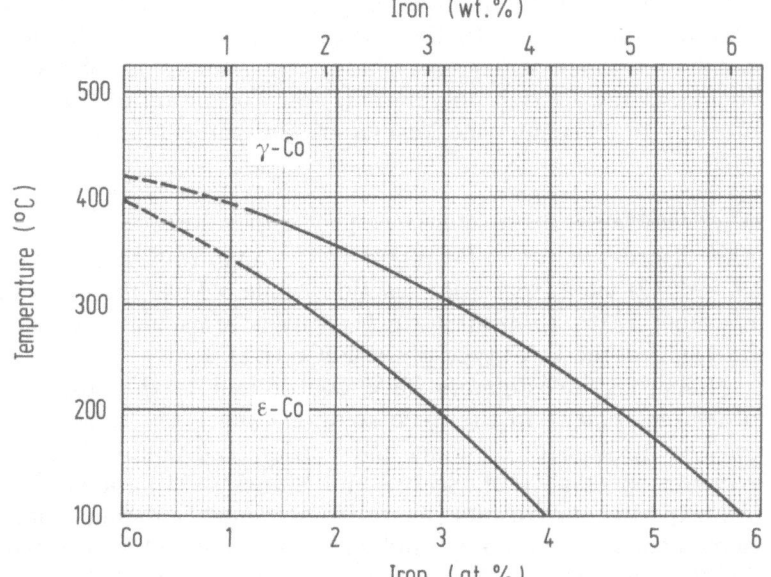

Fig. 14. Fe-Co

the effect of iron on the fcc/hcp transformation of pure cobalt. The calculated T_0 line in comparison with the data observed by [5] shows satisfactory agreement.

α/γ/δ transformations (ferrite/austenite)

Temperatures for the (austenite) γ(Fe)/δ(Fe) transformation have been reported among others by Harris and Hume-Rothery [7] and Fischer et al. [8] who used thermal analysis and susceptibility measurements respectively. Data quoted by [8], which are somewhat lower than those reported in the literature, have been combined with previous results in Fig. 13.

The addition of cobalt to iron hardly raises the α/γ transformation temperature between 0 and 8 at.% Co [8]. The temperature then rises very slowly to a maximum at about 970°C and ~ 45 at.% Co. Further additions of Co decrease the transformation temperature rapidly whilst increasing the (α + γ) two phase field.

This transformation has been investigated intensively in the 0–70 at.% Co concentration range but reliable information for the higher Co concentration region is still required. Here again, as with the γ/ε(Co) transformation, relatively low differences in Gibbs energy coupled with low diffusion rates owing to the relatively low temperature make equilibration a slow process resulting in substantial discrepancies between the various observations.

The phase boundaries indicated by [A] were taken from thermal and X-ray diffraction measurements (0–70) and (70–90 at.% Co) Ellis and Greiner [9]. Re-investigations by [8] (1.8–17.8 at.% Co) and Normanton et al. [10] (1.8–70 at.% Co) using susceptibility and calorimetric studies above 850°C respectively, confirmed the already established course of the α/γ(Fe) transformation temperature in the 0–70 at.% Co range. The maximum in Fig. 13 was placed at about 970°C, a compromise between various results reported in the literature but lower than the one in [A]. Data for the 70–90 at.% Co region below 850°C have been supplied by [9] (X-ray diffraction), Masumoto [4] (dilatometric) and Masumoto and Watanabe [3] (X-ray diffraction, thermal expansion and microscopy). Observations by [4] and [3] indicate an extremely narrow $(\alpha + \gamma)$ range (77–78.6 at.% Co at room temperature) in contrast to the lattice parameter data obtained by [9] which point to a much wider two phase field namely 75–88.5 at.% Co at 600°C. In Fig. 13 the Fe-rich polymorphic transformation of the $(\alpha + \gamma)$ range is an attempt to correlate the results of the three teams mentioned above [3, 4, 9].

Order/disorder

The disordered fcc γ-Fe phase (A1) transforms to a disordered bcc α-Fe phase (A2) via a first order reaction, while α-Fe at the maximum temperature of about 733°C ($\sim$48 at.% Co [11]) transforms via a second order reaction into an ordered CsCl type phase FeCo (B2).

Numerous investigators have contributed to the study of the order/disorder phenomena in the α-Fe solution range. The more recent publications may be mentioned: Castanet and Ferrier [12], Orehotsky and Schröder [13], Normanton et al. [10], Seehra and Silinsky [14], Inden [15], Oyedele [11]. All experimental results, apart from minor deviations, are in satisfactory agreement with each other and with the calculated critical temperature curve $T^{A2/B2}$ established by Inden [15] who applied the BWG model and also calculated interchange energies. His results are shown in Fig. 13 as hatched lines, indicating the second order reaction, where no abrupt physical changes take place and the transition is continuous.

Fe-Co anomaly

It is known that ordered iron-cobalt alloys exhibit anomalous behaviour near 550°C. Kuroki et al. [16] investigated the kinetic behaviour of the "550°C" change in ordered FeCo alloys of equiatomic composition by electrical resistivity and specific heat measurement on annealed samples prepared from 99.9% pure Fe and 99.9% pure Co respectively. The 550°C change appears in both types of measurements. The heating and cooling rate dependence of the electrical resistivity studies is in good agreement with calculated curves of [17] and [18], thus confirming the kinetic interpretation for the 550°C anomaly discussed by [17] and [18]. For a calculated phase diagram, the reader may be referred to a publication by Kaufman and Nesor [19].

Curie temperatures

The Curie temperature of Fe-Co alloys were extensively investigated by Masumoto [4] and to a more limited extent by Ellis and Greiner [9] Stuart and Ridley [20], Normanton et al. [10] and Inden and Meyer [21]. The data reported agree resonably well. However, values obtained by [10] (calorimetric study) at the magnetic coincidental point $(\alpha + \gamma)$Fe, are higher; about 22 at.% in contrast to 15–16 at.% quoted by other investigators. The results by Inden [21] who determined the Curie temperature of bcc Fe-Co alloys by using magnetization and Mössbauer measurements have been used in Fig. 13. [21] also estimated the Curie temperature of the hypothetical bcc Co from an extrapolation, a value of 1,151°C was suggested.

The high Curie point of Co (1,115°C) is an outstanding feature in the properties of metals. Measurements of the Curie temperature in fcc γ-Fe,Co alloys have been undertaken by [4] and [10]. Results in the 80–100 at.% Co range agree well, but a discrepancy is noted at lower Co concentrations. In view of rather scattered data found for this range by [4], data by [10] have been accepted in Fig. 13.

1   Viting, L.M.: Zh. Neorg. Khim. 2 (1957) 845
2   Predel, B.; Mohs, R.: Arch. Eisenhüttenwes. 41 (1970) 143
3   Masumoto, H.; Watanabe, K.: J. Jpn. Inst. Met. 42 (1978) 256
4   Masumoto, H.: Sci. Rep. Tohoku Univ. 15 (1926) 469, Trans. ASST 10 (1926)
    491
5   Hashimoto, U.: Nippon Kinzoku Gakkaishi 1 (1937) 177
6   Ishida, K.; Nishizawa, T.: Trans. Jpn. Inst. Met. 15 (1974) 225
7   Harris, G.B.; Hume-Rothery, W.: J. Iron Steel Inst. 174 (1953) 212
8   Fischer, W.A.; Lorenz, K.; Fabritius, H.; Schlegel, D.: Arch. Eisenhüttenwes.
    41 (1970) 489
9   Ellis, W.C.; Greiner, E.S.: Trans. ASM 29 (1941) 415
10 Normanton, A.S.; Bloomfield, P.E.; Sale, F.R.; Argent, B.B.: Met. Sci. J. 9 (1975)
    510
11 Oyedele, J.A.: McMaster Univ. Diss. Abstr. Int., Mar. (1979) 39 (9)
12 Castanet, R.; Ferrier, A.: Compt. Rendus C 277 (1971) 15
13 Orehotsky, J.; Schröder, K.: J. Phys. F: Met. Phys. 4 (1974) 196
14 Seehra, M.S.; Silinsky, P.: Phys. Rev. B 13 (1976) 5183
15 Inden, G.: Z. Metallkd. 68 (1977) 529
16 Kuroki, H.; Matsuda, H.; Eguchi, T.: Trans. Jpn. Inst. Met. 19 (1978) 211
17 Matsuda, H.; Kuroki, H.; Eguchi, T.: J. Jpn. Inst. Met. 35 (1971) 774
18 Matsuda, H.; Kuroki, H.; Eguchi, T.: Trans. J. Inst. Met. 12 (1971) 390
19 Kaufman, L.; Nesor, H.: Z. Metallkd. 64 (1973) 249
20 Stuart, H.; Ridley, N.: Brit. J. Appl. Phys. 2 (1969) Ser. 2, 485
21 Inden, G.; Meyer, W.O.: Z. Metallkd. 66 (1975) 725

Iron–Chromium

Fe–Cr

Figs. 15, 16

The sigma phase is the prominent feature of the important system iron-chromium. Early investigators (e.g. Adcock [1] and many others) failed to detect this phase, which forms only slowly, until Wever and Jellinghaus [2], after lengthy annealing found a new structure denoted σ. X-ray and microscopic investigations by Cook and Jones [3] on samples annealed up to 60 days at 600°C confirmed σ, which is formed below a temperature of 832°C: Hansen [A], Elliott [B] and Spencer [4]. The crystal structure was described by Bergman and Shoemaker [5].

Because of the slow equilibration rate and the small driving force (i.e. the small difference in Gibbs energy σ/α), this system is a typical example where the phase boundaries can only be assessed by thermochemical calculation. This was performed first by Kubaschewski and Chart [6] and eventually by Müller and Kubaschewski [7] from experimental heats and Gibbs energies of formation and transformation using the upper phase boundary of Elliott [B]. The then surprising result was the observation that the σ phase closes at the lower temperature of ca. 440°C (Fig. 15) below which the solid solutions in iron and chromium are in equilibrium, thus forming a miscibility gap which metastably closes at about 575°C (shown as a dotted curve in Fig. 15).

These equilibria have recently been re-assessed by Bernard [8] whose

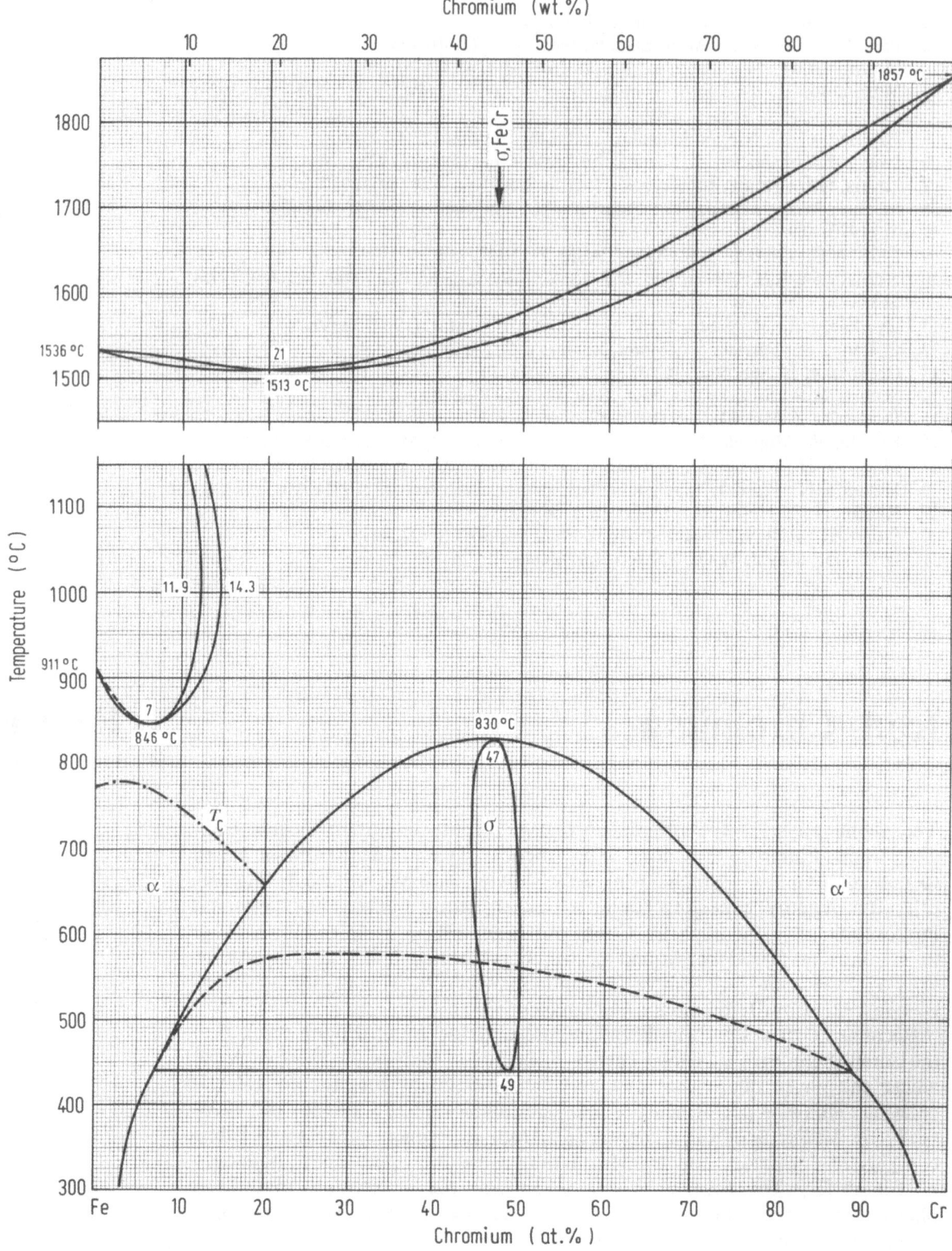

Fig. 15. Fe-Cr

results hardly differ from those of [7]. Williams and Paxton [9, 10] and other investigators [11, 12] have presented experimental evidence for the existence of the metastable miscibility gap, primarily from the changes in physical properties, the conventional equilibrium techniques largely failing because of the slow rates. In a more recent study, Williams [13] summarized and discussed the earlier work and proposed a calculated miscibility gap which agrees with the experimental results of [9, 10] and [11]. Vintaykin and Barkalaya [12] confirmed the results of [13] in a paper on the miscibility gap in the ternary Fe-Co-Cr system. They concluded that the boundary of the binary gap approaches 85 at.% Cr at 450°C corresponding to Williams' 86 at.%, that is slightly lower than the boundary calculated by [7] and [8].

This miscibility gap is responsible for the so-called 475°C embrittlement of ferritic chromium steels. An assumption confirmed by hardness tests, magnetic measurements and electron microscopy. More recently, Ettwig and Pepperhoff [22], using the Mössbauer spectren were able to prove the mechanism of the reaction "stable-metastable", namely the decomposition of the solid solution phase into a Fe-rich and a Cr-rich, paramagnetic phase.

The Curie temperatures versus concentration curves were measured by several investigators. A review is given in Hansen [A]. With increasing Cr-content the Curie points appear to reach a maximum before being lowered.

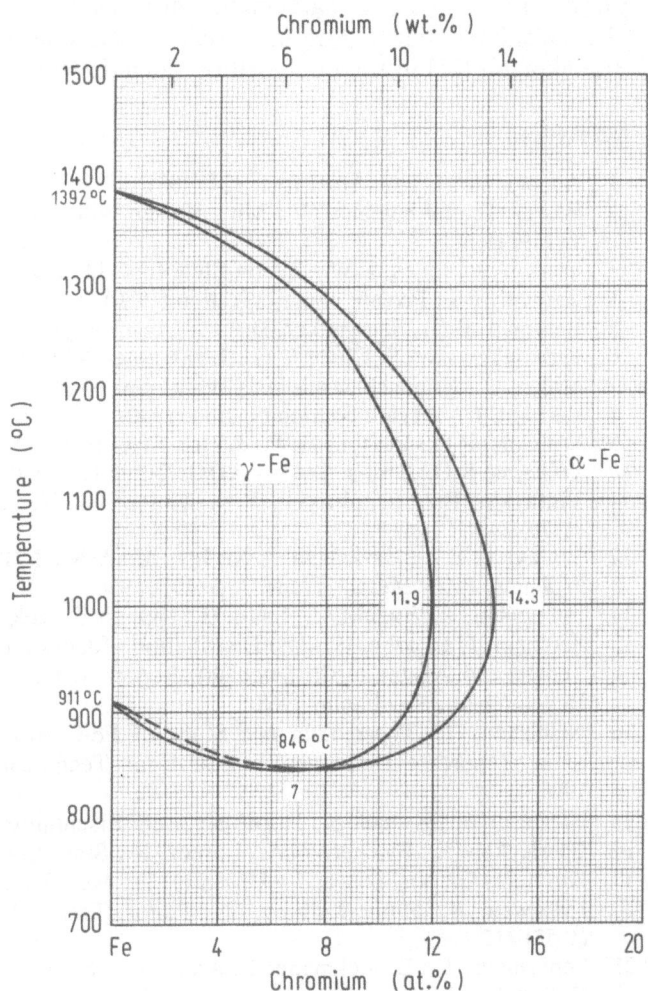

Fig. 16. Fe-C

Because of the sluggish formation of the phase, equilibrium has not been reached below 650°C and data for the magnetic transformation has been accepted down to this temperature only.

The concentration dependence of the Néel temperature was measured by Fukamichi et al. [14] up to 5 at.% Fe. The temperature decreases from 310–245 K with increasing concentration.

The γ-loop was thoroughly investigated by Kirchner and Larbo [15] at 850°C and Nishizawa [16] at 900–1,100°C who both used measurements by Bungardt et al. [17] and Baerlecken et al. [18] confirming earlier results [1]. More recently, Kirchner et al. [19] assessed all relevant experimental information available including thermodynamic data using the regular solution model. The calculated two-phase field ($\alpha + \gamma$) is in good agreement with most of the experimental evidence. The minimum temperature was found to be 846°C at 7.0 at.% Cr. The maximum solubility of austenite is 11.9 at.% Cr and no austenite is formed at normal pressure in alloys with more than 14.3 at.% Some uncertainty remains regarding the width of the ($\alpha + \gamma$) field between zero Fe and the concentration of the minimum [19] (Fig. 16).

The liquidus and solidus temperatures were determined up to 40 at.% Cr by Hellawell and Hume-Rothery [20] who found the minimum at 21 at.% Cr and 1,513°C. With this information and the thermochemical properties of the α solutions, [7] calculated the remainder of the liquidus and solidus curves, the former in agreement with [1], the latter substantially higher. The calculated curve has essentially been confirmed by Schürmann and Brauckmann [21] in the composition range investigated (0–35 at.% Cr).

References

1 Adcock, A.: J. Iron Steel Inst. 124 (1931) 99
2 Wever, F.; Jellinghaus, W.: Mitt. Kaiser Wilhelm Inst. Eisenforsch. Düsseldorf 13 (1931) 143
3 Cook, A.J.; Jones, F.W.: J. Iron Steel Inst. 148 (1943) 217
4 Spencer, P.J.: Private communication (1980)
5 Bergman, B.G.; Shoemaker, D.P.: J. Chem. Phys. 19 (1951) 515
6 Kubaschewski, O.; Chart, T.G.: J. Inst. Met. 93 (1965) 329
7 Müller, F.; Kubaschewski, O.: High Temp. High Press. 1 (1969) 543
8 Bernard, C.: Private communication (1980)
9 Williams, R.O.; Paxton, H.W.: J. Iron Steel Inst. 185 (1957) 358
10 Williams, R.O.: Trans. TMS – AIME 212 (1958) 497
11 Vintaykin, E.Z.; Dimitriyev, V.N.; Kolontzov, V.Yu.: Phys. Met. Metallogr. 29 (1970) 141
12 Vintaykin, E.Z.; Barkalaya, A.A.: Izv. Akad. Nauk SSSR Met. (1977) 6, 192
13 Williams, R.O.: Metall. Trans. A 5 (1974) 966
14 Fukamichi, K.; Suzuki, Y.; Saito, H.: Nippon Kinzoku Gakkaishi 37 (1973) 927
15 Kirchner, G.; Larbo, G.: The Distribution of Chromium and Manganese Between Ferrite and Austenite (in Swedish). Report to the Swedish Board of Technical Development, Stockholm 1968
16 Nishizawa, T.: Thermodynamic Study of Fe-C-Mn, Fe-C-Cr and Fe-C-Mo Systems. Report to the Swedish Board for Technical Development, Stockholm, 1966
17 Bungardt, K.; Kunze, E.; Horn, E.: Arch. Eisenhüttenwes. 29 (1958) 193
18 Baerlecken, E.; Fischer, W.A.; Lorenz, K.: Stahl Eisen 81 (1961) 768
19 Kirchner, G.; Nishizawa, T.; Uhrenius, B.: Met. Trans. 4 (1973) 167
20 Hellawell, A.; Hume-Rothery, W.: Philos. Trans. R. Soc. London, Ser. A: 249 (1957) 417
21 Schürmann, E.; Brauckmann: J., Arch. Eisenhüttenwes. 48 (1977) 3
22 Ettwig, H.; Pepperhoff, W.: Arch. Eisenhüttenwes. 41 (1970) 471

Iron–Copper

A thermodynamic analysis of phase relationships in the Fe-Cu system has been carried out by Kubaschewski et al. [1]. Available thermodynamic data have been perused to develop a set of analytical expressions which describe the Gibbs energies of phase formation for the liquid, the fcc and bcc phases in the Fe-Cu system as functions of composition and temperature. These functions predict phase equilibria in close agreement with the experimental

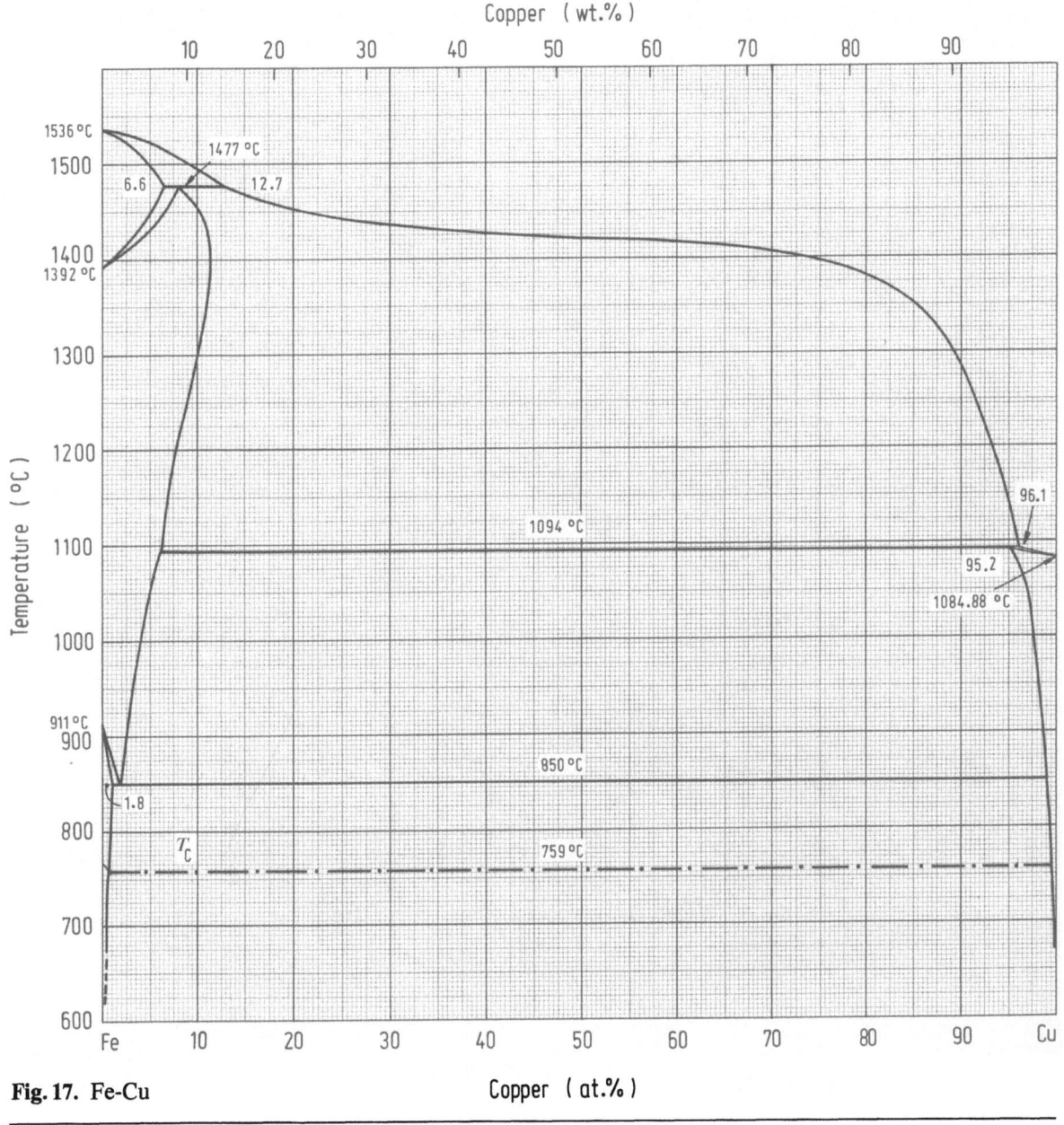

Fig. 17. Fe-Cu

temperature-composition diagram and are represented in Fig. 17. Detailed information is shown in Figs. 18 and 19. During the determination of diffusion coefficients in the Fe-Cu system with an electron probe microanalyser [2] the solubility limits were also analysed and published after the thermodynamical evaluation. The solubility of Cu in high purity Fe measured by [2] in the temperature range 650–1,050°C is recorded in Fig. 20. The results agree well with those used in the publication by [1].

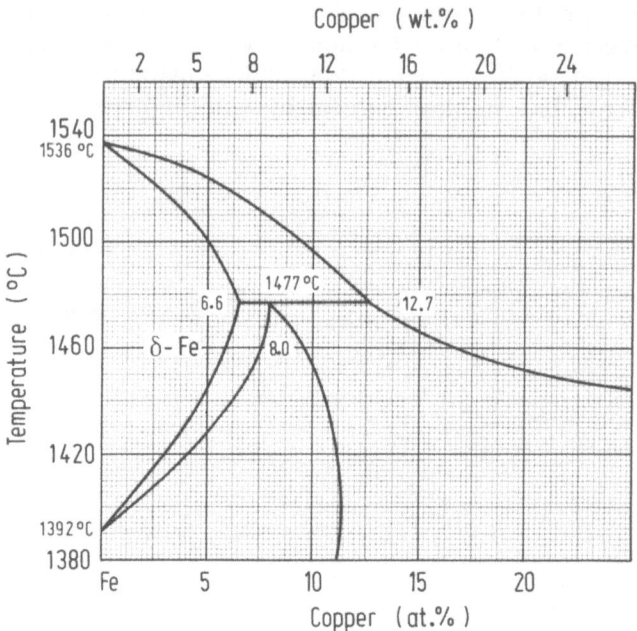

Fig. 18. Fe-Cu

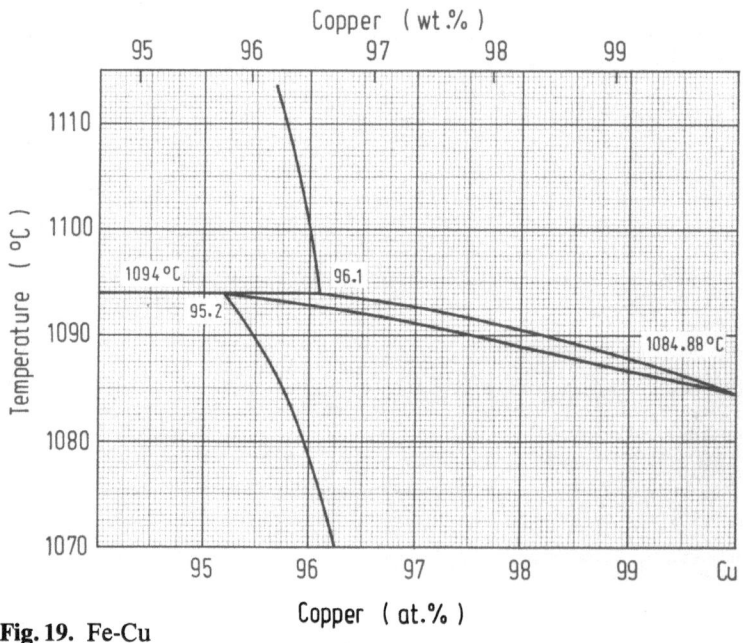

Fig. 19. Fe-Cu

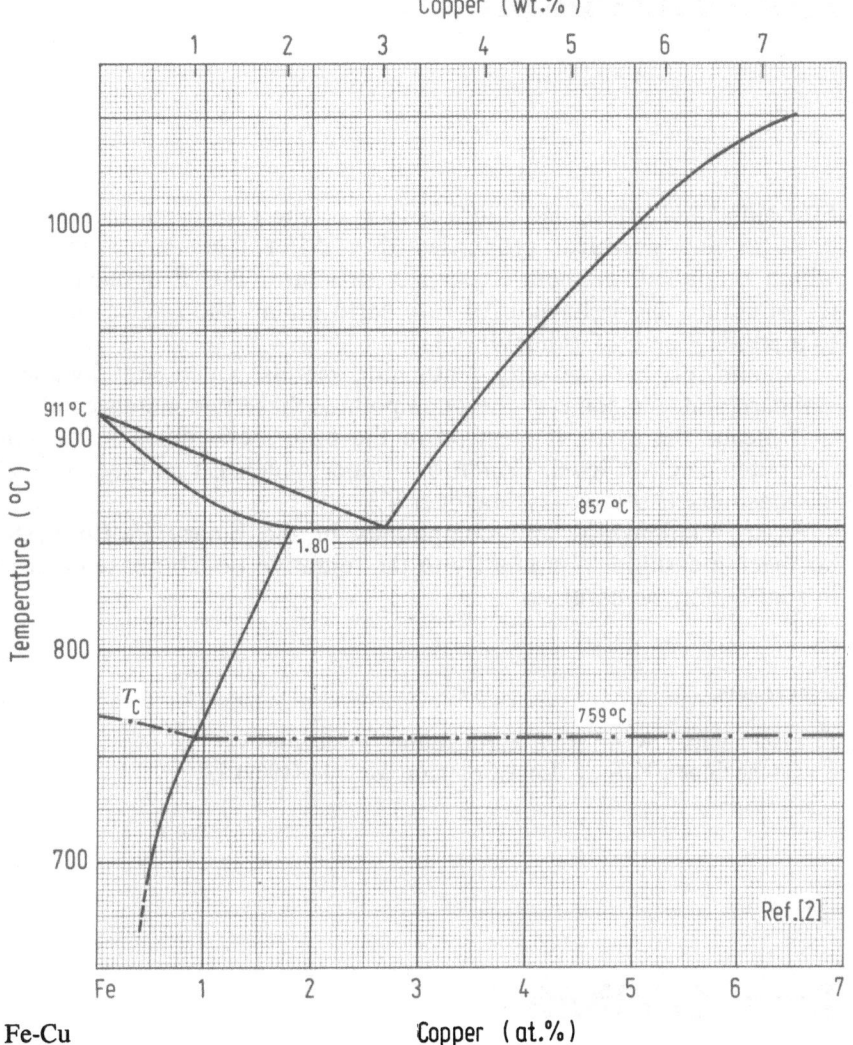

Fig. 20. Fe-Cu

References

1 Kubaschewski, O.; Smith, J.F.; Bailey, D.M.: Z. Metallkd., 68 (1977) 495. See also NBS Spec. Publ. 496, Vol. 2.
2 Salje, G.; Feller-Kniepmeier, M.: Z. Metallkd. 69 (1978) 167

Iron-Dysprosium, see Fe-R: Iron-Rare Earth Metals, Fig. 64

Fe–Dy

Iron-Erbium, see Fe-R: Iron-Rare Earth Metals, Fig. 66

Fe–Er

Fe–Eu   Iron–Europium

Europium and ytterbium (vide) belong to the rare earth metals but exhibit physical and chemical properties that differ from the other lanthanides which originates in their electronic structure. For a detailed summary of the structure and behaviour of rare earth metals the reader may be referred to a publication by Gschneidner [1].

Eu and Yb are the only lanthanide metals which are divalent in their standard state. In some of their compounds these two metals can be trivalent, that means the extra 4f electron in the pure metal must be "promoted" to the outer electron level. Conditions under which Eu may acquire trivalency has been discussed by Gschneidner [2].

Phase relationships and compounds in the binary Fe-Eu system have not been measured. A similarity to the tentative Fe-Yb phase diagram shown in Fig. 99 may be assumed.

References

1 Gschneidner, K.A.: Rare Earth Alloys. Princeton: van Nostrand 1961
2 Gschneidner, K.A.: J. Less Common Met. 17 (1969) 13

Fe–Ga   Iron–Gallium

Figs. 21–24

The Fe-Ga phase diagram shown in Fig. 21 is based primarily on a publication by Köster and Gödecke [1, 2] who investigated the complete system but concentrated on the 10–50 at.% Ga range and incorporated information reported by Wachtel and Maier [3], Dasarathy and Hume-Rothery [4], Meissner and Schubert [5] and others.

The Fe-Ga system is of the γ-loop type and entails four stable compounds, namely: Fe_3Ga, Fe_6Ga_5 (in two modifications) Fe_3Ga_4 and $FeGa_3$. Like Fe-Si and Fe-Al it is characterized by a large body-centred cubic solution phase which is stable in both the disordered (A2) and ordered (B2 and D0$_3$) forms.

Conclusive results for the range 21–26 at.% Ga and the temperature interval of 580–680 °C have not been obtained owing to the formation of metastable phases and the sluggishness of diffusion. An attempt to clarify the phase equilibria in this region has been undertaken by [1, 2], Bras et al. [6] and Tiemann and Schmand [7], see Figs. 23, 24.

The liquidus and solidus curves have been adopted from magnetothermal measurements on very pure alloys (Fe = 99.99 % with 0.0036 % O_2 and 0.005 %

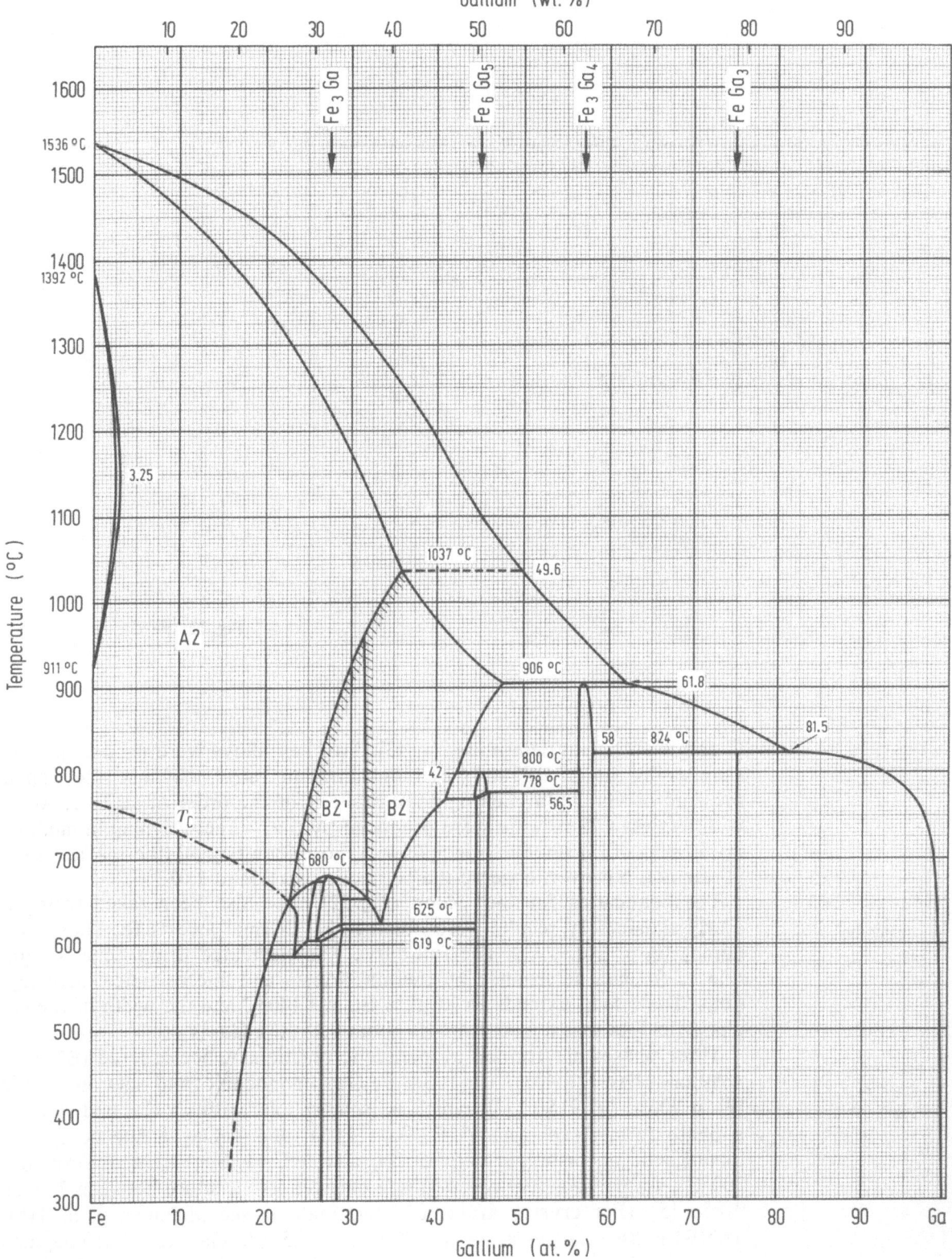

Fig. 21. Fe-Ga

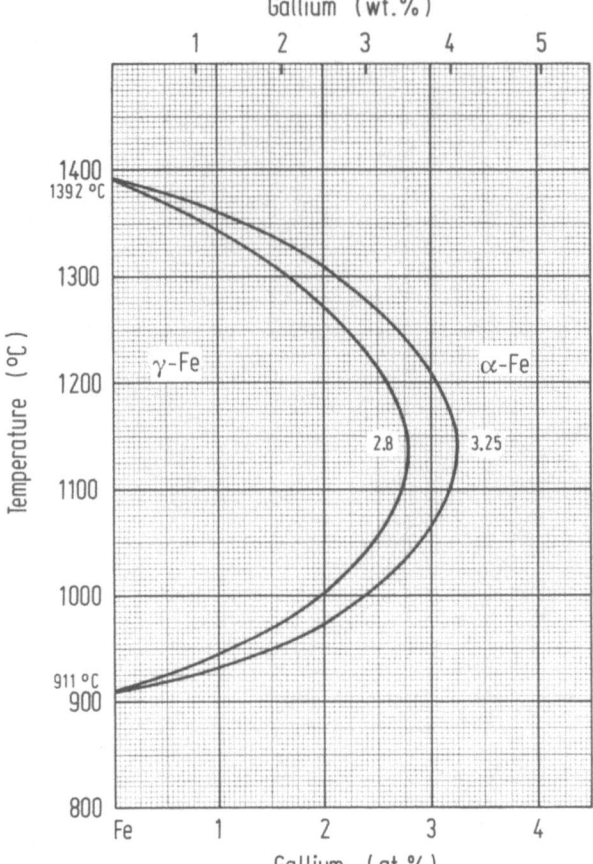

Gallium (wt.%)

Fig. 22. Fe-Ga

C and 99.99% Ga) determined by [3]. Two key alloys were used to prepare samples in the 25–100 at.% Fe concentration range, because the vapour pressure of Ga at the melting point of Fe is of the order of 10 Torr. Alloys with less than 25 at.% Fe were produced by the "directweighing-in method" [8]. Results of [3] agree in principle with those found by [4] (thermal analysis). [4] did not, however, observe the peritectic reaction at 1,037°C.

The boundaries for the γ-loop (Fig. 22) have also been taken from [4]. Fairly satisfactory results were obtained by thermal arrests for alloys containing up to 2 at.% Ga. The full curve was calculated using Zener's equation to fit data determined by thermal arrests and microscopical investigations. The γ-loop extends to 2.8 at.% Ga at 1,140°C and the width of the two phase field—at that temperature—is about 0.45 at.% Ga. [4].

Three phases are formed by peritectic reaction. A bcc ordered (B2) solid solution phase at 1,037°C [1–3], Fe_3Ga_4 (monoclinic and meta-magnetic) [9] at 906°C [4] 915°C [3], which is isomorphous with Cr_3Ga_4 and replaces a phase formerly disignated Fe_8Ga_{11} [5, 3]; and $FeGa_3$, a line compound, diamagnetic and isotypic with $CoGa_3$ at 824°C [4, 3]. Fe_6Ga_5 is formed at 800°C by a peritectoidal reaction and changes its structure between 770 and 778°C [5]. (For crystal structure determinations see Malaman et al. [15].) The B2 phase changes at about 680°C and 27.5 at.% Ga into a ferromagnetic compound Fe_3Ga which has two polymorphic ordered structures, DO_{19} (hexagonal) and $L1_2$ (fcc). The transition occurs between 605 and 619°C [2]. The transition from $L1_2$ to DO_{19} and from B2 are first order reactions (discontinuous) and are accompanied by a decrease in volume and increase in heat content [1, 2].

The homogeneity ranges of the Ga-rich phases were determined by [3] and [4] using thermal and magneto-thermal analyses supported by microscopical examinations. Findings of [3] confirmed for Fe$_3$Ga$_4$ [2] have been adopted.

The bcc Fe-Ga alloys exist in both, the ordered and disordered forms, over a wide area. [1, 2] employed thermal analysis, dilatometric, X-ray and microscopical investigations, [6] X-ray diffractions and thermal analysis as well as theoretical considerations based on the Bragg/Williams model and [7] Mössbauer effect data, to study phase stabilities and the nature of order/disorder transformations.

All teams agree on the formation of Fe$_3$Ga at lower temperatures. 680°C [1], 720°C [6], 630°C [7]. The results are in fairly good agreement. However, the interpretation is different. A "change of order" of the B2 phase, probably a difference in their defect structure [1] which takes place between 31 and 33 at.% Ga and 680–950°C was observed by [1]. A division into a B2 and a B2′ field was suggested. However, X-ray investigations carried out by Pitsch and Kudielka [10] on alloys with 25.3 and 33 at.% Ga in the temperature interval 550–850°C showed a B2 type structure only, which agrees with find-

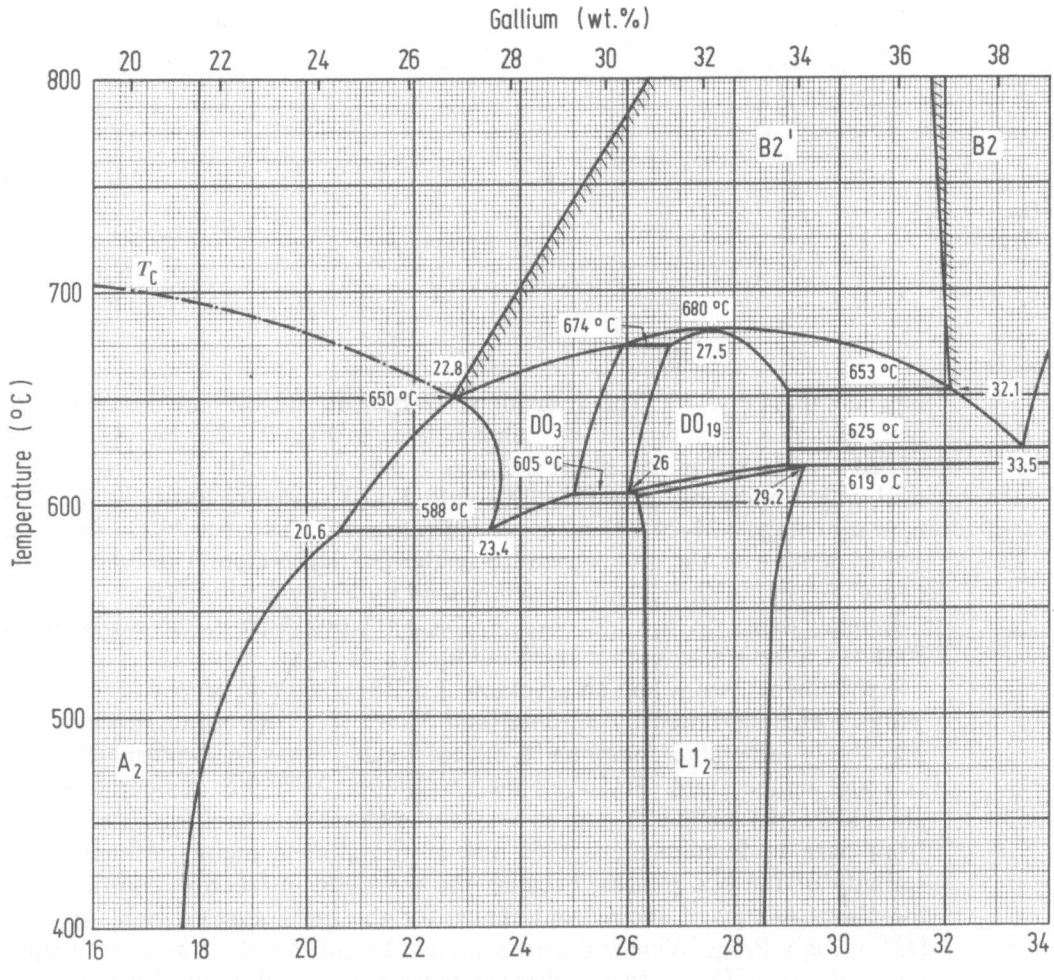

Fig. 23. Fe-Ga. Part phase diagram, Ref. [2]

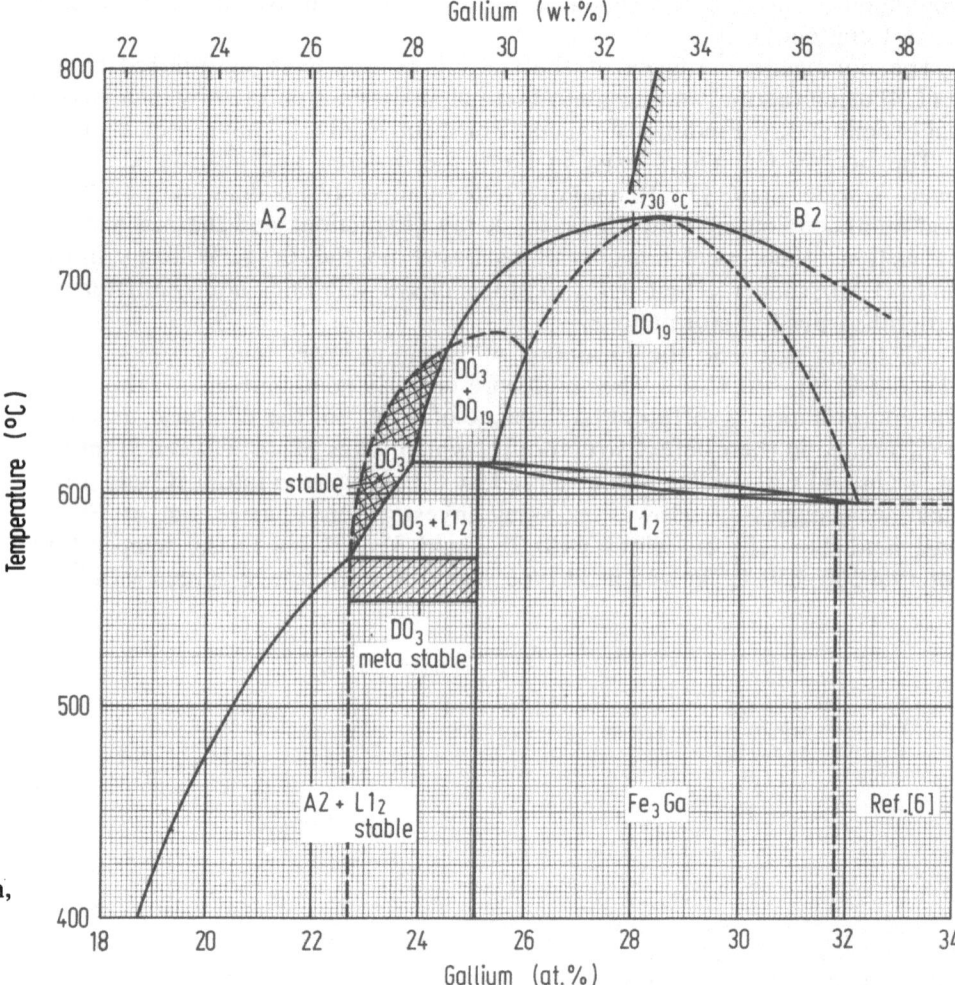

Fig. 24. Fe-Ga. Part phase diagram, Ref. [6]

ings obtained by [6]. The transition A2/B2 and B2/B2' are second order reactions and are indicated by hatched lines in Figs. 21, 23, 24.. Results obtained by [1, 2] for the phase boundary between A2/B2 agree reasonably well with finding by Gust [11]. D0₃ with Fe₃Al structure exists in the temperature range 674—588°C at a concentration of 22.8—26 at.% Ga. It is unstable relative to the D0₁₉ structure in the Ga-rich region. The limits of the two phase field D0₃ + L1₂ were determined by X-ray studies [1].

Investigations on the meta-stable phase relationships in the region up to 50 at.% Ga have also been carried out by [1].

Experimental work and an assessment of the magnetic properties of Fe-Ga alloys was undertaken by Wachtel and Maier [12]. According to their findings the intermetallic compounds, except $FeGa_3$, show ferromagnetic effects at room temperature. Fe_3Ga_4 is metamagnetic. At higher temperatures the phases are paramagnetic and have positive Curie temperatures. The change from para to ferromagnetism of one of the alloyed metals influences the stability of the bcc phases, because the magnetic interaction contributes to the Gibbs energy of the system. Thus, for the Fe-Si system, Schlatte et al. [13], using a Bragg/Williams/Gorsky model, found that the transition temperature of the A2/B2 phases decreases more rapidly below the Curie temperature.

References

1 Köster, W.; Gödecke, T.: Z. Metallkd. 68 (1977) 582
2 Köster, W.; Gödecke, T.: Z. Metallkd. 68 (1977) 661
3 Wachtel, E.; Maier, J.: Z. Metallkd. 58 (1967) 761
4 Dasarathy, W.; Hume-Rothery, W.: Proc. R. Soc. London A 268 (1965) 141
5 Meissner, H.-G.; Schubert, K.: Z. Metallkd. 56 (1965) 523
6 Bras, J.; Couderc, J.J.; Fagot, M.; Ferre, J.: Acta Metall. 25 (1977) 1077
7 Tiemann, K.; Schmand, J.: Z. Naturforsch., 33a (1978) 644
8 Damm, R.; Scheil, E.; Wachtel, E.: Z. Metallkd. 53 (1962) 196
9 Philippe, M.J.; Malaman, B.; Roques, B.: Acta Crystallogr. B31 (1975) 477
10 Pitsch, W.; Kudielka, H.: See Ref. [1]
11 Gust, W.: Private communication [14]
12 Wachtel, E.; Maier, J.: Z. Metallkd. 58 (1967) 885
13 Schlatte, G.; Inden, G.; Pitsch, W.: Z. Metallkd. 65 (1974) 94
14 Predel, B.; Vogelbein, W.: Thermochim. Acta 13 (1975) 133
15 Malaman, B.; Philippe, M.J.: Roques, B.: Acta Crystallogr. B30 (1974) 2081

Iron-Gadolinium, see Fe-R: Iron-Rare Earth Metals, Fig. 62   **Fe–Gd**

Iron–Germanium   **Fe–Ge**

Fig. 25

Recent studies of the iron germanides have shown the complexity of the Fe-Ge system. The phase diagram in Fig. 25 is a combination of evidence presented by Übelacker [1], Predel and Frebel [2], Chessin et al. [3], Lecocq [4], Richardson [5] and Kanematsu and Ohoyama [6] for the Fe-rich region, whilst the Ge-rich range has been adopted from Maier and Wachtel [7].

The phase diagram is characterized by a γ-loop and six intermediary phases, namely, Fe_3Ge, β, η, $\chi(Fe_6Ge_5)$, $FeGe$ and $FeGe_2$. A revision of the phase equilibria in the Fe-rich region by [2] led to the exact determination of the extent of the α phase which shows at a Ge concentration of over 10 at.% a superstructure of bcc ordered distribution of atoms. It is denoted α_1 and stems from a second order reaction (see Fe-Si) shown as hatched line in Fig. 25. The solubility of Ge in α_1 decreases with decreasing temperature (1,130–900 °C) but increases below 900 °C. This result confirms observations reported by [6].

The intermetallic compound Fe_3Ge, formed by continuous precipitation from α_1 [2], is dimorphic. The high-temperature modification ε (700–1,000 °C) crystallizes with the hexagonal $D0_{19}$ structure and the low-temperature modification, ε', (<700 °C) with the $L1_2$ (Cu_3Au) type structure. ε has a homogeneity range of 23.8–26.0 at.% Ge and ε' decomposes at 400 °C in α_1 and β [2]. The β phase melts congruently at 1,170 °C at a composition of 37.5 at.% Ge [8]. It has a homogeneity range, and the structure was found to be of the B8 (NiAs) type.

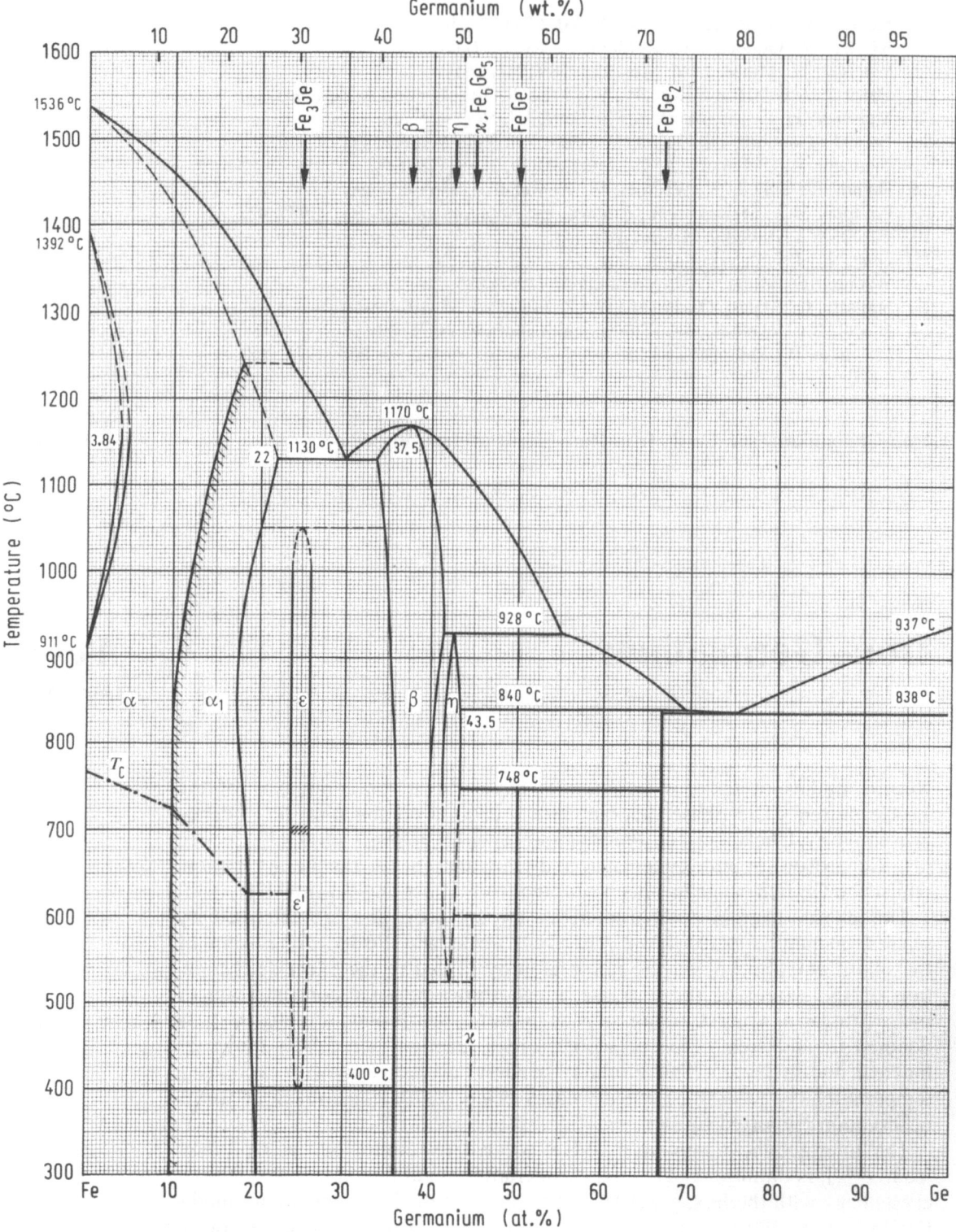

Fig. 25. Fe-Ge

η has been identified as a compound ($B8_2$ type) with a homogeneity range and a eutectoidal decomposition at about 520°C [7]. Fe_6Ge_5 (χ) is formed by peritectoidal reaction between η and FeGe. The structure has been determined by [15], and [5] reported a stability up to 740°C. Maier and Wachtel [7] using magnetothermal, X-ray and differential thermal analyses concluded that FeGe is formed as a line compound by peritectoidal reaction at 748°C. It has hexagonal B35 type structure. The different modifications of FeGe mentioned by [5] could not be confirmed by [6] and [7], but it shows a pronounced Néel point at 127°C.

$FeGe_2$, the compound richest in Ge, is formed at the stoichiometric composition by a peritectic reaction of η and liquid at 840°C according to [7] contrary to findings of [5, 6] and [10]. According to [7], $FeGe_2$ is a line compound whereas [11] and [12] reported a narrow homogeneity range.

Bugai et al. [13] measured the solid solubility of Fe in Ge using radioactive Fe. The results are as follows:

Temp. °C	925	900	850	800	750
at.% Fe	1.8×10^{-6}	2.5×10^{-6}	3.3×10^{-6} (max)	2.8×10^{-6}	1.6×10^{-6}

A comprehensive survey of the magnetic properties of the Fe-Ge alloys is given in the publication of Maier and Wachtel [7] and the magnetic structure of hexagonal FeGe at low temperature by Forsyth et al. [14].

References

1 Übelacker, E.: Thesis, Université de Paris 1966, Mém. Sci. Rev. Métall. 64 (1967) 183
2 Predel, B.; Frebel, M.: Z. Metallkd. 63 (1972) 393
3 Chessin, H.; Arajs, S.; Colvin, R.V.; Miller, D.S.: J. Phys. Chem. Solids 24 (1963) 261
4 Lecocq, P.: Ann. Chim. (Paris) 8 (1963) 85
5 Richardson, M.: Acta Chem. Scand. 21 (1967) 2305
6 Kanematsu, K.; Ohoyama, T.: J. Phys. Soc. Jpn. 20 (1965) 236
7 Maier, J.; Wachtel, E.: Z. Metallkd. 63 (1972) 411
8 Shtolts, A.K.; Geld, P.V.; Zagryazhskii, V.L.: J. Inorg. Chem. 9 (1964) 76
9 Laves, F., Wallbaum, H.J., Z. angew. Mineral. 4 (1941/42) 17
10 Samsonov, G.V., Bondarev, V.N., 'Germanides', Primary Sources Publishers, New York 1970
11 Ohoyama, T.; Kanematsu, K.; Yasukochi, K.: J. Phys. Soc. Japan 18 (1963) 589
12 Shtolts, A.K.; Geld, P.V.; Zh. Fiz. Khim. 36 (1962) 2400
13 Bugai, A.A.; Kosenko, V.E.; Miseliuk, E.G.: Sov. Physics — Tech. Phys. 2 (1957) 183
14 Forsyth, J.B.; Wilkinson, C.; Gardner, P.; J. Phys. F: Metal Phys. 8 (1978) 2195
15 Malaman, B; Philippe, M.J; Roques, B.: Acta Crystallogr. 30B (1974) 2081

The solubility of hydrogen in solid and liquid Fe has been reported by a number of investigators [1–14].

Measurements of Da Silva et al. [12, 13] have shown, that the temperature dependence of the solubility of hydrogen in annealed Fe at atmospheric pressure is essentially independent of purity and of the presence of grain boundaries and dislocations in the different polymorphic formes of Fe.

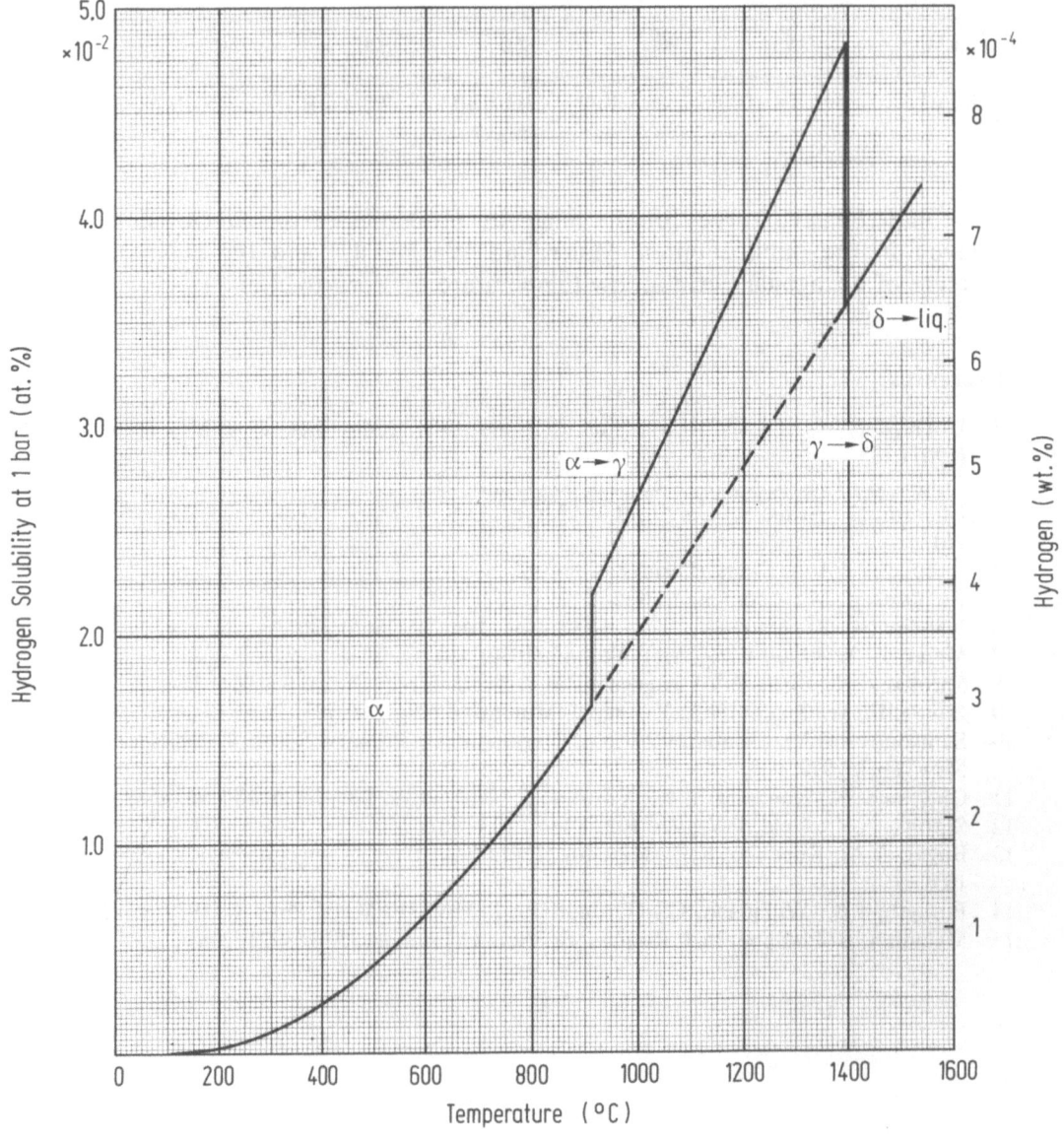

Fig. 26. Fe-H. Hydrogen solubility at 1 bar in α(Fe), γ(Fe) and δ(Fe)

Solution in α(Fe). Sieverts law is obeyed at concentrations up to the solubility limit, i.e. solubility of hydrogen = $k\sqrt{p_{H_2}}$ where k is a constant, at 573–1,184 K. The solubility of H found by investigators [1–13] for temperatures between 500 K and the α/γ(Fe) transition temperature of 1,184 K shows good agreement between all the different studies. The solubility of hydrogen in α(Fe) can be expressed by the following formula:

$$\lg(\text{at.\% H}) = -1,376\ T^{-1} - 0.665.$$

Solutions in γ(Fe). As in case of solutions in α(Fe), there is good agreement between the experimental results of the different investigators. Their findings for the solubility of hydrogen in γ(Fe) can be described by the following formula:

$$\lg(\text{at.\% H}) = -1,411\ T^{-1} - 0.468.$$

Solutions in δ(Fe). There is poor agreement between the results given by the investigators [2, 3, 11, 13]. The values obtained by [3, 11] and [13] join with the solubility data for the γ phase whereas the measurements obtained by [2] indicate an abrupt decrease in the solubility on passing from the γ phase to the δ phase. The latter values lie close to the extrapolated solubility curve for hydrogen in α(Fe) and therefore seem to appear to be more reasonable in view of the fact that α(Fe) and δ(Fe) have the same structure. The resulting curve (dependence of temperature vs at.%) can be calculated from the formula given for the solubility in α(Fe) and agrees very well with Oriani's equation [15].

Solutions in liquid Fe. Data reported by numerous investigators have been critically evaluated by Spencer [14]. The somewhat scattered findings for

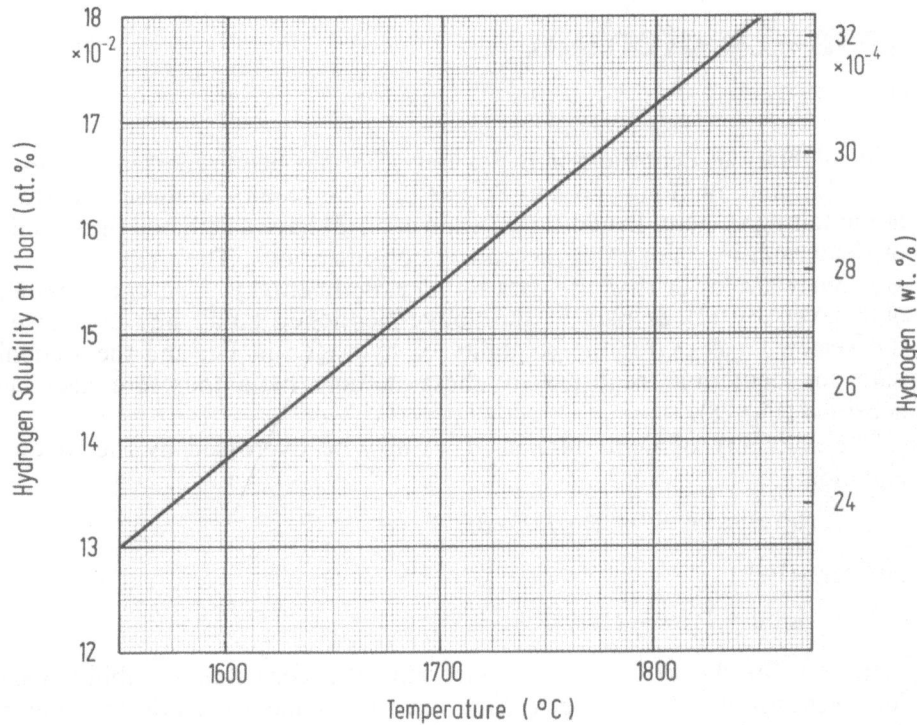

Fig. 27. Fe-H. Hydrogen solubility at 1 bar in *liquid* Fe

Fe–H

the solubility of hydrogen in liquid Fe can be summarized and expressed by the following formula:

$$\lg(\text{at.\% H}) = -1{,}820\,T^{-1} + 0.112.$$

Figures 26 and 27 represent the critically evaluated findings, expressed by the above mentioned formulae for the solubility of hydrogen in solid Fe (Fig. 26) and in liquid Fe (Fig. 27).

References

1 Martin, E.: Arch. Eisenhüttenwes. 3 (1929–1930) 407
2 Luckemeyer-Hasse, L.; Schenck, H.: Arch. Eisenhüttenwes. 6 (1933) 209
3 Sieverts, A.; Zaph, G.; Moritz, H.: Z. Phys. Chem. 183 (1938) 19
4 Pihlstrand, F.: Jernkontorets Ann. 121 (1937) 219
5 Andrew, J.H.; Lee, H.; Quarrell, A.C.: J. Iron Steel Inst. 146 (1942) 181
6 Armbruster, M.H.: J. Am. Chem. Soc. 65 (1943) 1043
7 Eichenauer, W.; Künzig, H.; Pebler, A.: Z. Metallkd. 49 (1958) 220
8 Hill, M.L.; Johnson, E.W.: Trans. AIME 221 (1961) 622
9 Heumann, Th.; Primas, D.: Z. Naturforsch. A 21 (1966) 260
10 Saliy, V.I.; Ryabov, R.A.; Gel'd, P.V.: Fiz. Net. Metalloved. 35 (1973) 119
11 Schenck, H.; Lange, K.W.: Arch. Eisenhüttenwes. 37 (1966) 739
12 Da Silva, J.R.G.; Stafford, S.W.; McLellan, R.B.: J. Less Common Met. 49 (1976) 407
13 Da Silva, J.R.G.; McLellan, R.B.: J. Less Common Met. 50 (1976) 1
14 Kubaschewski, O.; Spencer, P.J.; von Goldbeck, O.: Dok. 6210-CA-1/107 Düsseldorf: VDEh 1975–78
15 Oriani, R.A.: Conf. Fundamental Aspects Stress Corrosion Cracking NACF Houston, Texas, 1969

Fe–D Iron–Deuterium

Fig. 28

The solubility of deuterium in α(Fe) and γ(Fe) has been reported by Sieverts et al. [1], and by Heumann and Primas [2]. The results of these two studies are in fair agreement. In addition, Demin et al. [3] have calculated the solubility of deuterium in γ(Fe) from theoretical considerations. Their reported values are somewhat lower than the data found experimentally. Hydrogen solubility data given by [2] as part of the above-mentioned study lie very close to the selected values for the solubility of hydrogen in Fe, and the solubility data for deuterium reported by these authors have therefore also been accepted (Fig. 28).

The solubility of deuterium in α(Fe) may be expressed by the following equation:

$$\lg(\text{at.\% D}) = -1{,}376\,T^{-1} - 0.633$$

and for γ(Fe)

$$\lg(\text{at.\% D}) = -1{,}377\,T^{-1} - 0.51.$$

In view of the observations concerning the consistent solubility values for *hydrogen* in α- and δ(Fe) (which have the same structure) the equation for the solubility of deuterium in α(Fe) may readily be applied to that in δ(Fe).

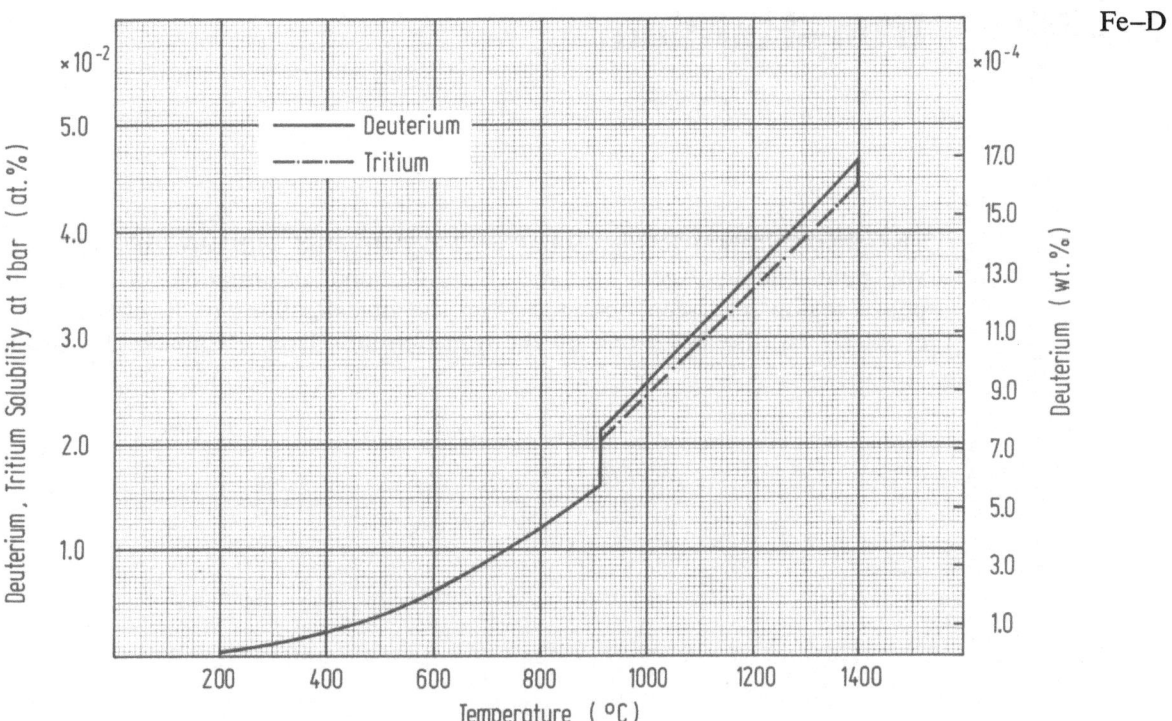

Fig. 28. Fe-D. Deuterium solubility at 1 bar and Fe-T. Tritium solubility at 1 bar

References

1 Sieverts, A.; Zapf, G.; Moritz, H.: Z. Phys. Chem. 183 (1938) 19
2 Heumann, Th.; Primas, D.: Z. Naturforsch. A 21 (1966) 260
3 Demin, V.B.; Vykhodets, V.B.; Gel'd, P.V.; Men', A.N.; Fishman, A.Ya.; Chufarov, G.I.: Izv. Akad. Nauk SSSR Met. (1972) 2, 201

Iron–Tritium

Fe–T

Fig. 28

The solubility of tritium in γ(Fe) has been calculated by Demin et al. [1]. Although the calculated solubility values for deuterium reported by these authors are somewhat lower than the experimental values, the difference in solubility of D_2 and T_2 in iron that they report has been used in conjunction with the selected solubility data for deuterium (vide) to produce corresponding values for tritium in γ(Fe). They may be represented by the equation

$$lg\,(at.\%\,T) = -1{,}400\ T^{-1} - 0.52.$$

Reference

1 Demin, V.B.; Vykhodets, V.B.; Gel'd, P.V.; Men', A.N.; Fishman, A.Ya.; Chufarof, G.I.: Izv. Akad. Nauk SSSR Met. (1972) 2, 201

Fe–Hf   Iron–Hafnium

Fig. 29

The phase diagram in Fig. 29 has been constructed with reference to a survey published by Ageev in 1968 [E]. Additional information on the Laves phase region has been obtained from the investigations of Kocherzhinskiy et al. [1] and Ikeda [2] and on the iron-rich area from Reinbach [3], Svech-

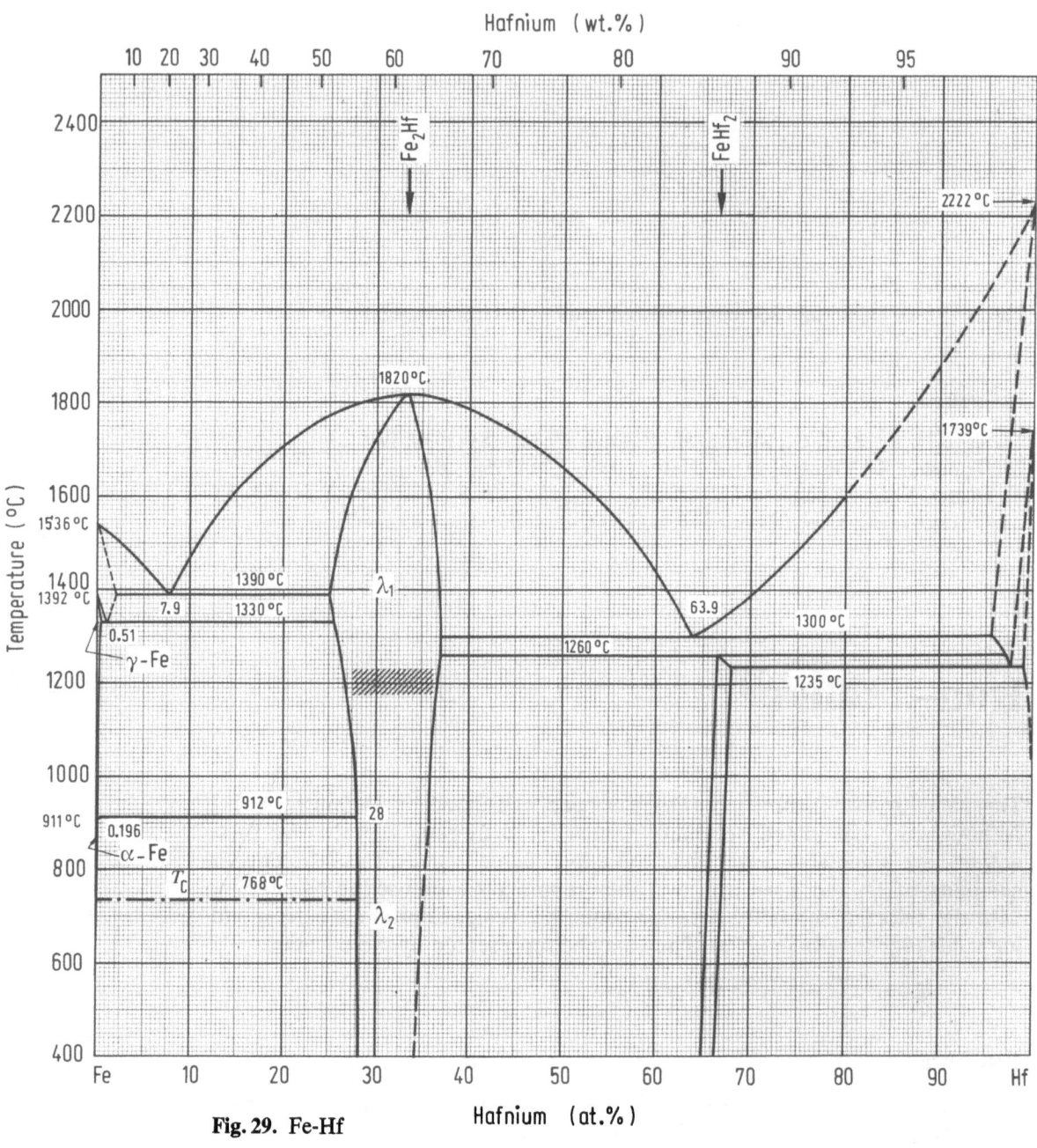

Fig. 29. Fe-Hf

nikov and Shurin [4], Kocherzhinskiy [1] and Abrahamson and Lopata [7]. There are two stable intermediary phases, $FeHf_2$ and Fe_2Hf. The latter is a Laves phase with a considerable range of homogeneity, at least at higher temperatures, and melts congruently at 1,810 °C [5] or 1,820 °C [1]. Hf_2Fe is formed by a peritectic reaction at 1,260 °C from $\beta(Hf)$ and $HfFe_2$ [5].

The melting point of hafnium 2,222 °C (IPTS-68), has been taken from measurements by Ackermann and Rauh [8], the transformation temperature of Hf containing 3 wt.% Zr, $(1,739 \pm 10)$ °C (IPTS-68), from Cezairliyan and McClure [9]. See also [10].

$HfFe_2$ seems to exist in three modifications of Laves type structure ([6] confirmed by [1]): λ_1 ($MgZn_2$ type), λ_2 ($MgCu_2$ type) and λ_3 ($MgNi_2$ type), λ_1 being the high-temperature and λ_2 the low-temperature phases; λ_3 is said to be stable between 1,200 and 1,640 °C. As is common with compounds of Laves structures in two or three modifications, the transformation temperatures cannot be precisely assessed because of the very small differences in Gibbs energy involved. In Fig. 29 therefore only the λ_1 and λ_2 modifications are indicated, separated by a shaded area. The formation of λ_1 and λ_3 has been confirmed independently by X-ray investigation and comparison with the similar patterns of the Hf-Cr system [1]. Ikeda [2] found the hexagonal λ_1 structure above 1,400 °C and the cubic λ_2 structure below 1,000 °C.

The magnetic properties of the Laves phase compound have been studied by [2]. The two modifications λ_1 and λ_2 were found to be ferromagnetic, showing only a difference in the Curie temperatures owing probably to the different crystal structures. For more detailed information on crystal structures and magnetic properties of this compound, the paper by Ikeda [2] may be consulted.

Solid solubilities have been determined by [3, 4] and [7]. The solubility of Hf in $\alpha(Fe)$ is 0.036, 0.065, 0.106 and 0.196 at.% at 600, 700, 800 and 900 °C respectively [7]. Data reported by [3] and [4] for the peritectoid temperature are somewhat lower, namely <0.1 and 0.06 at.%. In $\gamma(Fe)$ it is ca. 0.5 at.% Hf. The approximate solubility curves for Fe in Hf in Fig. 29 are from [E].

References

1 Kocherzhinskiy, Yu.A.; Markiv, V.Ya.; Pet'kov, V.V.: Russ. Metall. (1973) 134
2 Ikeda, K.: Z. Metallkd. 68 (1977) 195
3 Reinbach, R.: Z. Metallkd. 51 (1960) 292
4 Svechnikov, V.N.; Shurin, A.K.: Proc. Acad. Sci. USSR, Chem. Sect. 139 (1961) 774
5 Shurin, A.K.; Alfintseva, R.A.: Coll. Metal Physics "Naukova dumka", Kiev 2 (1968) 172
6 Elliott, R.P.: Trans. ASM 53 (1961) 321
7 Abrahamson, E.P.; Lopata, S.L.: Trans. Met. Soc. AIME 236 (1966) 76
8 Ackermann, R.J.; Rauh, E.G.: High Temp. Sci. 4 (1972) 272
9 Cezairliyan, A.; McClure, J.L.: High Temp. High Press. 8 (1976) 461
10 'Hafnium', At. Energy Rev. Special Issue No. 8, IAEA, Vienna 1981

Fe–Hg   Iron–Mercury

Fig. 30

Information on the Fe-Hg phase diagram has been supplied by a number of investigators who determined the solubility of Fe in Hg. On the basis of vapour pressure measurements, Jangg and Steppan [1] concluded that no stable compound is formed in this system. They used an improved isoteniscope, in the form developed by Biltz and Meyer [2], and found that the vapour pressure over the amalgams virtually corresponded to the one of pure Hg.

Reported values of the solubility of Fe in Hg differ widely (see Hansen [A: Refs. 1–13]). The findings by Palmaer [3], Irvin and Russell [4], Marshall et al. [5], de Wet and Haul [6], Weeks [7] and Parkman [8] agree at least in the order of magnitude. Jangg and Palman [9] pointed out that the discrepancies are mainly due to the methods of measurements (mostly "indirect"). They therefore selected a direct method for the study of metals in mercury, which is based on the analytical determination of the metal content of filtered

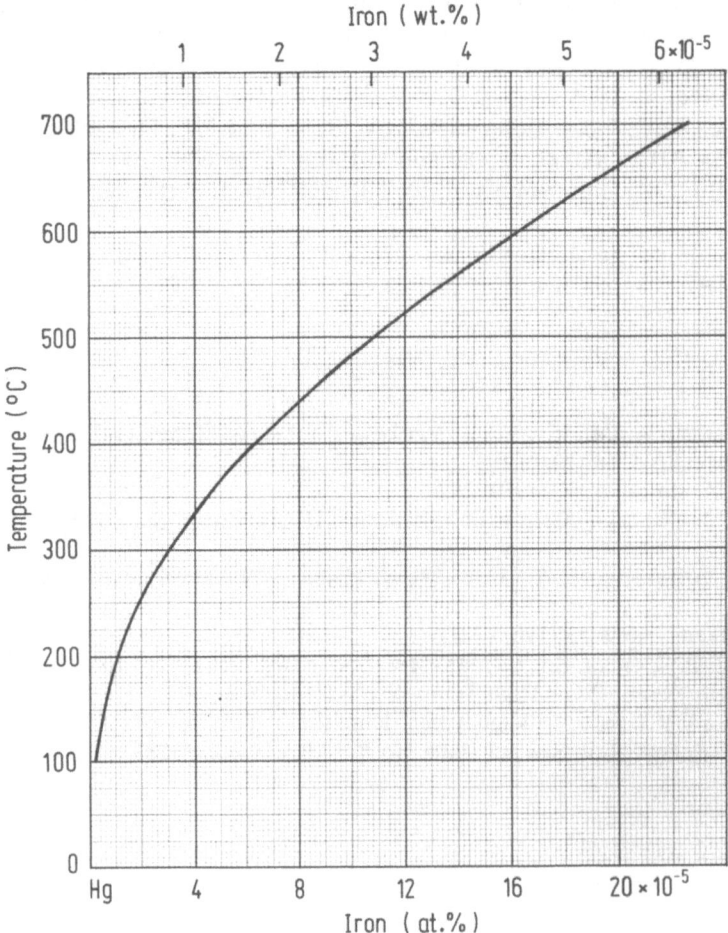

Fig. 30. Fe-Hg

saturated amalgams. They suggested that the heat of solution is about equal for all metals. An average value of about 23 kJ/mol was found empirically. On this basis, values for the solubility of Fe in Hg quoted by [5] have been critically reviewed by [9]. The findings of [5] Fig. 30 may be expressed by the following formula:

$$\lg(\text{at.}\%\,\text{Fe}) = -1{,}200\,T^{-1} - 2.41 \quad (273{-}973\,\text{K}).$$

The data reported by [7] and [8] are somewhat lower.

References

1 Jangg, G.; Steppan, F.: Z. Metallkd. 56 (1965) 172
2 Biltz, W.; Meyer, F.: Z. Anorg. Chem. 176 (1928) 23
3 Palmaer, E.: Z. Elektrochem. 38 (1932) 70
4 Irvin, N.M.; Russell, A.S.: J. Chem. Soc. (1932) 891
5 Marshall, A.L.; Epstein, L.F.; Norton, F.J.: J. Am. Chem. Soc. 72 (1950) 3514
6 De Wet, J.F.; Haul, R.A.W.: Z. Anorg. Chem. 277 (1954) 96
7 Weeks, J.R.: NASA Spec. Publ. NASA-SP-41 (1963) 21
8 Parkman, M.F.: Extended Abstracts of Electrothermics and Metallurgy Division, Electrochemical Soc., Fall Meeting Washington, 1964. Abstract No. 177
9 Jangg, G.; Palman, H.: Z. Metallkd. 54 (1963) 364

Iron-Holmium, see Fe-R: Iron-Rare Earth Metals, Fig. 65

Fe–Ho

Iron–Indium

Fe–In

Fig. 31

Phase relationships of the system Fe-In have been studied by Stadelmaier and Fiedler [1], Dasarathy [2–5] and Malingih and Logorely [6]. Authors [1] and [2–5] concentrated their investigations on the Fe-rich region and on the extent and feature of the miscibility gap [1, 5], while [6] established the In-rich region (1–8 at.% Fe). Their findings are summarized in Fig. 31.

The system is characterized by a range of liquid immiscibility ($\sim$ 3.4–91 at.% In), the absence of compounds and low mutual solid solubility.

The melting point of Indium, 156.634 °C is a secondary reference point on the International Practical Temperature Scale (IPTS-68 revised 1975).

Fe-rich region

In their investigation, [1] used metals of purity 99.9% Fe and 99.99% In, [2–5] BISRA "H" Fe and "pure" In. They applied thermal [1, 2], magnetic [1] and X-ray analyses [1, 2] supplemented by microscopical investigations [1] and electron microscopy [2]. The results are in very good agreement

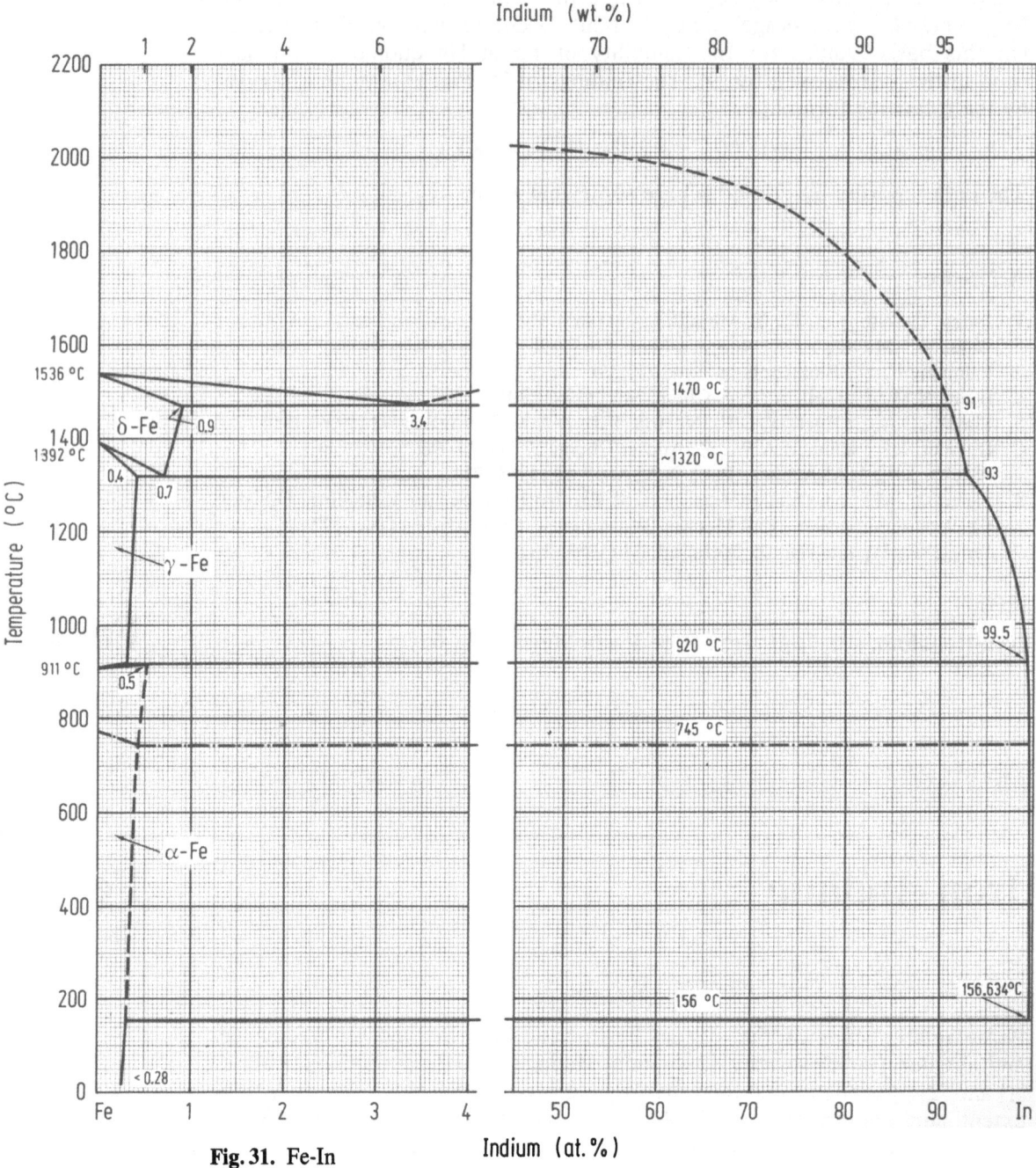

Fig. 31. Fe-In

except for the composition on the Fe-rich side of the miscibility gap, 3.4 and 7.7 at.% In respectively Addition of In to Fe contracts the γ field but expands the α- and δ-Fe region. The lattice parameter of Fe and In are little affected by the addition of the other metal. The Fe-rich region in Fig. 31 has been taken from the results obtained by [3], who determined especially the δ(Fe) liquidus and solidus most carefully. However, a temperature of 1,320 °C for the δ(Fe) eutectoid reaction has been chosen, which is lower than the one found by [3], to agree with results reported by [1] and [6].

In Fig. 31 the solubility of Fe in liquid In has been taken from data published by [6]. A temperature of 800, 900, 1,000, 1,100, 1,200 and 1,300°C corresponds to a composition of 0.27, 0.51, 1.0, 1.83, 3.53 and 5.98 at.% Fe.

In a more recent paper, [5] pointed out the similarity of the Fe-In and the Al-In systems and suggested that Fe in the presence of In behaves trivalently (at least up to 3 at.% In) analogous to Al. If these similarities are extended to the critical temperature of the miscibility gap (T_c), a temperature of ca. 2,000°C is estimated. The extent of the miscibility gap for the Al-In has been investigated and discussed by Predel [7]. A critical temperature of the miscibility gap of about 2,100°C is obtained if a mode of estimation by Wobst [8] is applied.

References

1 Stadelmaier, H.H.; Fiedler, M.L.: Z. Metallkd. 58 (1967) 633
2 Dasarathy, C.: Z. Metallkd. 58 (1967) 279
3 Dasarathy, C.: Trans. Metall. Soc. AIME 245 (1969) 1838
4 Dasarathy, C.: Z. Metallkd. 63 (1972) 209
5 Dasarathy, C.: Z. Anorg. Allg. Chem. 403 (1974) 173
6 Malingih, A.C.; Logorelyi, A.D.: Izv. Vyssh. Uchebn. Zaved. Tsvetn. Metall. 2 (1970) 107
   See also: Ageev [E] 1970, 38
7 Predel, B.; Z. Metallkd. 56 (1965) 791
8 Wobst, M.: Scr. Metall. 5 (1971) 583

Iron–Iridium   **Fe–Ir**

Fig. 32

Very little information on the Fe-Ir phase diagram is available. An attempt has been made to correlate existing data, and a tentative equilibrium diagram is suggested (Fig. 32). It is based mainly on experimental results obtained by Buckley and Hume-Rothery [1] (liquidus/solidus for the Fe-rich region) and Raub et al. [2], Fallot [3] (solid state reactions).

The system belongs to the expanded γ field type, it exhibits a continuous series of solid solutions between fcc γ(Fe) and Ir and shows an ordered ε phase (hcp) in the medium concentration range below 625°C.

The melting point of iridium, 2,247°C, is a secondary reference point on the International Practical Temperature Scale IPTS-68 (revised 1975).

[1] determined the liquidus/solidus and A_4 transformation temperatures of Fe-rich alloys (0–6 at.% Ir). They used iron of 99.95% purity with a melting point of 1,535°C and 99.99% iridium and employed thermal analysis. Their results show that δ-Fe is formed peritectically from the liquid and γ-Fe solid solution at 1,538°C. (δ-Fe (2.8 at.%) = liquid (1.98 at.%) + γ-Fe (2.84 at.%)).

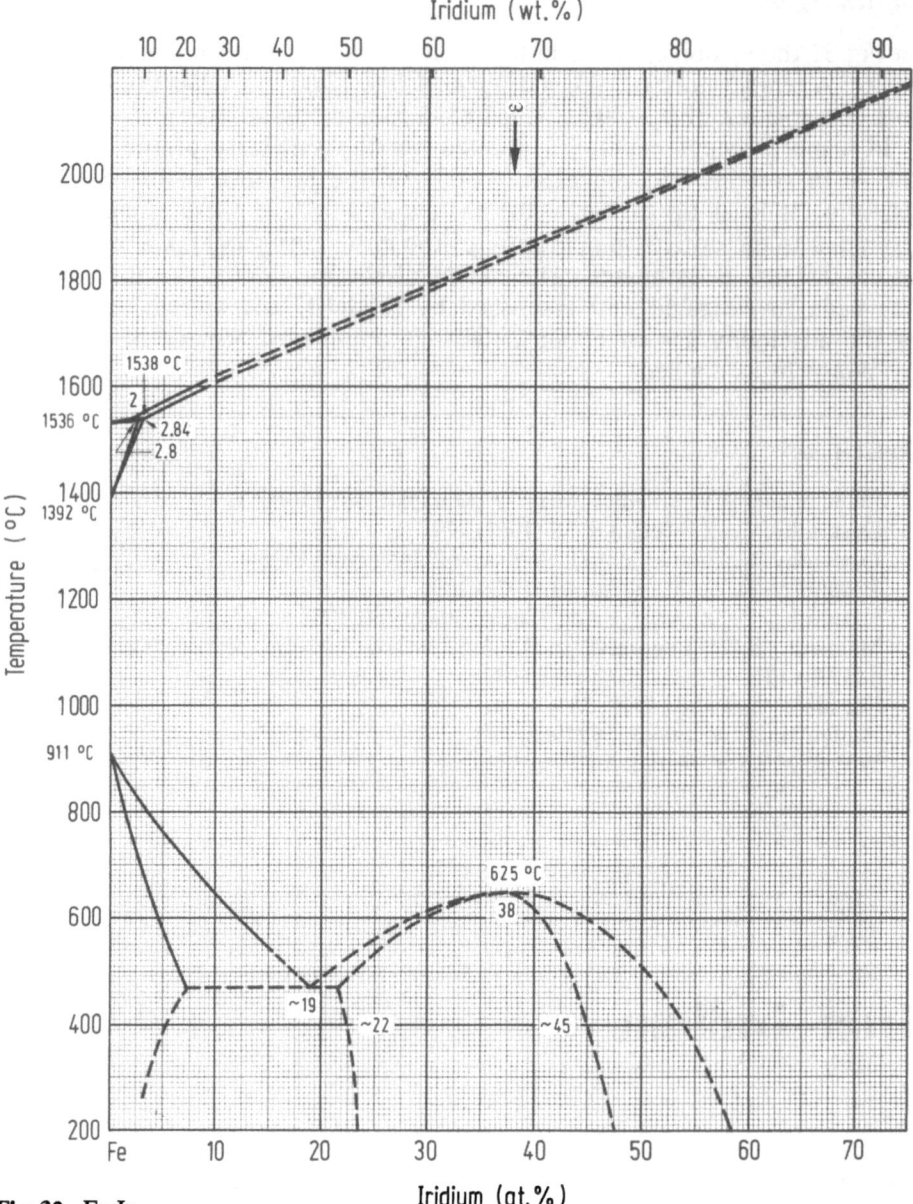

Fig. 32. Fe-Ir

The solid state equilibrium represented in Fig. 32 has been proposed by [2] who investigated the 15–80 at.% Ir range up to a temperature of 1,000°C and incorporated findings reported by [3] for the 0–15 at.% Ir concentration range. [2] carried out X-ray diffraction measurements at room and high temperatures on arc-melted alloys prepared from 99.96% Fe and 99.9% Ir and found a continuous range of fcc solid solution which transform into a hcp phase below 625°C and about 38 at.% Ir, thus confirming results previously quoted by Nemilov and Vidusova [4].

The sluggishness of the reactions at these low temperatures does not allow an exact determination of the homogeneity range which at 400°C is estimated to be ~22–45 at.% Ir [2].

Schwerdtfeger and Zwell [5] measuring the activities in solid Fe-Ir alloys

(0–75 at.% Fe) and using X-ray and microscopical analysis suggest a homogeneity range of 25–55 at.% Ir at room temperature. Miyagi and Wayman [6] on the basis of dilatometric, X-ray and metallographic analyses point out that the transformation of fcc phase into hcp phase can be explained by a martensitic transformation.

A systematic study of Fe-Ir alloys over the whole composition range using nuclear gamma resonance has been undertaken by Mössbauer et al. [7]. According to their findings, the fcc phase is paramagnetic at 4.2 K throughout the range of existence. For more detailed information, the reader may be referred to their publication.

References

1 Buckley, R.A.; Hume-Rothery, W.: J. Iron Steel Inst. London 201 (1963) 121
2 Raub, E.; Loebich, O.; Beeskow, H.: Z. Metallkd. 55 (1964) 367
3 Fallot, M.: Compt. Rendus 205 (1937) 517
4 Nemilov, V.A.; Vidusova, T.A.: Izv. Sekt. Platiny 20 (1947) 240
5 Schwerdtfeger, K.; Zwell, L.: Trans AIME 242 (1968) 631
6 Miyagi, M.; Mayman, C.M.: Trans. AIME 236 (1966) 806
7 Mössbauer, R.L.; Lengsfeld, M.; von Lieres, W.; Potzel, W.; Teschner, P.; Wagner, F.E.: Z. Naturforsch. 26 a (1971) 343

Iron-Potassium, see Fe-Alkaline Metals

Fe–K

Iron–Lanthanum

Fe–La

Fig. 33

The Fe-La phase diagram shown in Fig. 33 is essentially that of Gschneidner [1] who summarized and discussed previous investigations which were confirmed (1972) by Kepka et al. [10].

The system exhibits a eutectic at 91.5 at.% La and 780°C [3] a flattening of the liquidus in the ~8–19 at.% La range and an absence of intermediary phases.

Three teams, Haefling and Daane [2], Richerd [3] and Savitskii [4] have independently investigated the Fe-La system and have arrived at two different versions. According to [4] two compounds are formed, Fe_5La and Fe_2La, analogous to the Fe-Ce system, while [2, 3] and Wallace et al. [5, 6] agree that no stable compound exists in this system. The relative size of the component atoms and the electron concentration are involved in the stability of compounds. La is without 4f electrons and the largest in size of the lanthanons. These two factors might be the reason for the lack of a compound forming ability.

[4] gives no experimental details concerning this system; [2] and [3] used high purity materials (better than 99%) and based their results on data obtained by thermal analysis, X-ray and metallographic techniques. The liquidus anomaly between ~8 and 19 at.% La is "real" and not due to

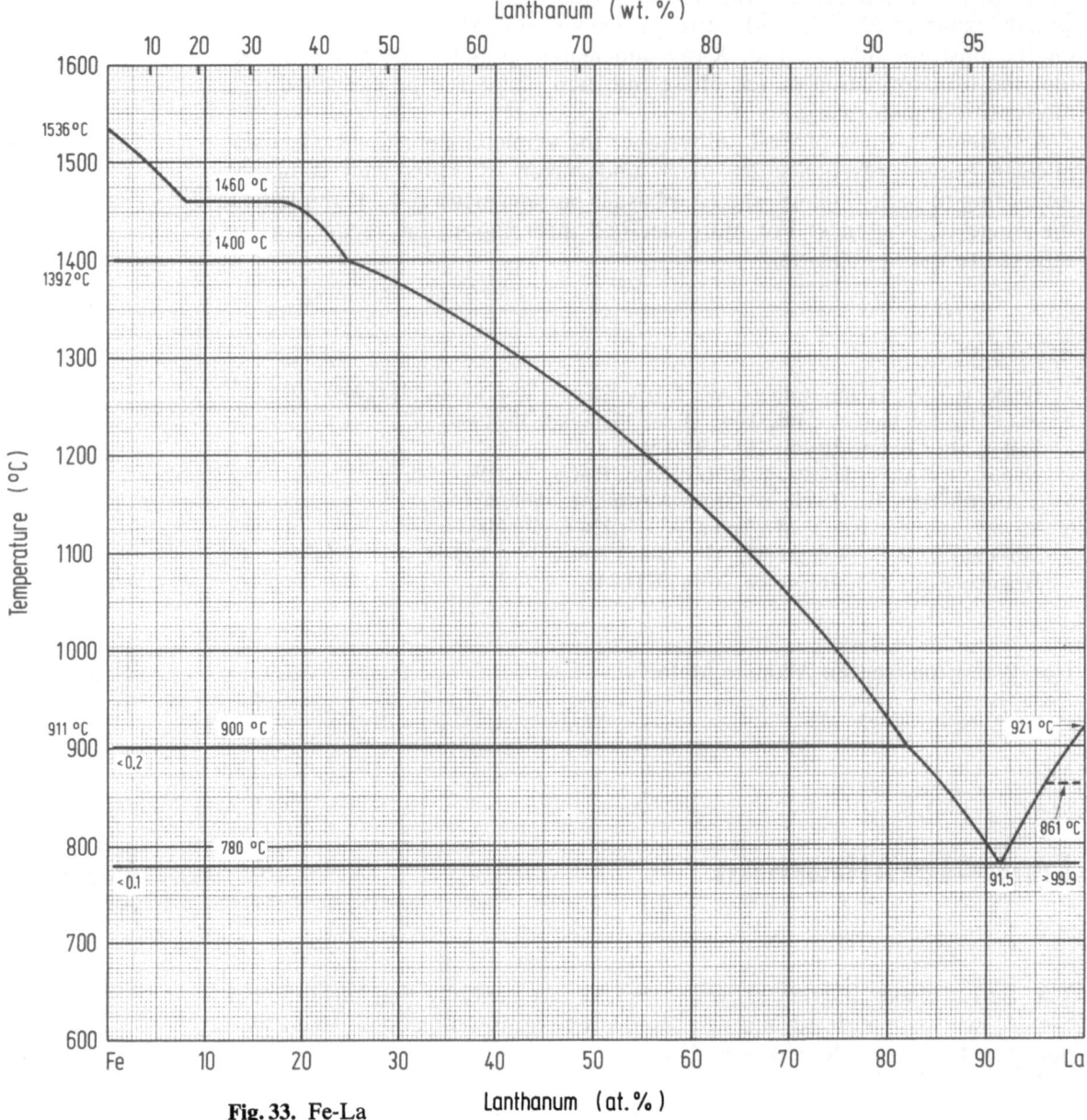

Fig. 33. Fe-La

the formation of two immiscible liquids [2]. ([3] did not investigate the Fe-rich region.) This phenomenon has also been reported by Mirgalovskaya and Strel'nikova [7], Iandelli [8] and Rolla and Iandelli [9] to occur in the Ce-Mn [7, 8] and La-Mn [9] system. An explanation of this flattening of the liquidus curve has not been proposed by [2, 7]. On the basis of microscopical examination, [2] and [7] discount the existence of a miscibility gap, [8, 9] however, support the miscibility gap version. In view of the small Gibbs energy differences obviously involved, influences such as impurity effects, grain boundary and strain energies may interfere, and an eventual decision is likely to be well nigh impossible.

The terminal solid solubilities are very small, less than 0.1 at.% at 780 °C [2, 3].

References

1 Gschneidner, K.A.: Rare Earth Alloys. Princeton: van Nostrand, 1961, pp. 187–188
2 Haefling, J.; Daane, A.: Inst. At. Res., Iowa State College; cited by [5]
3 Richerd, J.: Mem. Sci. Rev. Metall. 59 (1962) 539
4 Savitskij, E.M.: Redkie Metally i Splavy (Rare Metals and Alloys) Moscow: Dom Tekhniki 1959; as cited by [5] and by [2]
5 Wallace, W.E.: Private communication, see [1]
6 Nassau, K.; Cherry, L.V.; Wallace, W.E.: Phys. Chem. Solids 15–16 (1960) 123
7 Mirgalovskaya, M.S.; Strel'nikova, I.A.: Tr. Inst. Metall. Im. A.A. Baikova (2) 1957, 135
8 Iandelli, A.: Atti Accad. Naz. Lincei Rend. 13 (1952) 265
9 Rolla, L.; Iandelli, A.: Ber. Dtsch. Chem. Ges. 75 (1942) 2091
10 Kepka, M.; Skala, J.: Hutnik 1 (1972) 12

Iron-Lithium, see Fe-Alkaline Metals

Fe–Li

Iron-Lutetium, see Fe-R: Iron-Rare Earth Metals, Fig. 68

Fe–Lu

Iron–Magnesium

Fe–Mg

Fig. 34

The iron-magnesium system exhibits extensive miscibility gaps in the solid and liquid states. The determination of the phase boundaries boils down therefore to one of the mutual solubilities. After melting iron with only 0.057 wt.% C in the presence of magnesium at 1,600 °C under 25 bar of argon, Levchenko et al. [1] found a content of ca. 0.2 at.% Mg in the cooled sample. Taking this as the 'real' solubility, one may derive its temperature dependence using an empirical relationship suggested by Kubaschewski [2]:

$$\overline{H}_i = 3,400 \, \overline{S}_i^E \, (K)$$

where i is a solute element in a solvent metal. With this relationship and the above value, the solubility of *liquid* magnesium in *liquid* iron under 25 bar of argon may be expressed by the equation:

$$\lg (\text{at.\% Mg}) = -11,250 \, T^{-1} + 5.31.$$

The solubility of magnesium in *solid* iron is probably so small that it will elude experimental measurement.

Burylev [3] calculated the mutual solubilities by thermodynamic considerations, based mainly on experimental results of Trojan and Flinn [4] on Fe-Mg-C alloys. He obtained a symmetric miscibility gap in the liquid state, which is highly unlikely in view of the differences in atomic size and melting point of iron and magnesium. His solubility data for Mg in Fe are therefore substantially higher than those given by the equation above, which is preferred here. His calculated solubility data for iron in magnesium, however, are acceptable, in particular as they agree well with the experimental ones of Siebel [5] and reasonably well with those of Fahrenhorst and

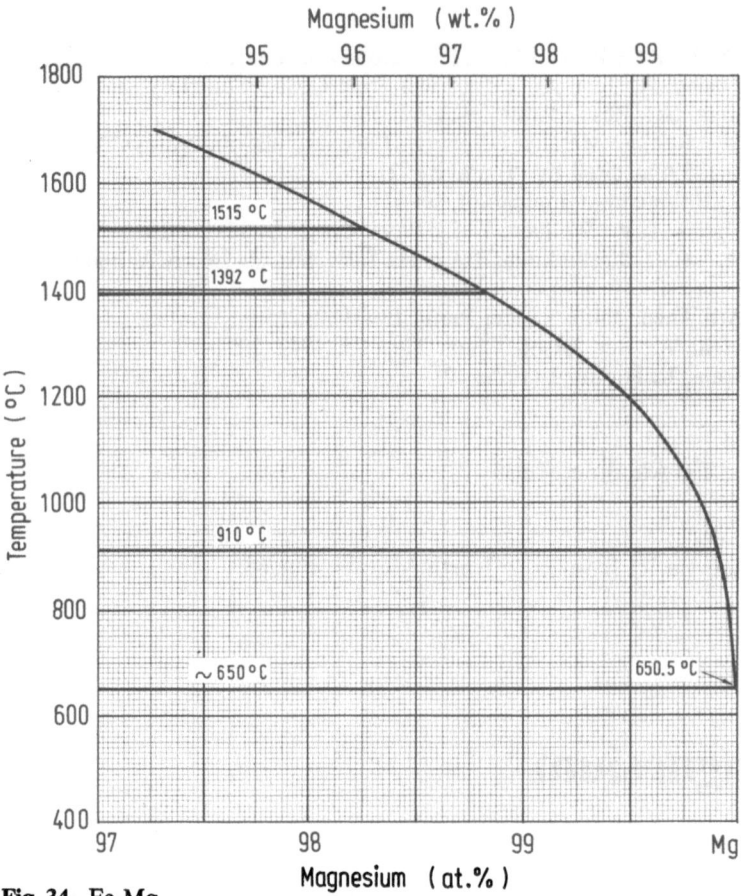

Fig. 34. Fe-Mg

Bulian [6] and Yue [7]. The selected solubilty curve is shown in Fig. 34. The values of [7], determined by segregation data from zone melting and directional freezing experiments, indicate a *solid* solubility at the eutectic temperature (~650°C) of 0.00043 at.% Fe and a eutectic composition of 0.0026 at.% Fe, which may be compared with the calculated 0.008 at.% Fe, (melting point of Mg = 650.5°C). Expressed in the form of equations, the curves of Fig. 34 are as follows [3]:

$$\lg(\text{at.\% } \alpha(\text{Fe})) = -4{,}478\ T^{-1} + 2.762,$$
$$\lg(\text{at.\% } \gamma(\text{Fe})) = -4{,}431\ T^{-1} + 2.722,$$
$$\lg(\text{at.\% } \delta(\text{Fe})) = -4{,}372\ T^{-1} + 2.687.$$

References

1 Levchenko, Y.M.; Khokholkov, V.M.; Gorshkov, A.A.: Dopov. Akad. Nauk Ukr. RSR (1963) 1602
2 Kubaschewski, O.: High Temp. High Press. 13 (1981) 435
3 Burylev, B.P.: Russian Castings Production 1966. In: Moffatt, W.G.: The Handbook of Binary Phase Diagrams. General Electric Comp. 1978, p. 466
4 Trojan, P.K.; Flinn, R.A.: Trans. ASM 54 (1961) 549
5 Siebel, G.: Z. Metallkd. 39 (1948) 22
6 Fahrenhorst, E.; Bulian, W.: Z. Metallkd. 33 (1941) 31
7 Yue, A.S.: J. Inst. Met. 91 (1963) 166

The system is characterized by an extensive solid solution range between fcc γ(Fe) and fcc γ(Mn) and by the stable solution phases, namely bcc α-Fe, δ-Fe, α-Mn, β-Mn and δ-Mn respectively.

The liquidus and solidus curves fall smoothly from the peritectic horizontal at 1,473°C to a minimum at 87 at.% Mn and 1,232°C. They then rise to the melting point of δ(Mn) at 1,246°C.

The phase diagram for this system is shown in Fig. 35. It is based on experimental results of Hellawell and Hume-Rothery [1] and Hume-Rothery and Buckley [2] and a critical assessment by Hellawell [3], calculations of the liquidus/solidus curve by Steiler [4] and Rao and Tiller [5] and a calculation of the α/γ equilibrium by Kirchner et al. [6]. A phase diagram consistent with the thermochemical properties has been calculated by Kaufman [7] using experimental thermochemical data combined with the analytical description of the liquid, bcc, fcc α-Mn and β-Mn phases.

The transformation temperatures of Mn have been taken from Hultgren and are as follows: (980 ± 20) K, $(1,361 \pm 10)$ K, $(1,412 \pm 5)$ K.

The liquidus-solidus curves have been established experimentally by [1]. A redetermination [2] of the 0–7 at.% Mn range confirmed earlier observations. The results in the 20–70 at.% Mn region agree well with earlier findings by Gayler [8]. Thermal analyses [1] indicate a minimum in the liquidus at 87 at.% Mn. This minimum is virtually indistinguishable from the peritectic horizontal at the same temperature (1,232°C). According to [3], the possibility of a eutectic reaction liquid ⇋ γ + δ-Mn cannot be excluded.

[5] and more recently [4] derived thermodynamic functions which adequately describe the solution behaviour of liquid and γ(Mn) over the entire range. Their results agree well and are represented in Fig. 35.

The α/γ transition in Fe-rich alloys shows a considerable temperature hysteresis. Results obtained in earlier investigations, which used thermal analysis and dilatometry, did apparently not reach equilibrium. The most reliable experimental results were obtained by Troiano and McGuire [9] who made precise determinations of lattice constants combined with microscopic investigations of alloys annealed in some cases for periods up to two and a half years. Alloys in the range 16–25 at.% Mn transform on cooling from fcc γ to hcp ε. [9] and Öhman [10] assume this to be a transition structure which eventually decomposes into (α + γ) solid solution. The α/γ equilibrium has been analysed thermodynamically by Hillert et al. [11] and more recently by [6]. The latter introduced a temperature dependence of the parameter A^γ. The calculated (α + γ) equilibrium [6] has been adopted in Fig. 35 and is in satisfactory agreement with the experimental data reported by [9]. The influence of high pressures on the (α + γ) phase field boundaries has been investigated by Claussen [12] and Ershova et al. [13]. Both agree that with increasing pressure (up to 50 kbar), the boundaries of the (α + γ) two phase area are shifted towards lower manganese concentrations.

The metastable diffusionless equilibrium, also called allotropic phase boundary, in the Fe-Mn system (0–700 K) has been calculated by Nishizawa [19].

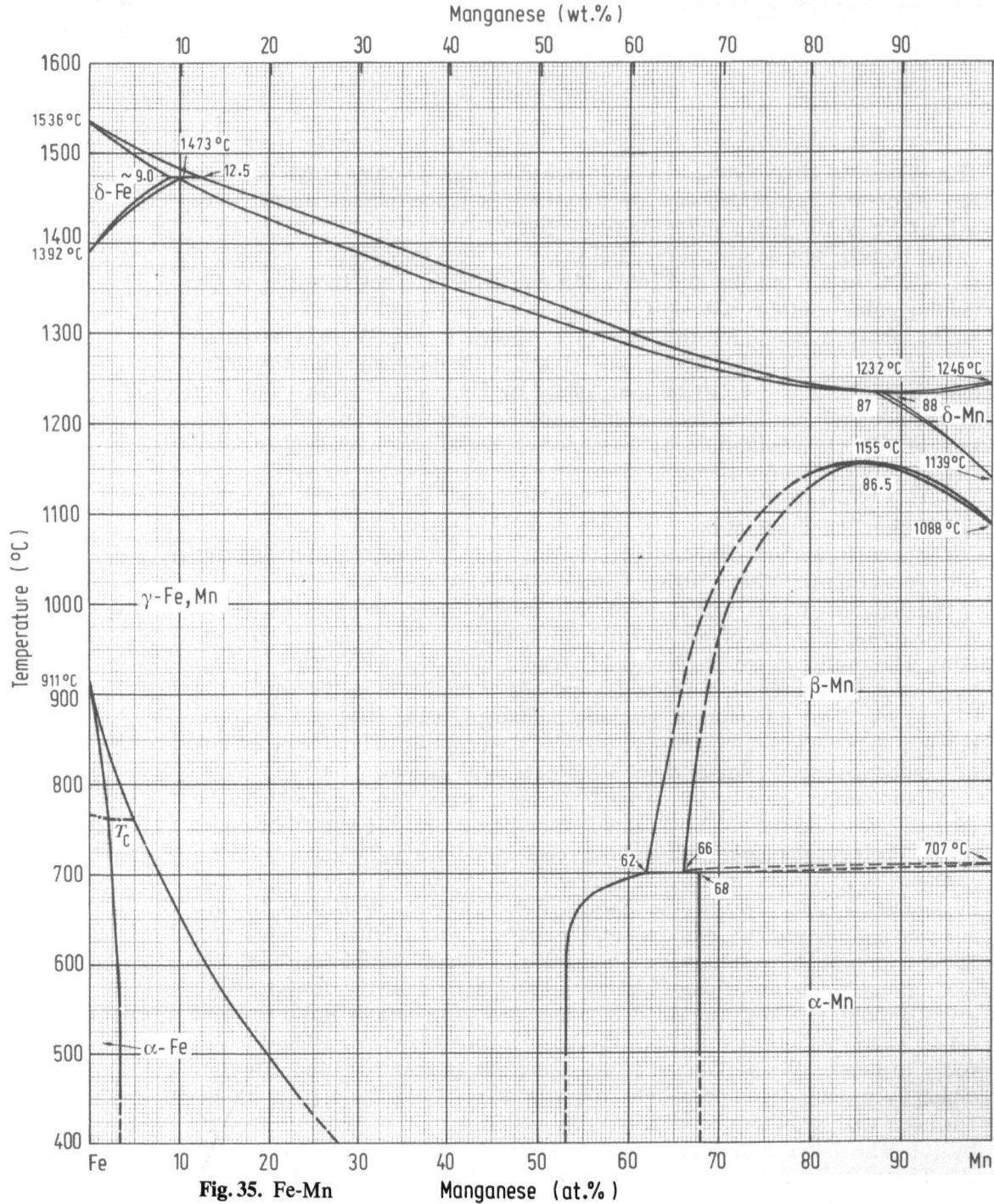

Fig. 35. Fe-Mn

In a thorough study to determine the structure and transformation behaviour of alloys containing 2–40 at.% Mn using dilatometry, X-ray diffraction and optical and electron metallography, Holden et al. [14], Gartstein and Rabinkin [20 23] Baličev and Tkačenko [21] and Gulyaev et al. [22] were able to show the effect of manganese on the metastable γ→α and γ→ε equilibria.

In the systems Mn-Co and Mn-Ni intermetallic superstructures occur. The possibility of a similar phenomenon in the solid state of Fe-Mn has been investigated by Sokolovskaya et al. [15]. X-ray, microstructural and thermal analyses supported by microhardness and electrical resistivity measurements have been carried out on very pure alloys (0.14 % total impurity) annealed in argon between 950 and 600°C for 600 hours. They found indications, not yet clearly defined, of superstructures (FeMn and Fe_2Mn) which require closer identification. An anomalous effect at about 150–200°C in the 25–55 at.% Mn region has been attributed to a paramagnetic ⇌ antiferromagnetic transformation.

The $(\delta + \gamma)$Mn two phase field is very narrow, namely about 1 wt.% at 1,232°C [3]. The β/γ(Mn) equilibrium in the range 80–100% Mn is based on thermal anysis [1]. The width of the two phase region has not been determined accurately owing to temperature hysteresis but is considered to be very small in the 85–100% region. The $(\gamma + \beta)$Mn phase widens below a maximum at 1,155°C [1, 8, 16].

The position of the β-Mn eutectoid (ca. 700°C and 62–68 at.% Mn) has been adopted from results reported by Walters and Wells [17] and confirmed by Gayler and Wainwright [18], but the temperature is lower than found by [10] (725°C). The width of the $(\gamma + \alpha)$Mn has been taken from the work published by [17] and [10].

References

1 Hellawell, A.; Hume-Rothery, W.: Philos. Trans. 249 (1957) 417
2 Hume-Rothery, W.; Buckley, R.A.: J. Iron Steel Inst. 202 (1964) 534
3 Hellawell, A.: Annotated Equilibrium Diagrams, No. 26, The Institute of Metals (1956)
4 Steiler, J.M.: Commission des Communautés Européennes CECA No. 7210-Ca/3/303, Nov. 1981
5 Rao, V.; Tiller, W.A.: Mater. Sci. Eng. 15 (1974) 87
6 Kirchner, G.; Nishizawa, T.; Uhrenius, B.: Metall. Trans. 4 (1973) 167
7 Kaufman, L.: CALPHAD 2 (1978) 117
8 Gayler, M.L.V.: J. Iron Steel Inst. 128 (1933) 293
9 Troiano, A.R.; McGuire, F.T.: Trans. ASM 31 (1943) 340
10 Öhman, E.: Z. Phys. Chem. 8 B (1930) 81
11 Hillert, M.; Wada, T.; Wada, H.: J. Iron Steel Inst. 205 (1967) 539
12 Claussen, W.F.: Technol. Rep. ASD-TDR-62-479, Part II (1963) 38
13 Ershova, T.P.; Ponyatovskiy, E.G.: Russ. Metall. No. 4 (1967) 81
14 Holden, A.; Bolton, J.D.; Petty, E.R.: J. Iron Steel Inst. 209 (1971) 721
15 Sokolovskaya, E.M.; Grigor'ev, A.T.; Altunin, Yu.F.: Russ. J. Inorg. Chem. 7 (1962) 1464
16 Yoshisaki, H.: Sci. Rep. Res. Inst. Tohoku Univ. Ser. A 3 (1951) 137
17 Walters, F.M.; Wells, C.: Trans. Am. Soc. Met. 23 (1935) 727
18 Gayler, M.L.V.; Wainwright, C.: J. Iron Steel Inst. 135 (1937) 269
19 Nishizawa, T.: CALPHAD 1 (1) (1977) 12
20 Gartstein, E.; Rabinkin, A.: Acta Metall. 27 (1979) 1053
21 Baličev, J.M.; Tkačenko, F.K.: Izv. Akad. Nauk SSSR Met. 21 (3) (1979) 169
22 Gulyaev, A.P.; Volynova, T.F.; Georgieva, I.Ya.: Metalloved. Termi. Obrab. Met. No. 3 (1978) 2
23 Rabinkin, A.: CALPHAD 3 (1979) 77

Fe–Mo   Iron–Molybdenum

Figs. 36–38

Since a diagram of the Fe-Mo system was assessed by Hansen and Anderko [A], quite a number of new data have been published. Most of these have been reviewed by Brewer and Lamoreaux [1] who paid particular attention to the work of Sinha et al. [2] and Heijwegen and Rieck [3]. In addition, Brewer et al. [1] Nüssler et al. [4] as well as Kaufmann and Nesor [5] have assessed the thermochemical properties of the various phases consistent with the phase diagram, though in a somewhat simplified form.

The selected phase boundaries are shown in Fig. 36. The limits of the individual phases may be discussed in turn.

a) The ferrite-austenite equilibria shown in the side diagram (Fig. 37) are due to Kirchner et al. [6] and are similar to the results of Alberry and Haworth [7]. The electron microprobe measurements of the former fix the maximum extent at 1,140°C with (1.69 ± 0.05) at.% Mo in fcc Fe and (2.26 ± 0.05) at.% Mo in bcc Fe. The results for the solubility of Mo in bcc Fe are on the whole in good agreement [2, 6, 8]; only [9] found a slightly higher solubility at the higher temperatures. The selected maximum solubility is 22.0 at.% at 1,450°C [8]. In equilibrium with liquid solutions, the bcc solutions form a minimum at 20 at.% Mo and 1,445°C [8].

b) At high temperatures, the bcc solutions are in equilibrium with the so-called R-phase formed peritectically at 1,488°C and about 37.4 at.% Mo [2]. [2] and [6] found the R phase to be stable above 1,245°C, whereas [3, 13] suggested a eutectoid temperature of 1,200°C. In a case like this where equilibration is very slow, one is inclined to select the lower value.

c) At low temperatures, the bcc solutions are in equilibrium with the Laves phase (MgZn$_2$ type). MoFe$_2$ forms with a narrow homogeneity range below 950°C [2, 13]. The phase boundaries for Fe$_2$Mo in Fig. 36 have been taken from results obtained by electron microprobe analysis [13]. 34.1, 33.9, 33.4 and 33.1 at.% Mo correspond to a temperature of 900, 850, 800 and 750°C. However, Rawlings and Newey [9] and Heijwegen and Rieck [3] were unable to detect the phase in their diffusion couple studies — obviously a nucleation problem at the temperatures in question [1]. Vol [10] in 1962 suggested that Fe$_2$Mo is only stable in the presence of certain impurities, which really means the same.

d) Fe$_3$Mo$_2$(μ) is formed by pertectoidal reaction at 1,370°C and has a rhombohedral D8$_5$ structure, isotypic with Fe$_7$W$_6$ [2]. That the ideal composition is not covered by its homogeneity range is not unusual with intermetallic compounds the deviation depending on its Gibbs energy and that of the neighbouring phases. Experimental results pertaining to the homogeneity range differ somewhat: 39–40 at.% Mo [2], 38.5–40.1 at.% Mo [6] and 38.5–43 at.% Mo [3], respectively.

e) The sigma phase also is somewhat controversial. Its tetragonal D8$_b$ structure has been established [A] with $a = 0.922$, $c = 0.481$ nm at 50 at.% Mo [12], but the phase boundaries are difficult to fix exactly, again owing to nucleation problems. The temperature and the composition of the eutectoidal decomposition of σFeMo are adopted from the careful work of [3], that is ca. 1,235°C and 55 at.% Mo. The remaining phase boundaries are based on micrographic

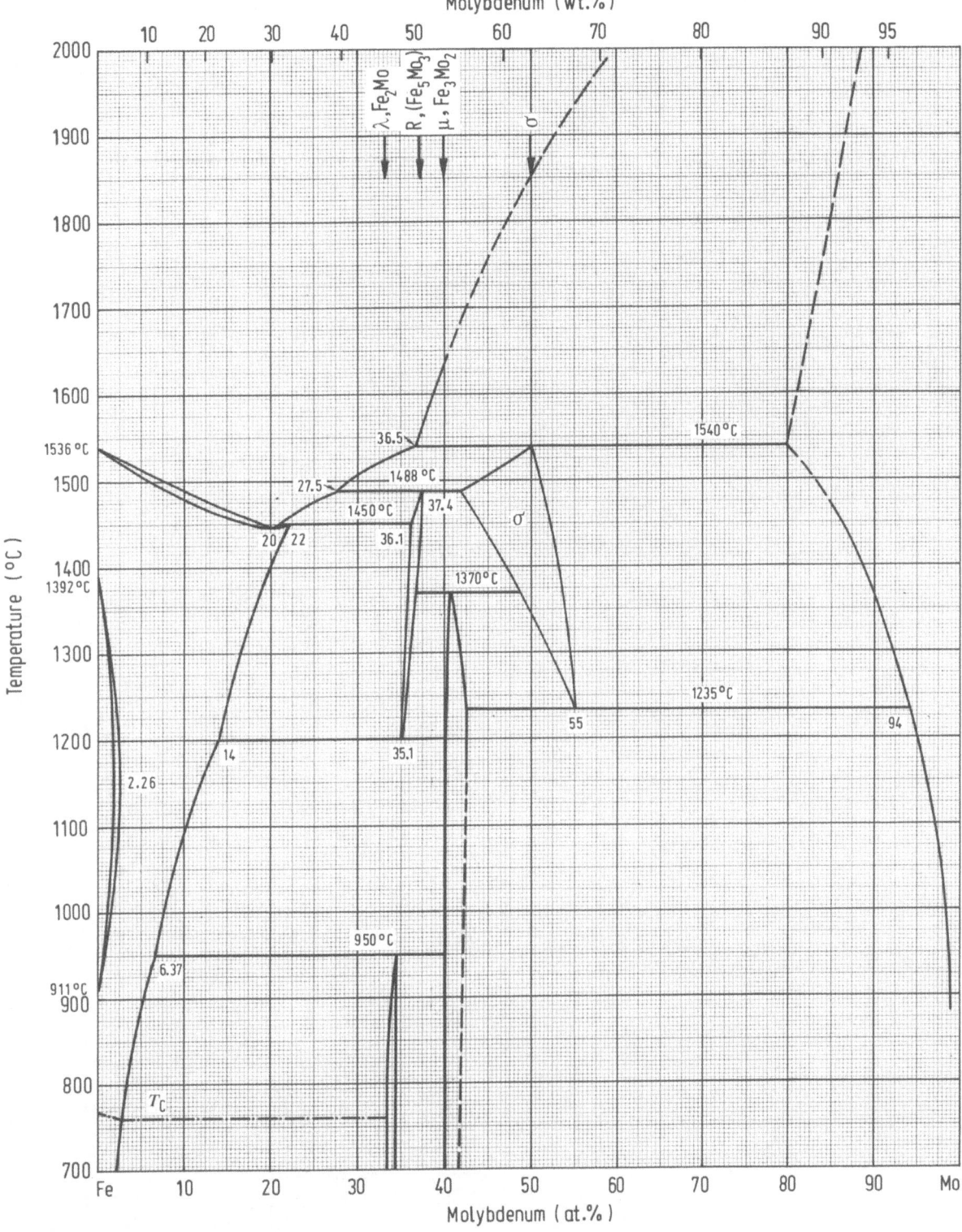

Fig. 36. Fe-Mo

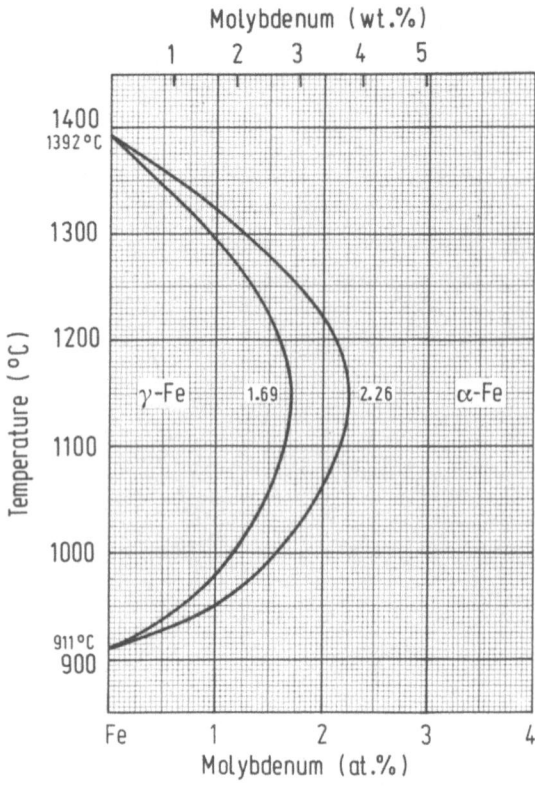

Fig. 37. Fe-Mo

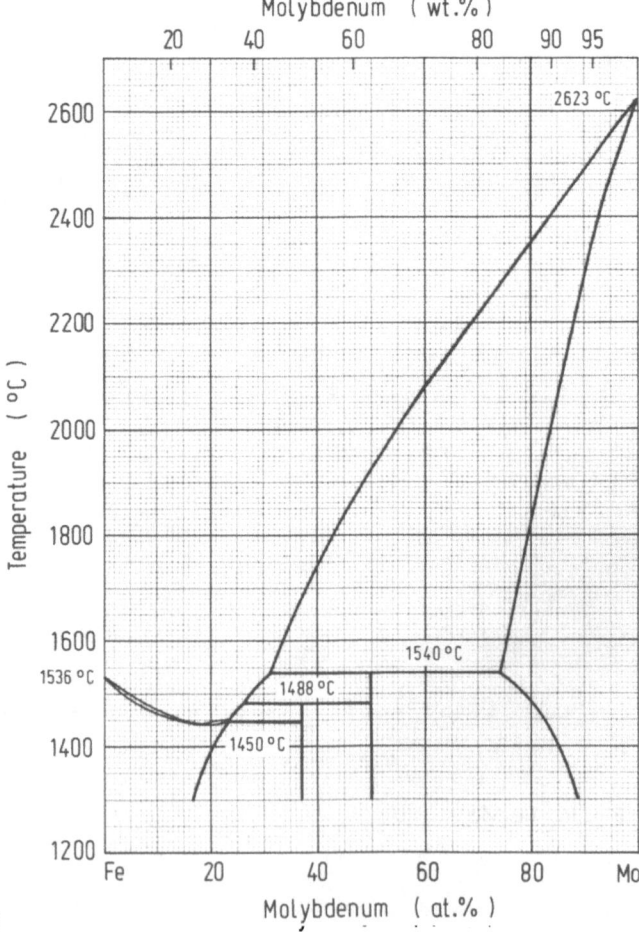

Fig. 38. Fe-Mo (calculated)

results (see [A]). Nüssler et al. [4] did not obtain sigma in the indicated range after annealing for two weeks at 1,250°C. The solid solubility curve shown in Fig. 36 represents a compromise of the experimental results of Heijwegen and Rieck [3] and Ham [11]. At 1,540°C, an uncertainty of about 5 at.% still remains.

The equilibrium diagram in Fig. 38 is the one calculated by Nüssler et al. [4] who for this purpose simplified the intermediary phases to "line compounds". For the corresponding thermochemical data, the original paper should be consulted.

References

1 Brewer, L.; Lamoreaux, R.H.: At. Energy Rev. Spec. Issue No. 7, IAEA, Vienna (1980) 244
2 Sinha, A.K.; Buckley, R.A.; Hume-Rothery, W.: J. Iron Steel Inst. 205 (1967) 191
3 Heijwegen, C.P.; Rieck, G.D.: J. Less Common Met. 37 (1974) 115
4 Nüssler, H.-D.; Hoster, T.; Kubaschewski, O.: Z. Metallkd. 71 (1980) 396
5 Kaufman, L.; Nesor, H.: CALPHAD 1 (1) (1977) 52; CALPHAD 2 (1) (1978) 66
6 Kirchner, G.; Harvig, H.; Uhrenius, B.: Metall. Trans. 4 (1973) 1059
7 Alberry, P.J.; Haworth, C.W.: Met. Sci. J. 2 (1975) 140
8 Gibson, W.S.; Lee, J.R.; Hume-Rothery, W.: J. Iron Steel Inst. 198 (1961) 64
9 Rawlings, R.D.; Newey, C.W.A.: J. Iron Steel Inst. 206 (1968) 723
10 Vol, A.E.: Handbook of Binary Metal Systems, Vol. II. Moskva-1962
11 Ham, J.L.: Trans. ASME 73 (1951) 723
12 Wilson, C.G.; Spooner, F.J.: Acta Crystallogr. 16 (1963) 230
13 Takayama, T.; Wey, M.Y.; Nishizawa, T.: Trans. Jpn. Inst. Met. 22 (1981) 315

Iron–Nitrogen

The iron-nitrogen diagram exhibits some uncommon features. For the iron-carbon system, as is well known, one usually presents two diagrams: a stable one for the equilibrium of Fe solid solution with graphite and a metastable one for its equilibrium with cementite (Fe_3C). Iron and nitrogen form three metastable compounds which because of kinetic checks often occur at moderately elevated temperatures: Fe_8N, Fe_4N and Fe_2N.

The solid solution of nitrogen in Fe is stable in equilibrium with nitrogen gas (of 1 bar pressure) at all temperatures above room temperature. Fe_4N is metastable with repect to Fe and N_2, $Fe_8N(Fe_{16}N_2)$ is metastable with respect to Fe_4N and Fe.

The solubility of nitrogen in iron depends accordingly on the phase with which it is in equilibrium. The thermochemical background of these relationships has been outlined quantitatively by Kubaschewski and Catterall [1]. A more recent thermochemical assessment stems from Hillert and Jarl [2]. The phase diagram commonly associated with the Fe-N system, shown in Fig. 39, is a metastable one.

The equilibrium between nitrogen gas and *ferrite* (bcc structure) at temperatures between 973 and 1,173 K (α(Fe)) and between 1,665 and 1,809 K

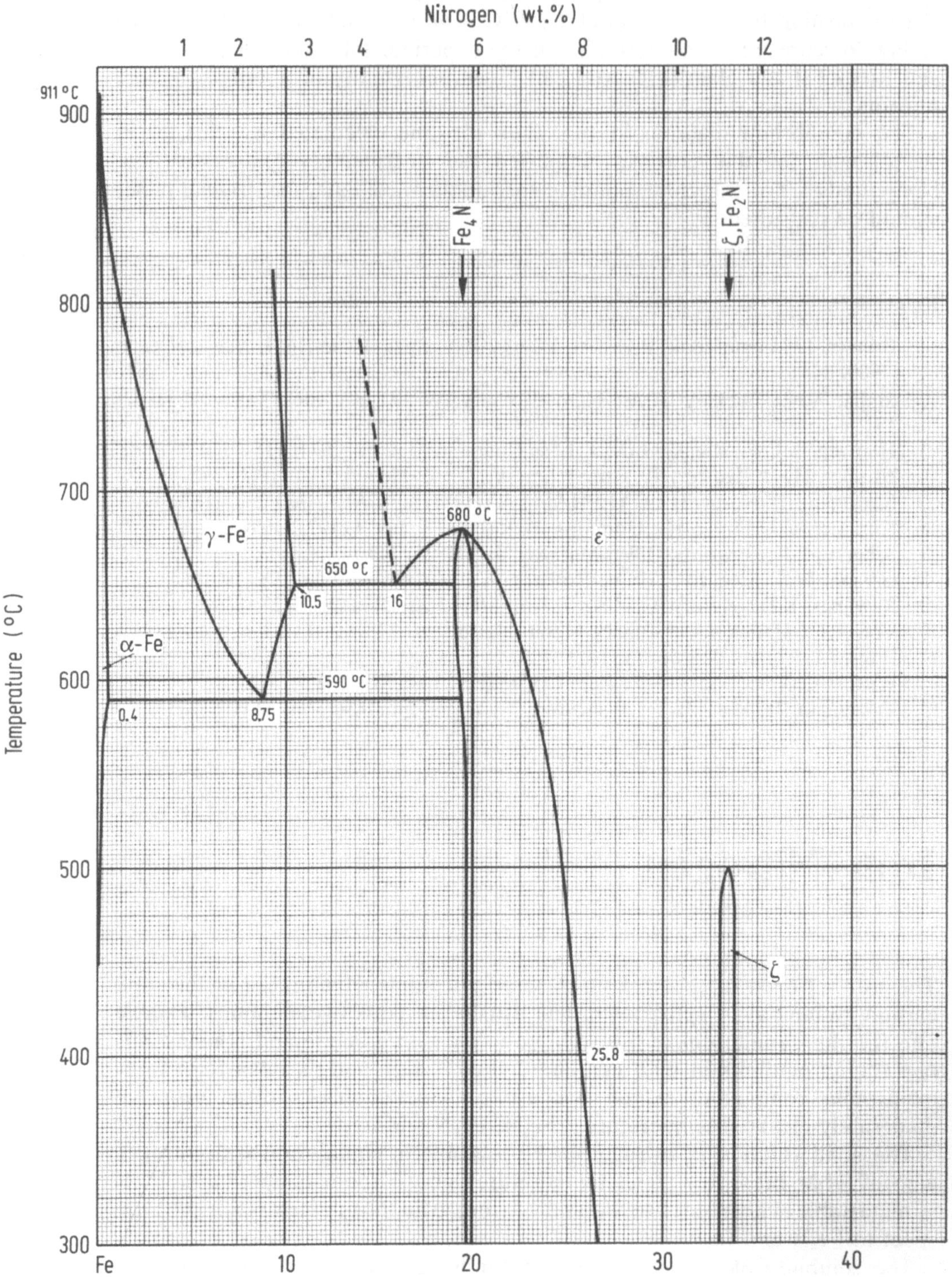

Fig. 39. Fe-N (metastable)   Nitrogen (at.%)

(δ(Fe)) has been measured by [6, 7, 12–16)] and their results are generally in good agreement. The solubility for the reaction $1/2(N_2) = [N]_{\alpha,\delta(Fe)}$ may be represented by the equation

$$\lg(\text{at.\% N}) = -1{,}700\, T^{-1} - 0.30 \quad (300\text{–}1{,}809\,\text{K}).$$

The solubility of nitrogen in $\gamma(Fe)$ in equilibrium with nitrogen gas at 1 bar has been measured by [7, 12–14, 16–22]. There are rather large differences between the results obtained by different investigators.

For the reaction $1/2(N_2) = [N]_{\gamma(Fe)}$, the solubility may be represented by the equation

$$\lg (at.\% N) = 155\, T^{-1} - 1.1.$$

Although this equation appears to be satisfactory for temperatures between 863 and 1,323 K or so, the nitrogen solubilities in equilibrium with N_2 at 1 bar are a little higher than those observed between 1,323 and 1,665 K.

The solubility of nitrogen in *liquid* iron in equilibrium with N_2 gas (1 bar) has been extensively reported and has been summarized by Hansen [A], Elliott [B], [1] and [23]. The results, on average, may be represented by the equation

$$\lg (at.\% N) = -457\, T^{-1} - 0.5.$$

The solubility of nitrogen in $\alpha(Fe)$ in equilibrium with Fe_4N has been determined by [3–11] (Fig. 40). The maximum solubility is 0.4 at.% at 590 °C. The lattice parameter is practically similar to that of $\alpha(Fe)$.

The γphase shows a fcc structure and is isomorphous with the γphase (austenite) in the Fe-C system.

The 'Fe_4N' or γ'phase is not a simple stoichiometric compound with ordered fcc structure but exists over a narrow range of compositions which according to measurements by Grabke [24] and by Wriedt [25] lie between ca. 19.6 and 19.95 at.% N at 500 °C for example.

Investigations by Jack [26] of the ε and $\zeta(Fe_2N)$ phase confirmed that both phases show an ordered arrangement of nitrogen atoms, hcp (D_3^1–C 312) and orthorhombic respectively.

In the $\zeta(Fe_2N)$ structure the Fe-atoms retain the same relative position as in $\varepsilon(Fe_2N)$, however, because of a rearrangement of the nitrogen atoms,

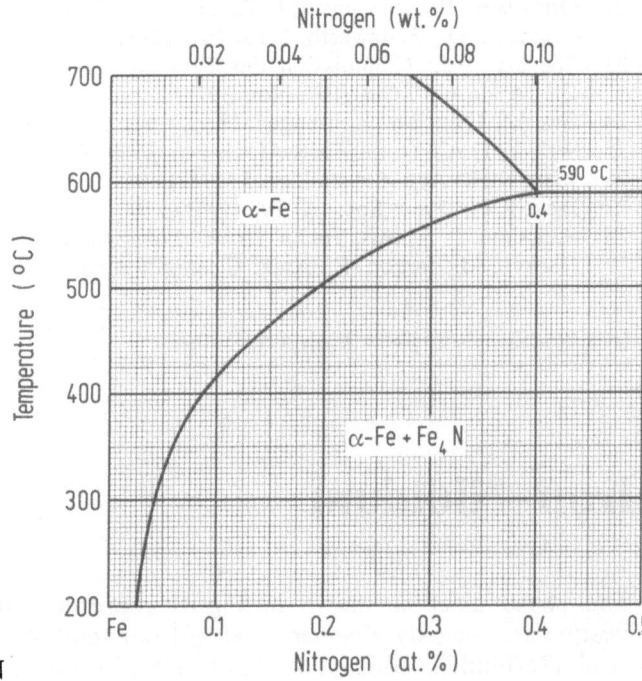

Fig. 40. Fe-N

Fe–N

which in ζ are more closely packed in one direction, the iron-atom lattice is anisotropically distorted. An explanation for the mechanism of the transition $\varepsilon(Fe_2N)/\zeta(Fe_2N)$ has been proposed by [26].

The $Fe_{16}N_2$ phase, as has been mentioned, is unstable with respect to Fe solutions and Fe_4N. However, its metastable existence and its structure have been determined by Jack [27]. The solubility of nitrogen in equilibrium with $Fe_{16}N_2$ has been measured with fairly good agreement by [3, 6, 28, 29].

References

1 Kubaschewski, O.; Catterall, J.A.: Thermochemical Data of Alloys. London, New York: Pergamon 1956
2 Hillert, M.; Jarl, M.: Metall. Trans. 6A (1975) 553
3 Dijkstra, L.J.: Trans. AIME 185 (1949) 252
4 Paranjpe, V.G.; Cohen, M.; Bever, M.B.; Floe, C.: Trans. AIME 188 (1950) 261
5 Aström, H.U.; Borelius, G.: Acta Metall. 2 (1954) 547
6 Fast, J.D.; Verrijp, M.B.: J. Iron Steel Inst. 180 (1955) 337
7 Corney, N.S.; Turkdogan, E.T.: J. Iron Steel Inst. 180 (1955) 344
8 Burdese, A.: Metall. Ital. 47 (1955) 357
9 Rawlings, R.; Tambini, D.: J. Iron Steel Inst. 184 (1956) 302
10 Darken, L.S.: The Physical Chemistry of Metallic Solutions and Intermetallic Compounds. NPL Symp. No. 9, 1958
11 Sakamoto, M.; Masumoto, T.; Imai, Y.: J. Jpn. Inst. Met. 37 (1973) 343
12 Sieverts, A.; Zapf, G.; Moritz, H.: Z. Phys. Chem. (A) 183 (1938) 19
13 Fountain, R.W.; Chipman, J.: Trans. Met. Soc. AIME 212 (1958) 737
14 Schenck, H.; Frohberg, M.G.; Reinders, F.: Stahl Eisen 83 (1963) 93
15 Grieveson, P.; Turkdogan, E.T.: Trans. Met. Soc. AIME 230 (1964) 1604
16 Zitter, H.; Habel, L.: Arch. Eisenhüttenwes. 44 (1973) 181
17 Darken, L.S.; Smith, R.P.; Filer, E.W.: J. Met. 3 (1951) 1174
18 Wriedt, H.A.; Gonzales, O.D.: Trans. Met. Soc. AIME 221 (1961) 532
19 Iwase, K., Fukusima, M.: Sci. Rep. Tohoku Univ. 27 (1939) 162
20 Mori, T.; Ichise E.: J. Jpn. Inst. Met. 29 (1965) 1001
21 Grieveson, P.; Turkdogan, E.T.: Trans. TMS-AIME 230 (1964) 407
22 Atkinson, D.; Bodsworth, C.: J. Iron Steel Inst. 208 (1970) 587
23 Spencer, P.J.: Commission des Communautés Européennes, Research project CECA No. 6210-Ca/1/107, Nov. 1977
24 Grabke, H.J.: Ber. Bunsenges. Phys. Chem. 73 (1969) 596
25 Wriedt, H.A.: Trans. TMS-AIME 245 (1969) 43
26 Jack, K.H.: Acta Crystallogr. 5 (1952) 404
27 Jack, K.H.: Proc. Soc. (A) 208 (1951) 216
28 Rahmann, J.: Thesis, Techn. Hochschule Aachen 1961
29 Nacken, M.; Jargon, F.: Arch. Eisenhüttenwes. 37 (1966) 989

Fe–Na

Iron-Sodium, see Fe-Alkaline Metals

Fe–Nb   Iron–Niobium

Fig. 41

The phase diagram shown in Fig. 41 represents the results of several investigators, namely Fischer et al. [1] (austenite-ferrite equilibrium), Gibson et al. [2] (liquidus-solidus, 0–15 at.% Nb), Ferrier et al. [3] (88.7–100 at.% Fe,

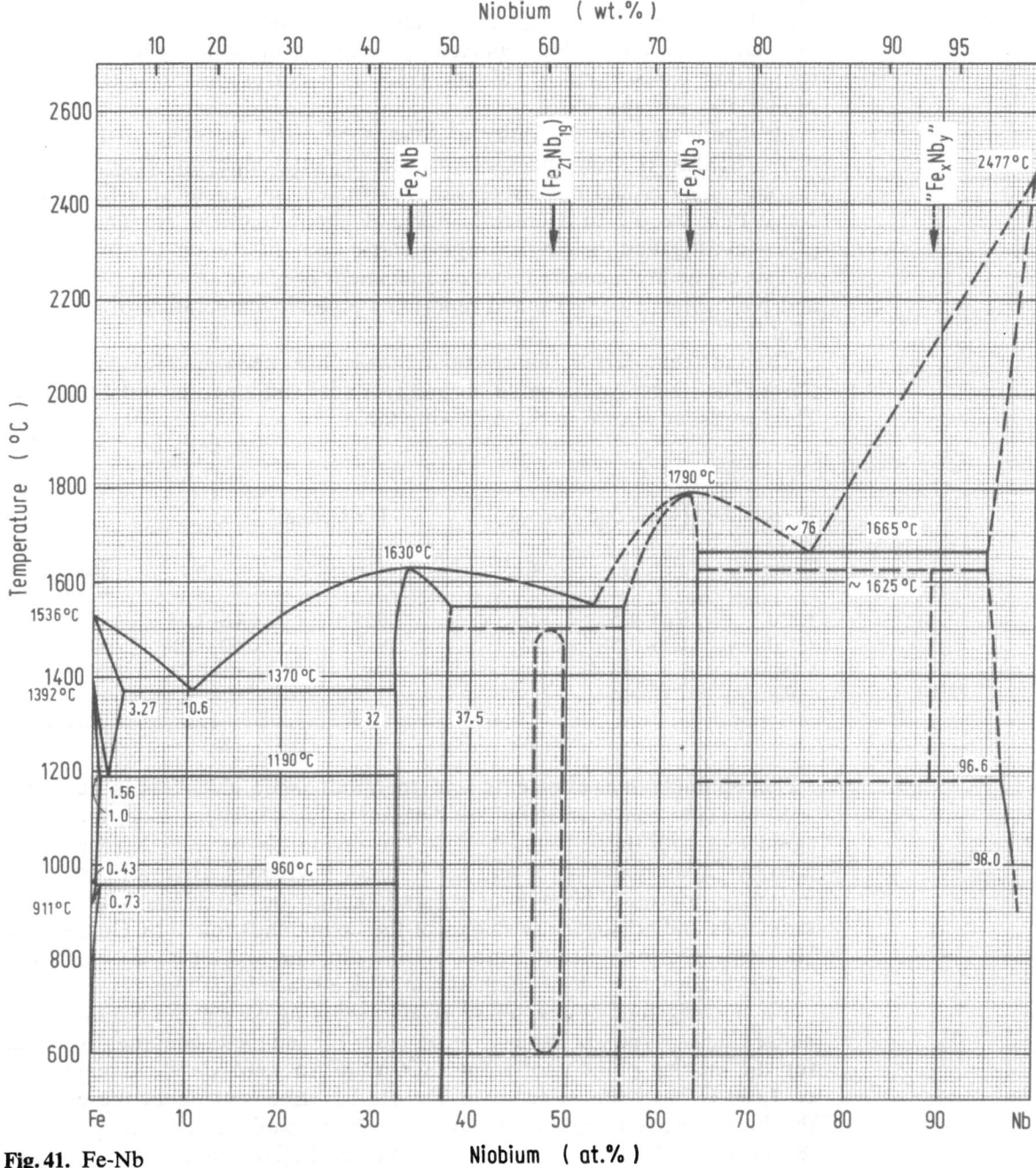

Fig. 41. Fe-Nb

>1,200 °C), Goldschmidt [4, 5] (0–80 at.% Fe), Elyutin et al. [6] (melting point of Fe₂Nb), Raman [7, 8] (Fe₂Nb and Fe₂₁Nb₁₉), Drobyshev and Rezukhina [9] (Fe₂Nb and Fe₂Nb₃). The melting point of Nb, 2,477 °C, is a secondary reference point (IPTS-68).

The Fe-Nb system displays four intermetallic phases, Fe_2Nb, $Fe_{21}Nb_{19}$ or Fe_7Nb_6, Fe_2Nb_3 and Fe_xNb_y (ca. 11 at.% Fe), and three eutectic reactions. The mutual solid solubility of the two metals is relatively small. The solubility of Fe in Nb at 1,000 °C and 1,200 °C is about 2.0 and 3.4 at.% respectively [4]. The most reliable values for the solubility of Nb are suggested to be: 0.73 at.% in

α(Fe) [1], 1.0 at.% in γ(Fe) [1] and 3.27 [3] or 2.8 at.% [2] in δ(Fe). Data for the solubility of Nb in α(Fe) in the temperature range 600–900°C (lattice parameter measurements) have been reported by Abrahamson and Lopata [15]. It was found that 0.095, 0.115, 0.185 and 0.330 at.% correspond to a temperature of 600, 700, 800 and 900°C.

[2] determined the liquidus and solidus curves in the range 0–15 at.% Nb using 'BISRA H' iron and spectroscopically pure niobium. They employed thermal analysis and microscopical examination and found that the liquidus curve falls steeply to a eutectic point at about 10.6 at.% Nb and 1,372°C. [3] re-investigating the Fe-rich region carried out differential thermal analysis and susceptibility measurements on specimens prepared from 99.99% pure Fe and Nb containing 1% Ta, and reported a eutectic temperature of 1,370°C at 12 at.% Nb.

The austenite-ferrite phase relationship has been investigated several times. The investigators agree on a eutectoidal and a peritectoidal reaction. The results are shown in the following table.

Eutectoidal (γ/eutectic compos.)			Peritectoidal (γ/α)		
Temp. °C	at.% Nb	Ref.	Temp. °C	at.% Nb	Ref.
1,220	0.60–1.50	10	965	0.48–1.08	10
1,220	1.20–1.56	11	989	0.60–1.08	11
1,190	1.0–1.56	1	960	0.43–0.73	1
1,208	1.1 –1.6	3			

Susceptibility measurements and thermodynamic calculations have been used by [1] to construct the phase diagram in Fig. 41. Their values are in good agreement with those found by [3]. The region 20–100 at.% Nb was examined by Goldschmidt [4, 5] and his findings have been used in Fig. 41. Later investigations have been concentrated mainly on establishing the crystal structures and homogeneity ranges of the compounds found by Goldschmidt. He used spectroscopically pure iron and niobium and based his evaluations on the disappearing phase method and lattice space variations.

All investigators of the phase relationships agree that Fe_2Nb is a Laves phase ($MgZn_2$ type), is Pauli-paramagnetic [12] and melts congruently. Values for the melting point vary from 1,630°C [6] to 1,665°C [9]. A temperature of 1,630°C has been selected in Fig. 41. Disagreement about the extent of the homogeneity range is, however, apparent. Re-investigations by [9] and [8] seem to be fairly realistic. The extent of the Fe_2Nb phase should not exceed 7 at.% according to the available data and because of the limited interchangeability of the atoms (size factor). [9] based their results on electrochemical and X-ray investigations and suggested that the Fe-rich boundary is close to 32–33 at.% Nb. [12] on the basis of X-ray measurements proposed a homogeneity range of 28–26 at.% Nb.

$Fe_{21}Nb_{19}$ identified by [4] was claimed to have the tetragonal $D8_b$ structure isotypic with CrFe. It is stable above ca. 600°C. [9] could not verify this superstructure but mention that the intensity of quenching was possibly insufficient to retain it. Raman [7, 8] supported by [13] maintains that this phase (with a 2 at.% homogeneity range) is not sigma but of the W_6Fe_7 type denoted μ. It may be pointed out that the structures of σ and μ are closely related.

Both are electron rather than size factor compounds wherein sigma exists at higher electron concentrations. The notation should thus be Fe_7Nb_6 rather than $Fe_{21}Nb_{19}$.

'Fe_2Nb_3' was found to possess the Ti_2Ni-type structure [4]. It melts congruently between 1,790 and 1,800°C and exhibits a homogeneity range. X-ray work carried out by [9] confirmed the characteristics of this phase whereas Raman could not establish evidence for its existence.

The diagram proposed by Goldschmidt includes a Nb-rich phase at about 89 at.% Nb existing possibly between 1,180 and 1,625°C. The structure has not been determined, and as far as the present compiler was able to find out, this compound has not been verified.

In a recent publication, Kaufman and Nesor [14] present a calculated equilibrium diagram. They give the experimental thermochemical and phase diagram data along with the analytical descriptions of the various phases and the computed diagram resulting from them.

References

1 Fischer, W.A.; Lorenz, K.; Fabritius, H.; Schlegel, D.: Arch. Eisenhüttenwes. 41 (1970) 489
2 Gibson, W.S.; Lee, J.R.; Hume-Rothery, W.: J. Iron Steel Inst. 198 (1961) 64
3 Ferrier, A.; Übelacker, E.; Wachtel, E.: Compt. Rendus 258 (1964) 5424
4 Goldschmidt, H.J.; J. Iron Steel Inst. 194 (1960) 169
5 Goldschmidt, H.J.: Research 10 (1957) 289
6 Elyutin, O.P.; Bokshitskii, I.Ya.; Rogova, I.V.: Sb. Tr. Tsentr. Nauchno Issled. Inst. Chern. Metall. No. 51 (1967) 169
7 Raman, A.: Z. Metallkd. 57 (1966) 301
8 Raman, A.: Proc. Indian Acad. Sci. A65 (1967) 250
9 Drobyshev, V.N.; Rezukhina, T.N.: Russ. Metall. No. 2 (1966) 85
10 Peter, W.; Fischer, W.A.: Arch. Eisenhüttenwes. 19 (1948) 161
11 Genders, W.; Harrison, R.: J. Iron Steel Inst. 140 (1939) 29
12 Nevitt, M.V.; Kimball, C.W.; Preston, R.S.: Proc. Int. Conf. 'Magnetism', Nottingham 1964, p. 137
13 Kripyakevich, P.I.; Gladyshevskii, E.I.; Pylaeva, E.N.: Sov. Phys. Crystallogr. 7 (1962) 165
14 Kaufman, L.; Nesor, H.: CALPHAD 2 (1978) 64
15 Abrahamson, E.P.; Lopata, S.L.: Trans. AIME 236 (1966) 76

Iron-Neodymium, see Fe-R: Iron-Rare Earth Metals, Fig. 59

Fe–Nd

Iron–Nickel

Fe–Ni
Figs. 42–44

Iron-nickel alloys have been investigated on numerous occasions, but there is still considerable uncertainty regarding the phase relationships at lower temperatures. Experimental difficulties are the main reason for this. Owing to the slow diffusion rates at these temperatures, very long times of annealing or electron or neutron irradiation are needed to accelerate diffusion.

Apart from this, X-ray and neutron diffraction methods are too insensitive unless the sensitivity is increased by the use of isotopes. Mössbauer effect spectroscopy (MES) may also be employed.

The system is characterized by a continuous series of solid solutions at high temperatures between fcc γ(Fe) and fcc Ni. The liquidus and solidus curves fall smoothly from the peritectic horizontal at 1,513 °C to a minimum between 60 and 70 at.% Ni at 1,425 °C. They then rise to the melting point of nickel at 1,455 °C (IPTS-68). At lower temperatures, three ordered phases exhibiting homogenous ranges have been established, FeNi$_3$ with the AuCu$_3$ superstructure (L1$_2$), FeNi with the AuCu superstructure and Fe$_3$Ni with an as yet undetermined structure.

Chamberod et al. [1] and van Deen [2] have recently (1979 and 1980) proposed a tentative phase diagram of the solid state for the binary system. These investigators, though assessing similar information, suggest different equilibria for the solid state reactions. In Fig. 42 the phase diagram of [2] has been adopted for the temperature range 200–950 °C, except for the α-Fe phase boundary. The complete diagram (Fig. 42) has in principle been based on the results of [3–7] for the liquidus-solidus relationships, on those of [8–21, 36] for the $(\alpha + \gamma)$ region and on those of [1, 2, 22–24] for the order-disorder behaviour at temperatures below 700 °C.

High temperatures (above 950 °C)

Measurements pertaining to the liquidus and solidus curves including the δ-Fe region carried out after the publication of the Fe-Ni system by Hansen [A: Fig. 379] confirmed the shapes of both curves. Additional data were obtained by thermal analysis: Schürmann and Brauckmann [3], Hellawell and Hume-Rothery [5] and Hume-Rothery and Buckley [6], confirmed by Knudsen-cell mass-spectrometry by Conard et al. [7].

The δ(Fe)/liquidus and -/solidus fall smoothly to a peritectic horizontal at 1,513 °C, corresponding to the reaction: liquid (5.3 at.% Ni) + δ(Fe) (4.1 at.% Ni) $\rightleftharpoons$ γ-Fe (4.5 at.% Ni) [3] (Fig. 43). All investigators agree that the Ni-rich part shows a minimum between 60 and 70 at.% Ni. However, they disagree about the temperature. Thermochemical measurements [7] support a temperature of the congruently melting alloy of 1,425 °C already suggested by Jenkins et al. [4], which is somewhat lower than that quoted by [A]. The chemical activity measurements [7] further suggest, also confirming findings by [4], that the liquidus-solidus gap extends to a maximum of 5–10 K. In view of the strongly exothermic nature of the liquid and solid solutions, however, one would expect the gap to be smaller.

Low temperatures (below 950 °C)

Heumann and Karsten [20] investigated the low-temperature equilibria in the 25–90 at.% Ni region and concluded that γ-(Fe, Ni) decomposes eutectoidally at 345 °C and 52 at.% Ni into α-Fe and FeNi$_3$. In order to obtain alloys that are in equilibrium at a given temperature, they applied two different methods: for the range 230–360 °C, they used the decomposition of Ni- and Fe-carbonyls and for alloys above 370 °C the transpiration method ("Aufdampfverfahren"). The alloys contained up to 0.2 wt.% C. The phase boundaries were determined by X-ray analysis and electron microscopy. Results for the position of the γ phase boundary above 350 °C agree reasonable well with those of Owen and Sully [9], Owen and Liu [10] Troiano and McGuire [11], Goldstein and Ogilvie

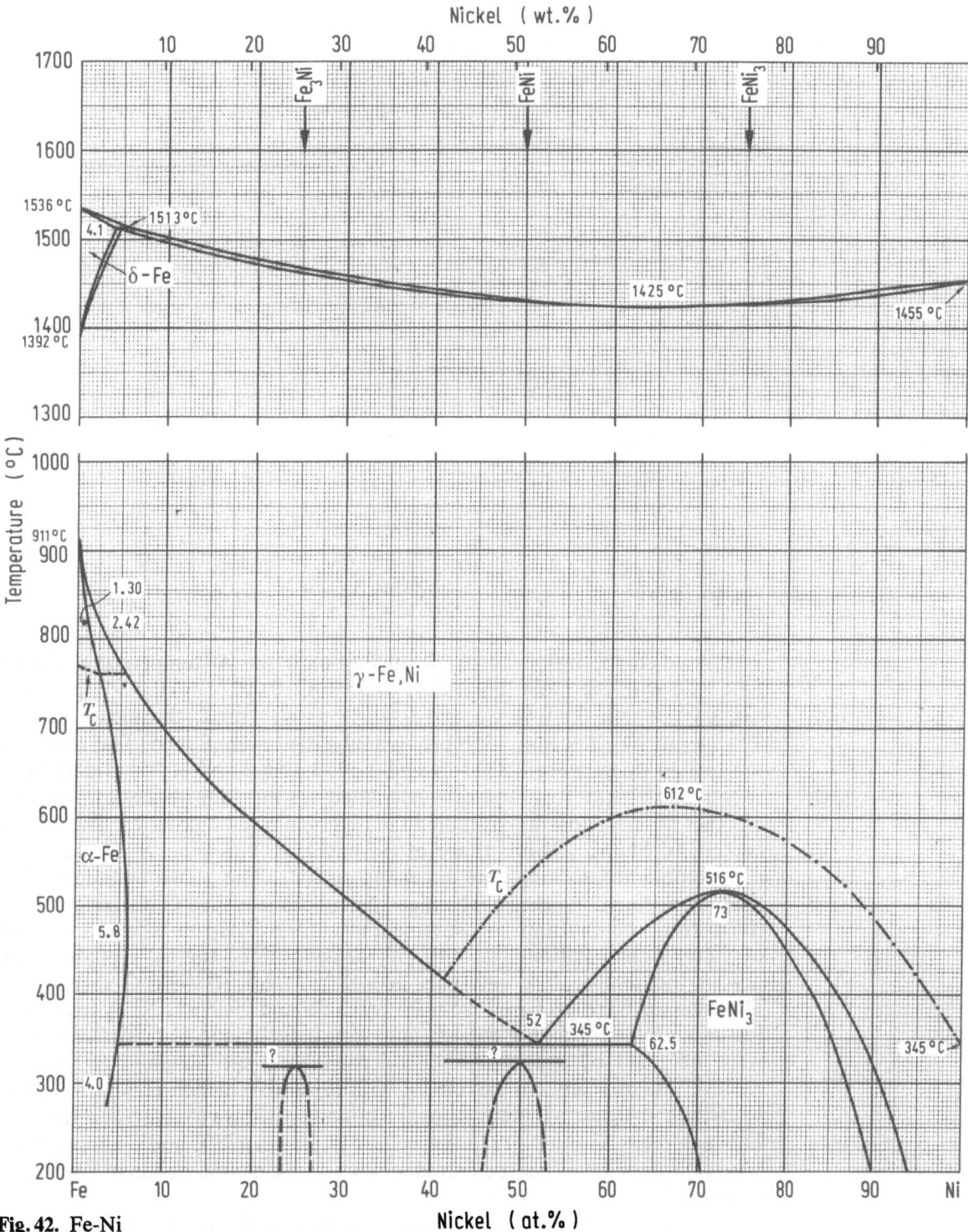

Fig. 42. Fe-Ni

[14] and Romig and Goldstein [36]. The temperature of the eutectoidal reaction was found to lie between 340 and 360°C [20] or at 350°C according to [21]. The latter studied Fe-Ni alloys prepared by decomposition of (Fe, Ni) oxalates by means of X-ray analyses. The temperature of 345°C, suggested by [20], has been accepted in Fig. 42.

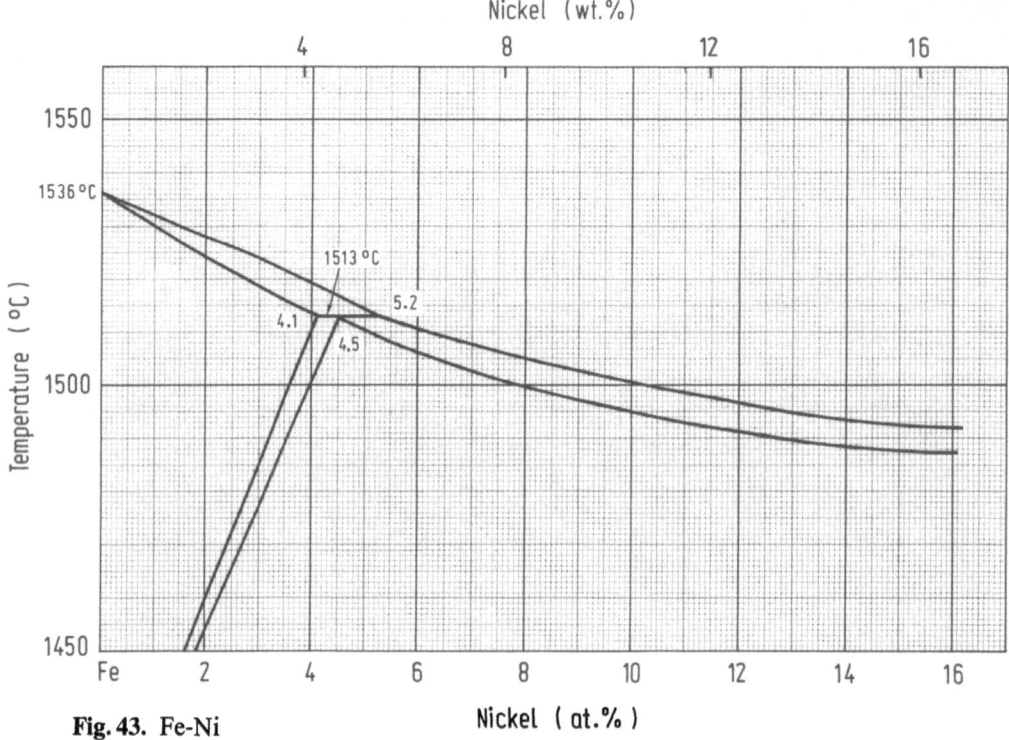

Fig. 43. Fe-Ni

[8] studied the influence of Ni on Fe below the A_3 temperature. Electrolytic iron with the main impurities: 0.035% C, 0.002% N, 0.009% S, 0.005% Co, 0.005% Ni, 0.001% Cu and 0.001% Si was used, the nickel being of 99.95% purity. Electron microprobe analysis was employed. The results for the $(\alpha + \gamma)$ range were 0.83–1.29 and 1.30–2.42 at.% Ni at 850 and 820°C respectively. The $(\alpha + \gamma)$ field was calculated via the Gibbs energies. A number of other investigators have studied the $(\alpha + \gamma)$ phase boundaries. The ones accepted by Shunk [C] are those of [9] and [10], based on X-ray studies on alloys subjected to long-time annealing periods. The phase boundaries have been confirmed by thermochemical calculations using Gibbs energies derived from equilibrium measurements with H_2O/H_2 mixtures and alloys of various compositions [12, 14]. [36] re-determined the α-Fe phase boundary in the temperature intervals 500–800°C and 300–600°C respectively. The samples were prepared from 99.999% Fe and 99.999% Ni by using a diffusion couple [14] and a quench and anneal technique [14, 36]. The results were obtained by electron-probe microanalysis and scanning-transmission electron-microscopy. In agreement with [8] and Winchell [15], it was found by both investigators that the $\alpha/(\alpha + \gamma)$ boundary shows a retrograde solubility. The maximum Ni solubility in the αphase occurs at 475°C and is about 5.8 at.% (Fig. 42). [8] believe that the corresponding experimental points of [9] and [10] be in error.

From the practical point of view, the metastable states occurring in the Fe-rich alloys are sometimes more important than the equilibrium states. The α/γ transformation on continuous heating and cooling has the features of a diffusionless, martensitic-type transformation. Ishida and Nishizawa [16] re-determined the temperatures of the beginning of the martensitic transformation using electrolytic iron (99.95%) and electrolytic nickel (99.95%), the total contents of C and N being less than 0.004%.

It has recently become possible to calculate the equilibrium diagrams of ordering fcc systems fairly accurately by means of a numerical method known as the cluster variation method [26]. Anomalous behaviour has been found for the thermochemical properties of fcc Fe-Ni alloys [25, 27, 28]. Alloys containing 20–50 at.% Ni have a tendency to split into two fcc phases even at high temperatures, which is consistent with the observation that the composition dependence of physical properties (elastic moduli, electrical resistivity, thermoelectric power, etc.) is also anomalous at high temperatures. Kaufman et al. [28] attribute this well known invar-effect to the two γ states of iron. Further information may be obtained from publications of Matsui and Adachi [29], Weiss [30], Shimizu [31] and Katsuki [32]. The latter has published a survey of the theories of the invar characteristics. Investigating the 28–50 at.% Ni alloys, [1] found with zone-melted iron and nickel employing electron irradiation to accelerate diffusion (by means of MES, X-ray and electron microscopy) that the invar character disappears after irradiation up to 250°C. The invar state has thus been shown to be a metastable state.

Order/disorder transformations

The occurrence of superstructures in the regions $FeNi_3$ [2], FeNi and Fe_3Ni [1] mentioned above has been confirmed and more accurately investigated recently. [2] determined the phase diagram in the $FeNi_3$ region by using MES on samples of 99.999% purity. Figure 44 represents his findings. The transition temperature of $FeNi_3$, which has been measured with an accuracy of 1 K, reaches a maximum at 73 at.% Ni and 516°C. It represents a first-order transformation, the two phase region, expected according to Gibbs' phase rule, being thus confirmed experimentally. For other publications pertaining to the ordering of $FeNi_3$, see [22–24].

Experiments on polycrystalline samples obtained from zone-melted iron and nickel using electron irradiation to accelerate diffusion were carried out

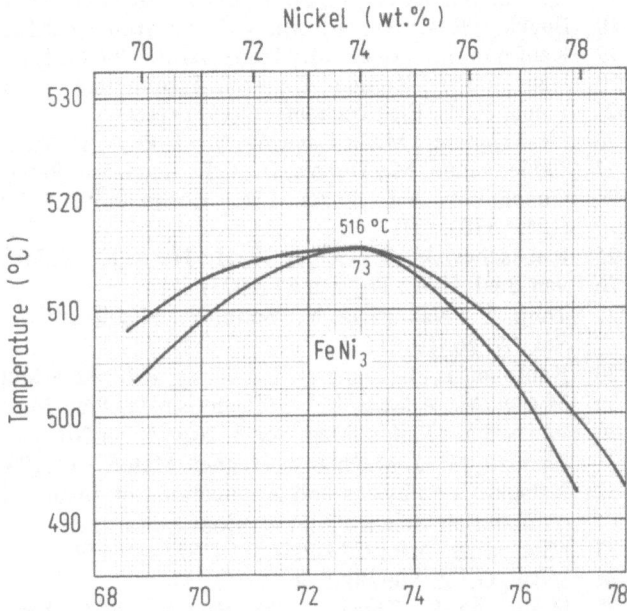

Fig. 44. Fe-Ni

by [1] in the concentration range 26–50 at.% Ni. The investigators believe that equilibrium was actually reached below 400°C. A phase with an ordered FeNi superstructure and an Fe-rich (probably ordered) Fe_3Ni compound were found. The homogeneity ranges and ordering temperatures in Fig. 42 are tentative only.

Magnetic transformations

The Curie points of the γphase lie on a curve which reaches a maximum at 64–68 at.% Ni and 612°C. Miodownik [33] compared the Curie temperatures fitted by the "Inden parameter" [34] with experimental values. Huthmann, to whose thesis [35] the reader may be referred, investigated the simultaneous influence of chemical and magnetic ordering processes in fcc iron. A part phase diagram is proposed.

References

1 Chamberod, A.; Laugier, J.; Penisson, J.M.: J. Magnetism Magnet. Mater. 10 (1979) 139
2 Van Deen: Thesis, University Groningen 1980
3 Schürmann, E.; Brauckmann, J.: Arch. Eisenhüttenwes. 48 (1977) 3
4 Jenkins, C.H.M.; Bucknall, E.H.; Austin, C.R.; Mellor, G.A.: J. Iron Steel Inst. 136 (1937) 188
5 Hellawell, A.; Hume-Rothery, W.: Philos. Trans. R. Soc. London, A 249 (1957) 59
6 Hume-Rothery, W.; Buckley, R.A.: J. Iron Steel Inst. 202 (1964) 531
7 Conard, B.R.; McAneney, T.B.; Sridhar, R.: Metall. Trans. B 9 (1978) 463
8 Hillert, M.; Wada, T.; Wada, H.: J. Iron Steel Inst. 205 (1967) 539
9 Owen, E.A.; Sully, A.H.: Philos. Mag. 27 (1939) 614
10 Owen, E.A.; Liu, Y.H.: J. Iron Steel Inst. 163 (1949) 132
11 Troiano, A.R.; McGuire, F.T.: Trans. ASM 31 (1943) 340
12 Kubaschewski, O.; von Goldbeck, O.: Trans. Faraday Soc. 322 (1949) 948
13 Zener, C.: Trans. AIME 167 (1946) 513
14 Goldstein, J.I.; Ogilvie, R.E.: Trans. Metall. Soc. AIME 233 (1965) 2083
15 Winchell, P.: Private communication, quoted by [8]
16 Ishida, K.; Nishizawa, T.: Jpn. Inst. Met. 15 (1974) 217
17 Oelsen, W.; Wever, F.: Arch. Eisenhüttenwes. 19 (1947) 97
18 Buckley, R.A.; Hume-Rothery, W.: J. Iron Steel Inst. 201 (1963) 227
19 Kaufman, L.; Cohen, M.: Trans. AIME 206 (1956) 1393
20 Heumann, T.; Karsten, G.: Arch. Eisenhüttenwes. 34 (1963) 781
21 Hund, F.: Z. Elektrochem. 56 (1952) 609
22 Desjonquères, M.C.; Lavagna, M.: J. Phys. F: Met. Phys. 9 (1979) 1733
23 Lebienvenue, M.; Dubois, B.: C.R. Acad. Sci. Paris Ser. C, 280 (1975) 1251
24 Gonam'kov, V.I.; Puzei, I.M.; Loshmanov, A.A.: Sov. Phys. Crystallogr. 10 (1965) 338
25 Kubaschewski, O.; Geiger, K.-H.; Hack, K.: Z. Metallkd. 68 (1977) 337
26 van Baal, C.M.: Physica 64 (1973) 571
27 Tanji, Y.; Nakagawa, Y.; Saito, Y.; Nishimura, K.; Nakatzuka, K.: Phys. Stat. Sol. (A) 56 (1979) 513
28 Kaufman, L.; Clougherty, E.; Weiss, R.J.: Acta Metall. 11 (1963) 323
29 Matsui, M.; Adachi, K.: J. Magnetism Magnet. Mater. 10 (1979) 152
30 Weiss, R.J.: Philos. Mag. Ser. 8, 26.1 (1972) 261
31 Shimizu, M.: J. Magnetism Magnet. Mater. 10 (1979) 231
32 Katsuki, A.: Physics and Applications of Invar Alloys, Honda Memorial Series on Materials Science 3 (Proc. Conf.), Maruzen
33 Miodownik, A.P.: CALPHAD 1 (1977) (2) 157
34 Inden, G.: Z. Metallkd. 66 (1975) 577
35 Huthmann, H.: Thesis, Techn. Hochschule Aachen 1977
36 Romig, A.D.; Goldstein, J.I.: Metall. Trans. 11A (1980) 1151

Iron–Oxygen

The iron-oxygen system exhibits the following stable phases: The solution of oxygen in $\alpha(\delta)$, γ and liquid Fe, "FeO" (wüstite), Fe_3O_4 (magnetite), Fe_2O_3 (hematite) and liquid iron oxide. A complete and consistent assessment of the phase boundaries and the thermochemical properties has recently been presented by Spencer and Kubaschewski [1] whose development of the properties of the various phases will here be adhered to, starting with the wüstite phase.

The results of many of the earlier experimental determinations are summarized in the compilations due to Hansen and Anderko [A], to Elliott [B] and to Shunk [C]. More recent experimental studies of wüstite have provided additional phase boundary information [2–6], and reviews of the thermodynamic properties made by Giddings and Gordon [7] and by Löhberg and Stannek [8] also contain summary information on the stability range of wüstite. Fig. 45 illustrates the results of a number of investigations [2–4, 6, 7, 9–17]. The phase limits of Darken and Gurry [9] have been accepted since their studies were self-consistent both with regard to their directly-observed phase boundaries and to the thermodynamic values determined as part of the same study. Thermodynamic values across the wüstite phase have been assessed from numerous experimental measurements by [1].

There has been some controversy in the literature regarding the existence of three allotropic varieties of wüstite, W_1, W_2, and W_3 above 1,184 K and three additional varieties W_1', W_2' and W_3' below this temperature. Carel and Gavarri [14] have recently summarized the structural evidence for varieties W_1, W_2, and W_3, while [7] have discussed the experimental problems associated with some thermodynamic studies which, by demonstrating changes in measured properties, also appear to reveal structural changes in wüstite.

Phase boundary studies relevant to the wüstite-magnetite equilibrium have been summarized by [A] whose selected values are based on the work of Darken and Gurry [15] and are shown in Fig. 45. Activity measurements across the *magnetite* phase are due to [15], Greig et al. [16] and Smiltens [17] (gas equilibration technique) and to Sockel and Schmalzried [18] (solid electrolyte emf).

There are considerable differences with regard to the range of stability of *hematite*. For example, White et al. [19] and Schmahl [20] report a solubility of Fe in Fe_2O_3 of 0.12–0.35 at.% at 1,573 K increasing to about 0.53 at.% at 1,723 K. On the other hand, [16] found the solubility to be less than 0.024 at.% Fe for all temperatures between 1,473 and 1,723 K, and Ruer and Nakamoto [21] report "practically zero" solubility at 1,423–1,473 K. The boundaries illustrated in Fig. 45 are based on the emf and static-oxygen-pressure measurements carried out by Komarov and Oleinikov [22] which produced the following values for the solubility of Fe in *hematite*.

T (K)	1,173	1,273	1,526	1,567	1,611	1,657
Fe (at.%)	0.014	0.024	0.054	0.066	0.086	0.123

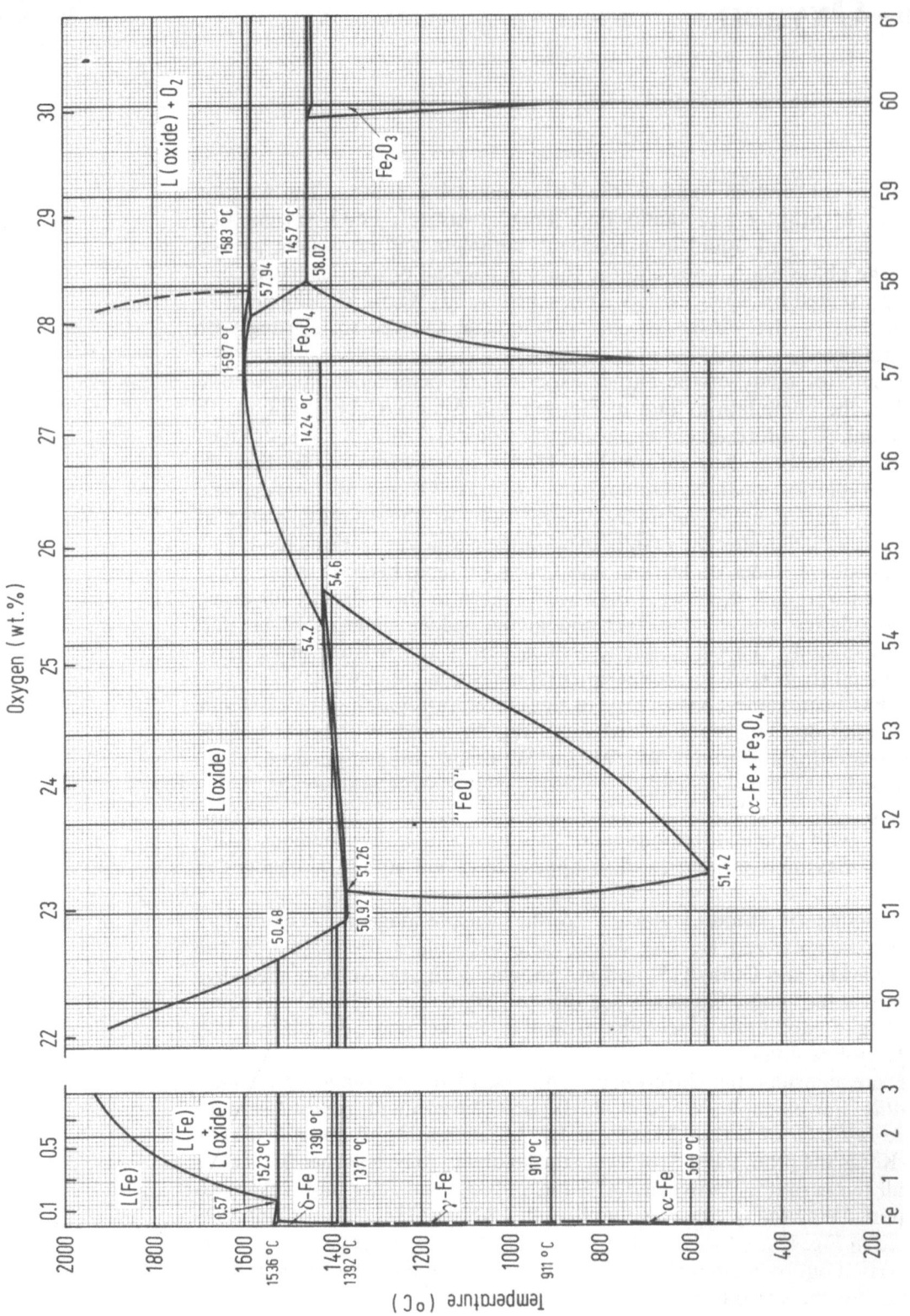

The equilibrium between *liquid* iron, oxygen and solid iron has been studied for instance by Chipman and Marshall [23] using H_2/H_2O equilibration, by [15] using CO/CO_2 equilibration, by Goto and Matsushita [24] and Fischer and Pateisky [25] both employing emf techniques. The Gibbs energies of the liquid oxide at the Fe-rich boundary obtained by [15, 23] and [24] are in good agreement whereas the values reported by [25] are rather different. It may be seen from Fig. 45 that between 1,663 and 1,796 K the liquid oxide phase is in equilibrium with *solutions of oxygen in δ(Fe)* and at temperatures above 1,796 K with *solutions in liquid Fe.* Summary reviews of the phase boundaries relevant to these equilibria have been presented by [A–C], by Sawamura and Sano [26] and Distin et al. [27], the last mentioned having defined the solubility limit at temperatures up to 2,230 K using an experimental method involving levitation of the specimen. Their results [27] are incorporated in Fig. 45.

More recent measurements by Fischer and Schumacher [32] based on the levitation melting method, are in fair agreement with data used by [1]. Their [32] results may be expressed by the following formulae:

$$\lg(\text{wt.\% O}) = -6{,}629\, T^{-1} + 2.939 \quad (1{,}508\text{--}1{,}850\,°C),$$
$$\lg(\text{wt.\% O}) = -9{,}830\, T^{-1} + 4.496 \quad (1{,}850\text{--}2{,}046\,°C).$$

Expressed in the form of actual solubility values:

Temp. °C	wt.%	at.%
1,600	0.25	0.86
1,700	0.38	1,31
1,800	0.55	1.89
1,850	0.73	2.52
1,900	0.94	3.20

The solubility of oxygen in δ(Fe) has been determined by Tankins and Gokcen [28] from studies of the equilibrium with gaseous H_2/H_2O mixtures in the temperature range 1,693–1,783 K.

The authors summarize their data and those obtained by previous investigators by the following equation:

$$\lg(\text{wt.\% O}) = -12{,}630\, T^{-1} + 5.51.$$

The solubility boundary for oxygen in δ(Fe) shown in Fig. 45 is based on this equation.

The solubility of oxygen in both γ(Fe) and α(Fe) is very uncertain and there are considerable discrepancies between the results obtained by many investigators. It is apparent that even very small amounts of impurities, in particular silicon and aluminium, can give rise to formation of oxide particles and lead to experimental values for the solubility of oxygen which are too high. Consequently it has been found that the purer the iron sample used, the lower is the oxygen solubility determined. It is also a general finding that the solubility of oxygen in γ(Fe) is 'much lower' than in α(Fe). Slowman [29] for instance recommended the value of Kitchener et al. [30] for the solubility in γ(Fe), namely 0.003 wt.% O, which obviously is not much of a help. In Sloman's assessment the unavoidable presence of surface oxygen was accounted for by deducting 0.004 wt.% due to this source. The very tentative solubilities selected here, based on the lowest reported values in α(Fe) [31], are the temperature invariant values of 2.4×10^{-4} at.% O in γ(Fe) and 2.4×10^{-3} at.% O in α(Fe).

References

1   Spencer, P.J.; Kubaschewski, O.: CALPHAD 2 (1978) 147
2   Ackermann, R.J.; Sandford, R.W.: U.S. At. Energy Comm. Rep. ANL-7250 (1966)
3   Swaroop, B.; Wagner, J.B.: Trans. AIME 239 (1967) 1215
4   Campserveux, J.; Boreau, G.; Picard, C.; Gerdanian, P.: Rev. Int. Hautes Temp. Refract. 6 (1969) 165
5   Lykasov, A.A.; Kuznetsov, Yu.S.; Pil'ko, E.I.; Shishkov, V.I.; Kozheurov, V.A.: Russ. J. Phys. Chem. 43 (1969) 1754
6   Rizzo, H.F.; Gordon, R.S.; Cutler, I.B.: J. Electrochem. Soc. 116 (1969) 266
7   Giddings, R.A.; Gordon, R.S.: J. Am. Ceram. Soc. 56 (1973) 111
8   Löhberg, K.; Stannek, W.: Ber. Bunsenges. Phys. Chem. 79 (1975) 244
9   Darken, L.S.; Gurry, R.W.: J. Am. Chem. Soc. 67 (1945) 1398
10   Schenck, R.; Dingmann, T.: Z. Anorg. Chem. 166 (1927) 113; and 171 (1928) 239
11   Barbi, G.B.: J. Phys. Chem. 68 (1964) 2912
12   Vallet, P.; Raccah, P.: Compt. Rendus 258 (1964) 3679, 4028
13   Fender, B.E.F.; Riley, F.D.: J. Phys. Chem. Solids 30 (1969) 793
14   Carel, C.; Gavarri, J.R.: Met. Res. Bull. 11 (1976) 745
15   Darken, L.S.; Gurry, R.W.: J. Am. Chem. Soc. 68 (1946) 798
16   Greig, J.W.; Posnjak, E.; Merwin, H.E.; Sosman, R.B.: Am. J. Sci. 30 (1935) 239
17   Smiltens, J.: J. Am. Chem. Soc. 79 (1957) 4877
18   Sockel, H.G.; Schmalzried, H.: Ber. Bunsenges. Phys. Chem. 72 (1968) 745
19   White, J.; Graham, R.; Hay, R.: J. Iron Steel Inst. 131 (1935) 91
20   Schmahl, N.G.: Z. Elektrochem. 47 (1941) 821
21   Ruer, R.; Nakamoto, M.: Rec. Trav. Chim. 42 (1923) 675
22   Komarov, V.F.; Oleinikov, N. N.: Izv. Akad. Nauk SSSR, Neorg. Mater. 3 (1967) 1064
23   Chipman, J.; Marshall, S.: J. Am. Chem. Soc. 62 (1940) 299
24   Goto, K.; Matsushita, Y.: Tetsu To Hagane 52 (1966) 827
25   Fischer, W.A.; Pateisky, G.: Arch. Eisenhüttenwes. 41 (1970) 661
26   Sawamura, H.; Sano, K.: Spec. Rep. No. 9, Sub-Committee for Physical Chemistry of Steelmaking, Jap. Soc. for Promotion of Science (1967)
27   Distin, P.A.; Whiteway, S.G.; Masson, C.R.: Can. Metall. Q. 10 (1971) 13
28   Tankins, E.S.; Gokcen, N.A.: Trans. ASM 53 (1961) 843
29   Slowman, H.A.: Private communication 1956
30   Kitchener, J.A.; Bockris, J.O'M.; Gleiser, M.; Evans, J.W.: Acta Metall. 1 (1953) 93
31   Sifferlen, R.: Compt. Rendus 247 (1958) 1608
32   Fischer, W.A.; Schumacher, J.F.: Arch. Eisenhüttenwes. 49 (1978) 431

Fe–Os   Iron–Osmium

Figs. 46, 47

The only data pertaining to the Fe-Os phase diagram stem from investigations undertaken by Buckley and Hume-Rothery [1] and by Fallot [2]. The former studied the liquidus-solidus relations in the Fe-rich region up to 7 at. % Os, while [2] carried out thermal analyses to establish the α/γ transformation temperatures.

However, osmium with a melting point of $(3,027 \pm 20)$ °C accepted here shows great similarity to ruthenium (both having the hcp, A3 structure) and one may assume that both metals are also similar in their alloying behaviour. The Fe-Ru system has been established experimentally in the region 0–40 at.% Ru [3] (*vide*). Therefore, the tentative Fe-Os phase diagram suggested by Moffatt [4], which is based on the Fe-Ru diagram has been accepted and is shown in Fig. 46.

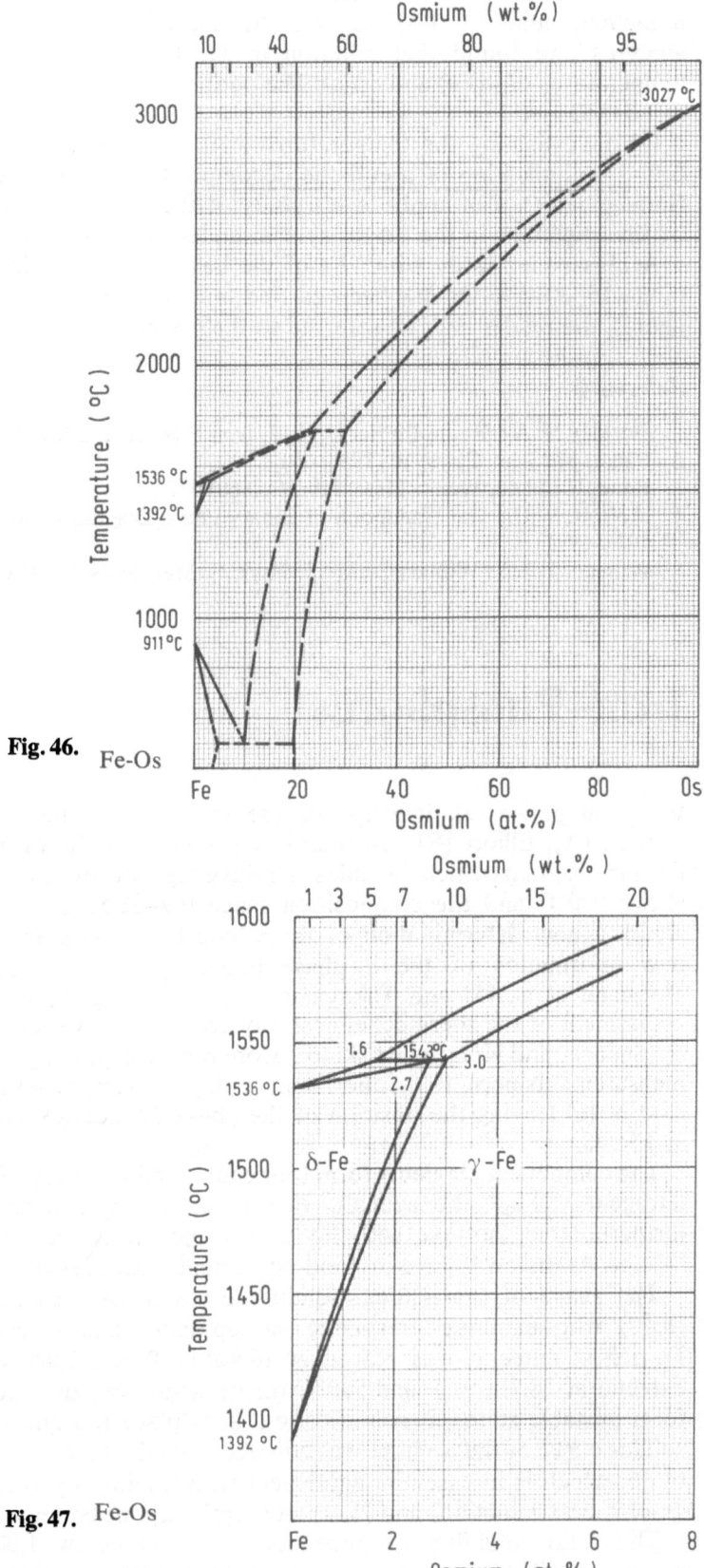

Fig. 46. Fe-Os

Fig. 47. Fe-Os

[1] prepared alloys from BISRA "H" iron with a purity of 99.95% and a melting point of 1,532°C (see 'Introduction'). [1] carried out thermal analyses and found that addition of Os to Fe raises the transformation temperatures of γ(Fe) and δ(Fe). The peritectic reaction $L + \gamma \rightleftharpoons \delta$ takes place at 1,543°C (see Fig. 47).

[2], investigating the temperature range 600–900°C in the Fe-rich region, found no indication of a eutectic decomposition of the γ solid solution. His findings which also apply to his study of the Fe-Ru system, are superseded by investigations in the Fe-Ru system undertaken by Raub and Plate [3].

A Mössbauer study with ^{57}Fe of the hexagonal phase in the concentration range 15–30 at.% Os has been carried out by Pearson and Williams [5]. For details, the reader may be referred to their paper.

References

1 Buckley, R.A.; Hume-Rothery, W.: J. Iron Steel Inst. 201 (1963) 121
2 Fallot, M.: Ann. Phys. 10 (1938) 291
3 Raub, E.; Plate, W.: Z. Metallkd. 51 (1960) 477
4 Moffatt, W.G.: The Handbook of Binary Phase Diagrams. General Electric Comp. 1976–1981
5 Pearson, D.I.C.; Williams, J.M.: J. Phys. F: Met. Phys. 9 (1979) 1797

Fe–P

Figs. 48, 49

Iron–Phosphorus

Work on phase relationships of the system Fe-P has been reviewed by Hansen [A], Elliott [B] and Shunk [C]. Figure 48 shows the modified equilibrium diagram which includes a reinvestigation of the temperature range above 900°C and the composition range 0.9–21.5 at.% P by Wachtel et al. [1, 2], a new determination of the γ-loop by Lorenz and Fabritius [3] and new information of the α phase boundary by Doan and Goldstein [4], Hofmann et al. [5] and Takayama et al. [11]. Fig. 48 is consistent with an assessment of the thermochemical properties of the various phases published by Spencer and Kubaschewski [6]. More recent high-temperature calorimetric measurements permitted Schürmann et al. [12] to improve the thermochemical data of [6] leaving the position of the phase boundaries essentially the same as Fig. 48.

The component elements form four compounds, namely: Fe_3P (body centred tetragonal), Fe_2P (hexagonal), FeP (orthorhombic, MnP type) and FeP_2 (orthorhombic, isotypic with FeS_2). The parameters reported by Rundqvist [7] and Aronsson [8] are in good agreement with previous investigations.

The shape of the liquidus/solidus curves in [A] has been confirmed by [1, 2] who employed magnetic susceptibility and thermal analysis. The Fe-richest eutectic was placed at 16.9 at.% P and 1,048°C, the metastable eutectic at 18.7 at.% P and 930°C (melts with more than 16.7 at.% P tend to form unstable states, if solidification takes place too quickly). In Fig. 48 the solidus curve of the α phase was adopted from the thermodynamic assessment by [6] which is in essential agreement with findings by Haughton [9] (micrographic determination) and [1, 2] (susceptibility measurements).

The solid solubility of phosphorus in iron below 1,000°C [5, 11] and below 1,100°C [4] has been studied more recently by microprobe and micro-

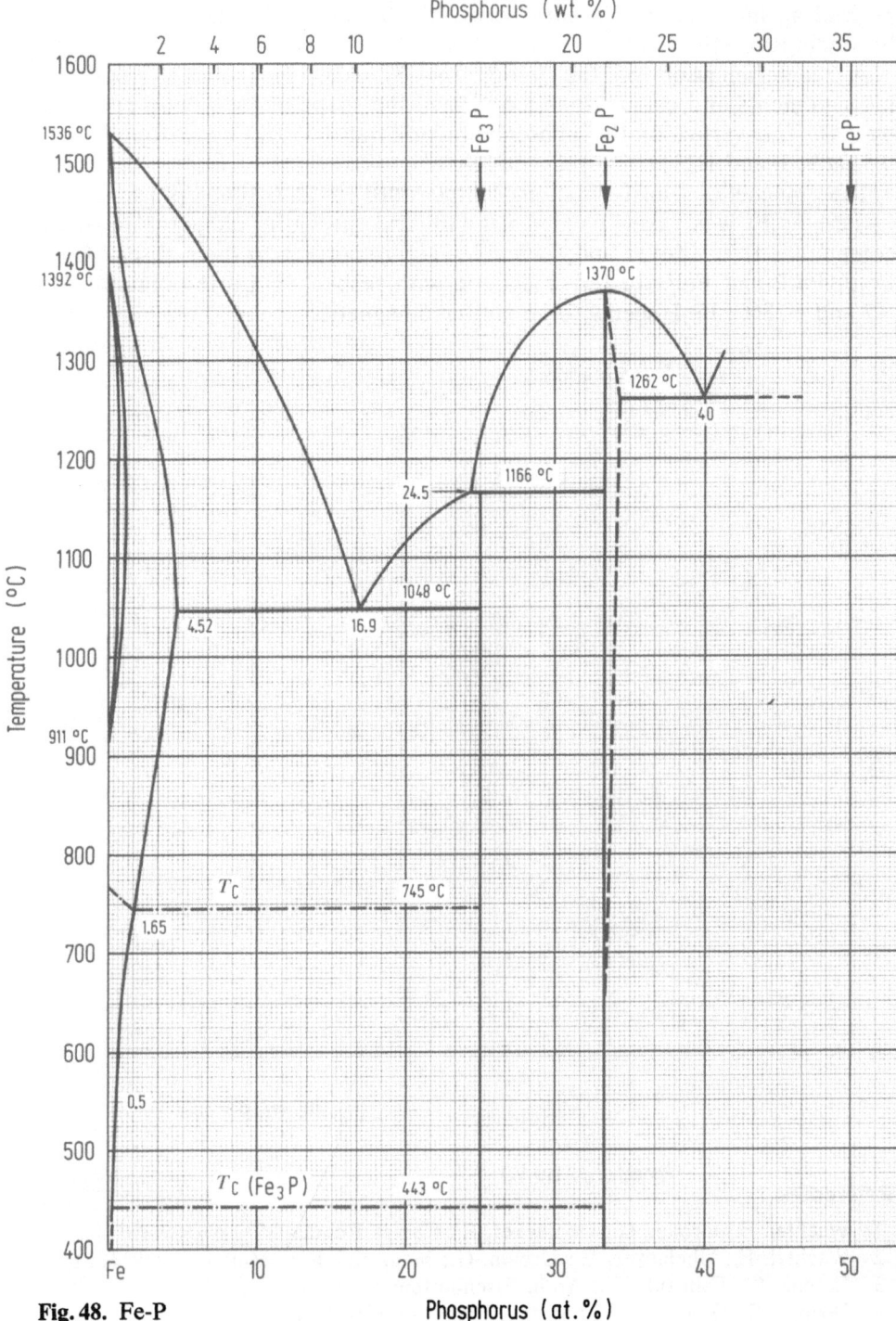

Fig. 48. Fe-P

hardness of "diffusion-couples" [5], by direct measurements (electron microprobe analysis) and extrapolation from the ternary system Fe-Ni-P [4] and by X-ray diffraction and electron probe micro-analyses [11]. The results of [4, 5] are in excellent agreement and confirm the maximum solubility of P in α(Fe) (4.52 at.%) already found by [1], while values of [11] point to a slightly lower solubility. The results indicate that the magnetic transition has a marked effect on the α-Fe phase boundary. It may be mentioned that [5] noticed an increase in solubility between 800 and 835°C which is supported by findings of [4].

A redetermination of the γ-loop by [3] using magnetic susceptibility measurements and calculations, confirmed the shape of the γ-loop published by [A]. The vertex of the phase boundaries at 1,150°C were found to be 0.55 at.% P and 1.23 at.% P respectively (Fig. 49).

The magnetic transformation temperature in the two phase region (α-Fe + Fe_3P) was taken from Lorig [10] and the Curie temperature of the compounds from Meyer and Cadeville [11]. According to their observations, Fe_3P, Fe_2P and FeP are ferromagnetic with respective Curie points at 443, −7 and −58°C. Fe_2P seems to be antiferromagnetic.

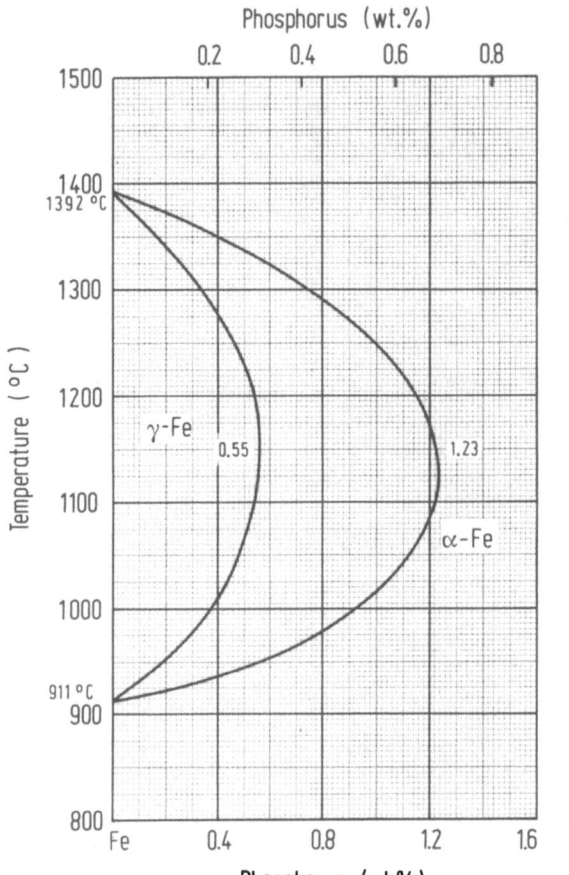

Fig. 49. Fe-P

References

1 Wachtel, E.; Urbain, G.; Übelacker, E.: Compt. Rendus 257 (1963) 2,470
2 Wachtel, E.; Übelacker, E.; Urbain, G.: Mem. Sci. Rev. Metall. 61 (1964) 515
3 Lorenz, K.; Fabritius, H.: Arch. Eisenhüttenwes. 33 (1962) 269
4 Doan, A.S.; Goldstein, J.I.: Metall. Trans. 1 (1970) 1,759
5 Hofmann, H.-P.; Löhberg, K.; Reif, W.: Arch. Eisenhüttenwes. 41 (1970) 975
6 Spencer, Ph.; Kubaschewski, O.: Arch. Eisenhüttenwes. 49 (1978) 225
7 Rundqvist, S.: Acta Chem. Scand. 16 (1962) 1
8 Aronsson, B.: Acta Chem. Scand. 9 (1955) 137
9 Haughton, J.L.: J. Iron Steel Inst. 115 (1927) 417
10 Lorig, C.H.: In: Metals Handbook, publ. ASM, 7th ed., 1948
11 Takayama, T.; Wey, M.Y.; Nishizawa, T.: Trans. Jpn. Inst. Met. 22 (1981) 315
12 Schürmann, E.: Arch. Eisenhüttenwes. 52 (1981) 51

Iron–Lead

According to Hansen and Anderko [A] liquid iron and lead are mutually virtually insoluble. However, later investigations revealed that mutual solubility is significant and can be measured. Lord and Parlee [1], Araki [2] and Morozov and Ageev [3] determined the solubility of *Pb in liquid* Fe. [1] and [2] found the monotectic temperature at 1,535°C and 1,530°C, respectively. The experimental solubility values may be seen in Fig. 50. The slope of the solubility vs. temperature curves of none of these results agrees reasonably

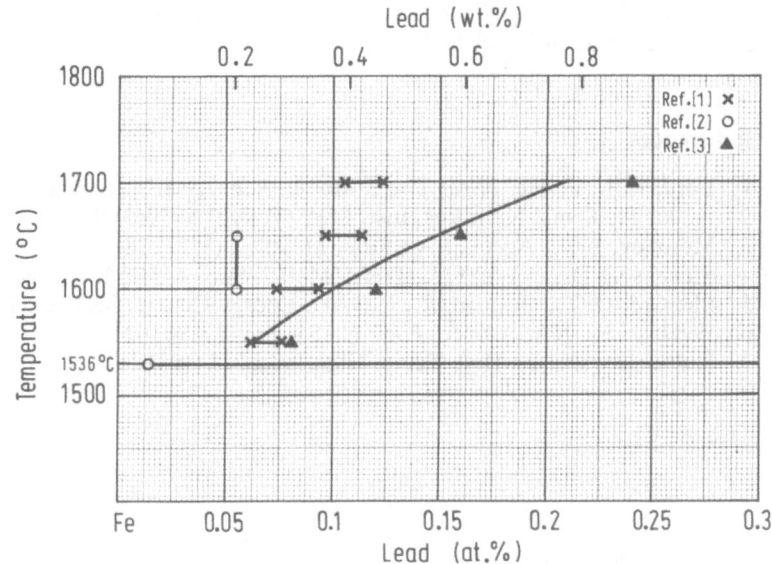

Fig. 50. Fe-Pb

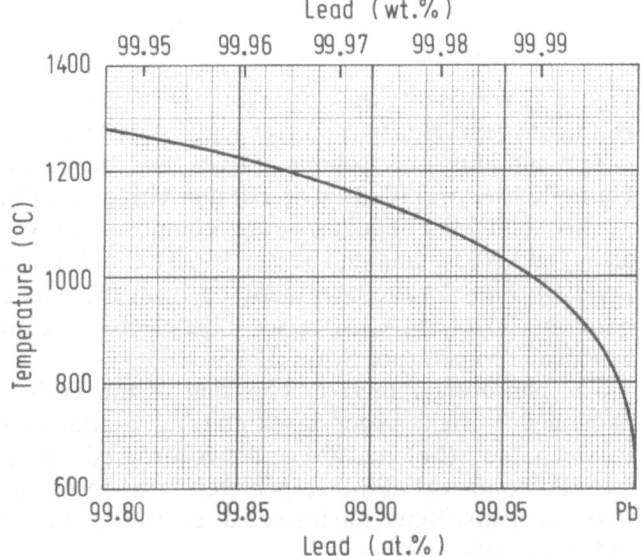

Fig. 51. Fe-Pb

well with a rule suggested by Kubaschewski [7], namely that the quotient of the excess heats and entropies of solution should be $\bar{H}_{Pb}/\bar{S}_{Pb}^E$ equal to about 3,400 K. A compromise curve is therefore drawn in Fig. 50, corresponding to the relationship:

$$\text{or}\quad \begin{aligned} -RT \lg N_{Pb} &= 61{,}300 - 19.0\,T \\ \lg (\text{at.\% Pb}) &= -13{,}400\,T^{-1} + 6.16. \end{aligned}$$

This, in the opinion of the present author, is the best that can be suggested on the basis of the available information. The equations may be used for an extrapolation to higher temperatures.

The solubility of Pb in *solid* iron at the monotectic temperature (ca. 1,530 °C) was found to be 2.7×10^{-4} at.% [2].

Results reported on the solubility of Fe in *liquid* Pb and represented in Fig. 51 are taken from a publication by Stevenson and Wulff [4] who surveyed earlier data and re-investigated the Fe solubility in the temperature range 700–1,300 °C. They agree with the results published by Shephard and Parkman [5] (700–900 °C) and by Weeks [6]. The latter found a solubility of 8.5×10^{-4} at.% Fe at 600 °C.

References

1 Lord, A.E.; Parlee, N.A.: Trans. Metall. Soc. AIME 218 (1960) 644
2 Araki, T.: Trans. Natl. Res. Inst. Met. (Tokyo) 5 (1963) 91
3 Morozov, A.N.; Ageev, I.A.: Russ. Metall. 4 (1971) 78
4 Stevenson, D. A.; Wulff, J.: Trans. Metall. Soc. AIME 221 (1961) 271
5 Shepard, O.C.; Parkman, R.: U.S. At. Energy Comm. ORO-38 (1950)
6 Weeks, J.R.: NASA Spec. Publ. NASA-SP-41 (1963) 21
7 Kubaschewski, O.: High Temp. High Press. 13 (1981) 435

Fe–Pd   Iron–Palladium
Fig. 52

Figure 52 represents the phase relationships in the system Fe-Pd constructed from data reported by the investigators [1] to [5].

The liquidus and solidus curves are based on results of Gibson and Hume-Rothery [1] (0–11 at.% Pd) and Grigorjew [2].

The solid state equilibria are composed of data obtained by Raub et al. [4] and Kussmann and Jessen [5]. The melting and γ/δ(Fe) transformation points in Fig. 52 have been adopted from the publication by [1]. The melting point of Pd (1,554 °C) is a secondary reference point on the International Practical Temperature Scale (IPTS-68).

The phase diagram resembles that of the Fe-Ni system. It belongs to the expanded γ field type. It exhibits a continous series of solid solutions between fcc γ(Fe) and Pd and two ordered phases at lower temperatures, γ_1(FePd) and γ_2(FePd$_3$) respectively. In contrast to the Fe-Ni and Fe-Pt systems, a miscibility gap instead of an ordered phase Fe$_3$Pd has been observed in the fcc solid solution.

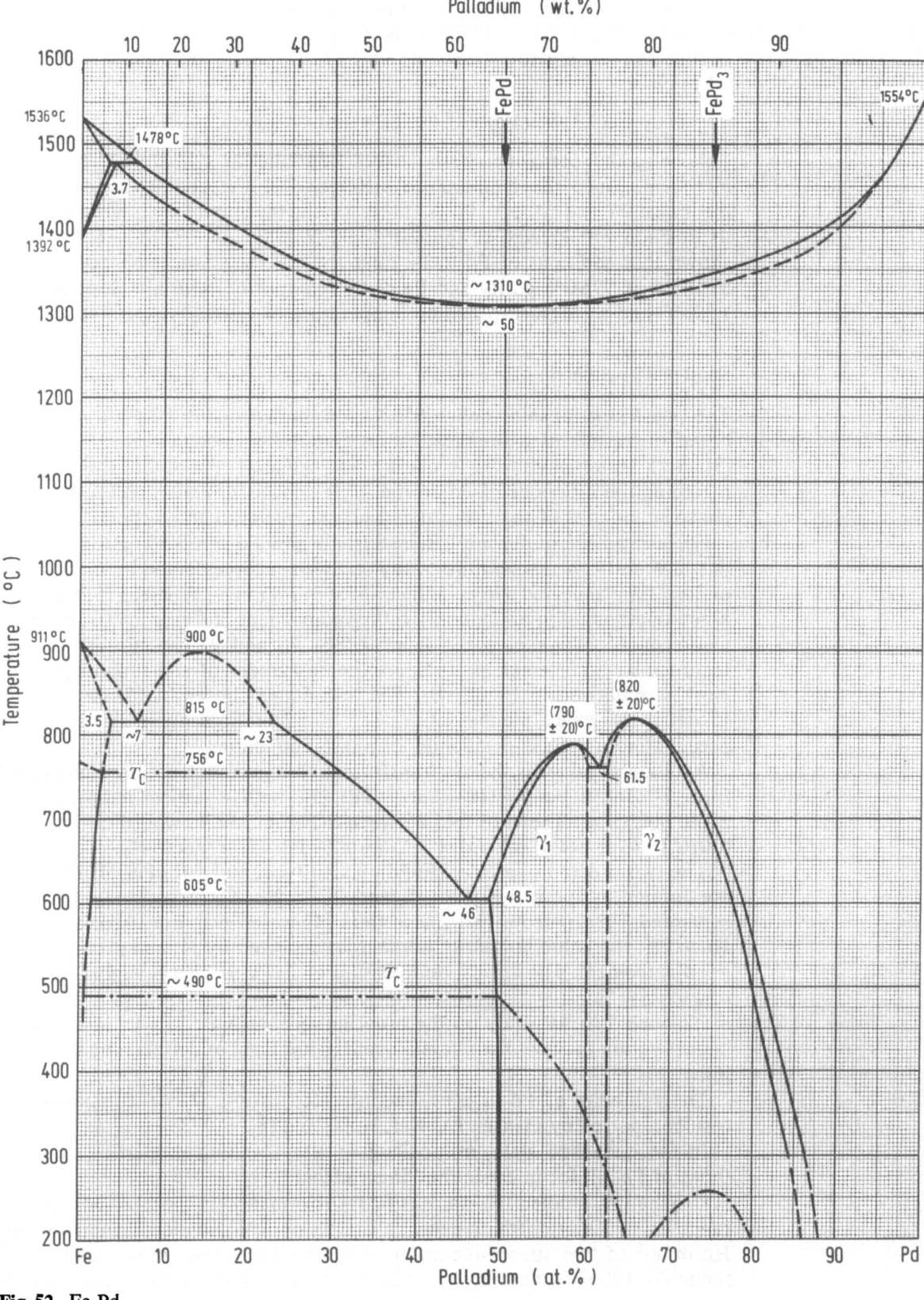

Fig. 52. Fe-Pd

The liquidus and solidus curves were determined by thermal analysis. Gibson and Hume-Rothery [1] re-investigated the 0–11 at.% Pd region and confirmed the results obtained by Grigorjew [2]. The δ(Fe)-liquidus and solidus fall steeply to a peritectic at 1,478 °C, where the following reaction takes place:

$$\delta\text{-Fe}(3.3\,\text{at.\% Pd}) + L(6.5\,\text{at.\% Pd}) \leftrightarrows \gamma\text{-Fe}(3.7\,\text{at.\% Pd}).$$

Following the peritectic, both curves continue to fall steeply towards a minimum. Since the temperature difference of the liquidus-solidus gap in the range 37 and 61 at.% Pd is rather small, the minimum is very flat, and its position was located at approximately 50 at.% and 1,310 °C. [1] have also shown that additions of Pd to Fe raise the γ/δ transition temperature. They used Pd of 99.96 % purity and Fe of 99.96 % purity (with a melting point of 1,533 °C and a γ/δ transformation temperature of 1,389 °C respectively).

Solid state equilibria

The phase relationships below 950 °C and shown in Fig. 52 are a composite based on results of a re-investigation undertaken independently by [4] and [5]. Both teams have tried to clarify discrepancies reported in the literature regarding the influence of Pd on the α/γ(Fe) transformation and to settle the phase boundaries of the two ordered phases γ_1(FePd) and γ_2(FePd$_3$).

[4], using high purity Fe ('Reinsteisen') and Pd of 99.9 % purity, carried out high and low temperature X-ray analyses, supplemented by microscopy, on long-time annealed alloys in the 6–85 at.% Pd range. Whereas [5] employed kinetic methods of investigation, i. e. thermomagnetic and dilatometric measurements on 25 alloys prepared from 99.8 % Pd and 99.99 % Fe. The results recorded by both teams are in approximate agreement and support previous work summarized in Elliott [B]. The techniques chosen by [5] do not allow to detail the exact number of phases in equilibrium compared to [4] but permit a more precise positioning of phase boundaries.

Addition of Pd leads the α/γ transformation to a monotectoid which extends from 7–23 at.% Pd [4] (10–25 at.% Pd) [5], with a critical temperature at about 900 °C [4]. The eutectoid between α(Fe) and FePd was placed at about 620 °C (~43 at.% Pd) [4] and 605 °C (46 at.% Pd) [5] respectively.

The two ordered phases, γ_1(FePd), CuAu-type, and γ_2(FePd$_3$), CuAu$_3$-type, are formed at about (790 ± 20) °C and (820 ± 20) °C, respectively. They are separated by a narrow two phase field [4] which was not detected by [5]. However, the solubility phase boundaries reported by [4] and [5] are in very good agreement. The critical temperature of FePd$_3$ is clearly higher than that of FePd [4, 5].

A detailed X-ray study on the ordering mechanism of Fe-Pd alloys in the concentration range of the phases γ_1 and γ_2 has been undertaken by Babanova [6].

The solubility of Pd in α(Fe) incorporated in Fig. 52 is a combination of results reported by Hansen [A: Ref. 3], [4] and [5].

In the quenched state, iron-palladium alloys exhibit properties similar to those of Fe-Ni, for instance the "invar effect".

Oshima [10] studied the martensitic transformation in the 23.6–30.6 at.% Pd range using X-ray diffractometry as well as optical and electron microscopy. He observed two successive martensitic transformations in the temperature range 40–196 °C in an alloy containing 29.7 at.% Pd:–firstly a thermoelastic transition from fcc austenite to fct martensite (shape memory effect), thus confirming results of Sohmura et al. [11] and secondly a typical non-thermo-elastic type transition from fct martensite to bct martensite.

The Curie temperatures indicated in Fig. 52 are based on findings quoted by [5]. Meier et al. [9] have measured the effect of pressure on the Curie temperatures of Fe-Pd alloys by thermal scanning of the ^{57}Fe Mössbauer resonance up to 170 kbar. The critical magnetic behaviour has also been studied by Kouvel and Comly [7] who confirmed among others the existence of the "giant moments" in dilute Pd-Fe alloys, previously discovered by neutron diffraction. Biehl and Flanagan [8], who found in this dilute Pd-Fe region (about 0.3 at.% Fe) an anomalous lattice parameter expansion, suggest a connection between this anomaly, the "giant moments" and the enhanced H_2 solubility.

References

1 Gibson, W.S.; Hume-Rothery, W.: J. Iron Steel Inst. 189 (1958) 243
2 Grigorjew, A.T.: Z. Anorg. Chem. 209 (1932) 295
3 Kuprina, V.V.; Grigorjew, A.T.: Russ. J. Inorg. Chem. 4 (1959) 297
4 Raub, E.; Beeskow, H.; Loebich, O.: Z. Metallkd. 54 (1963) 549
5 Kussmann, A.; Jessen, K.: Z. Metallkd. 54 (1963) 504
6 Babanova, E.N. et al.: Fiz. Met. Soedin., (Sverdlovsk) 6 (1978) 3, Met. Abstr. 12–1326 (1979)
7 Kouvel, J.S.; Comly, J.B.: Critical Phenomena in Alloys, Magnets and Super-conductors, (1971) 437, Met. Abstr. 72 0084
8 Biehl, G.; Flanagan, T.B.: Solid State Commun. 28 (1978) 751
9 Meier, J.S.; Christoe, C.W.; Wortmann, G.: Solid State Commun. 15 (1974) 485
10 Oshima, R.: Scr. Metall. 15 (1981) 829
11 Sohmura, T.; Oshima, R.; Fujita, F.E.: Scr. Metall. 14 (1980) 855

Iron-Promethium, see Fe-R: Iron-Rare Earth Metals, Fig. 60

Fe–Pm

Iron–Platinum

Fe–Pt

Fig. 53

The Fe-Pt system shown in Fig. 53 resembles in many respects the Fe-Ni and Fe-Pd systems, i. e. it exhibits a very narrow solidification interval, belongs to the open γ field type, entails three superstructures (Fe_3Pt, $FePt$, $FePt_3$), shows an invar effect in alloys containing 22–35 at.% Pt [12] and, like Fe-Pd, has the desirable ability to form thermoelastic martensites that underlie the shape-memory effect, which is linked to prior ordering in the austenite.

In Fig. 53 the liquidus and solidus curves are based on results of Buckley and Hume-Rothery [1] (0–10 at.% Pt) and Isaac and Tammann [2] (0–90 at.% Pt). The solid state equilibria have been constructed from results reported

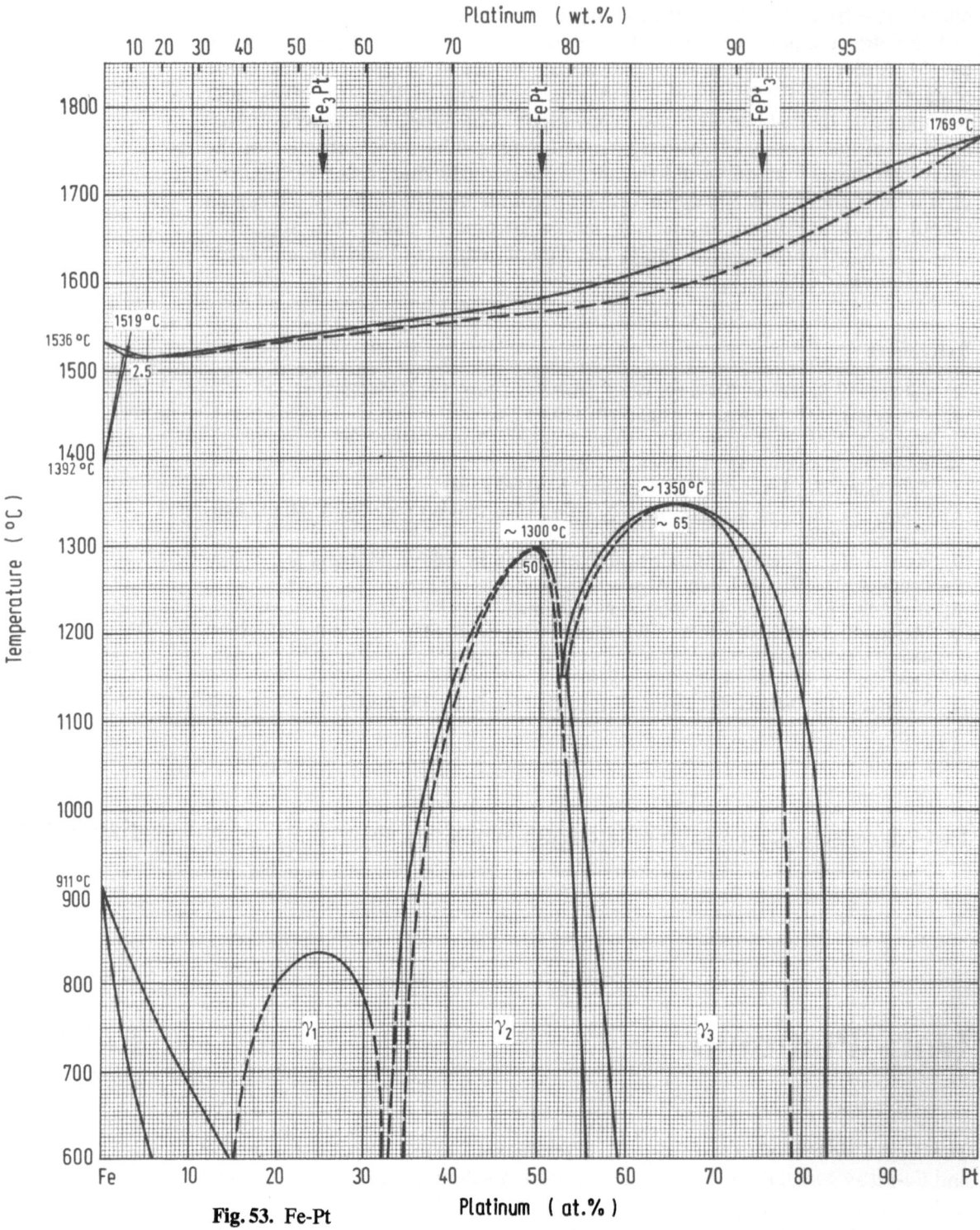

Fig. 53. Fe-Pt

by [15, 16] for the α/γ transformation temperatures, by [3–6, 8–10] for the order/disorder transformation of Fe₃Pt, by Hansen [A: Refs. 2, 7–12], [2] and [4] for the order/disorder transformation of FePt and by [4, 7] and [8] for the order/disorder transformation of FePt₃. The melting point of Pt, 1,769°C is a secondary reference point on the International Practical Temperature Scale (IPTS-68, revised 1975).

The liquidus and solidus curves in the 0–90 at.% Pt concentration range were established by thermal analysis [2]. [1] re-determined by thermal analysis the 0–10 at.% Pt region of alloys prepared from 99.99% Pt and high purity Fe (melting point 1,533°C, transformation point 1,389°C). They ascertained that addition of Pt to Fe lowers the δ(Fe) liquidus to a peritectic at 1,519°C, where the following reaction takes place:

$$\delta\text{-Fe}(2.3\,\text{at.\%\,Pt}) + L(3.6\,\text{at.\%\,Pt}) \leftrightharpoons \gamma\text{-Fe}(2.5\,\text{at.\%\,Pt}).$$

With further additions of Pt, the liquidus sinks to a very shallow minimum, the temperature of which is nearly coincidental with the peritectic temperature [1]. The transformation temperature of Fe is raised by addition of Pt.

Solid state equilibria

Investigations of the α/γ transformation, which is very similar to that of the corresponding Fe-Ni system, habe been summarized by Hansen [A: Refs. 1–5]. Fig. 53 shows results by Martelly [15] and Fallot [16]. Berkowitz et al. [6], by means of resistometric and magnetic measurements supported by X-ray and metallographic examinations, suggest an eutectoidic reaction between α-Fe and Fe_3Pt at about 550°C and 15 at.% Pt.

The Fe-Pt alloys have been extensively investigated in recent years because of the "shape memory effects" that can be produced by suitable heat treatment. Dunne and Wayman [3] applied metallography, cinematography and electrical resistivity measurements on alloys near the composition Fe_3Pt and confirmed the transformation from fcc austenite to bcc martensite and furthermore established that ordering of a ~25 at.% Pt alloy prior to transformation results in a reversible martensite transformation with a small temperature hysteresis. This transformation is a truly thermoelastic one, i. e. these alloys underlie shape memory effects. Kajiwara and Owen [9] determining the reversible martensite transformation of 25–27 at.% Pt alloys and Chang and Sastri [5] studying the ordering kinetics of Fe_3Pt agree with [3] and extend the knowledge underlying the varying degree of order. The tentative phase boundary for Fe_3Pt shown in Fig. 53 is based on magnetic, dilatometric and X-ray measurements by [4]. The homogeneity range was found to extend from about 19 to 33 at.% Pt. [6] and [8] mention a stability range from about 600 to 740°C.

The order/disorder transformation of FePt has been extensively investigated. A summary of this work is given by Hansen [A]. The phase boundaries in Fig. 53 are in principle those of [4] who studied alloys up to 1,100°C and clearly identified a two phase region between γ and FePt and between FePt and $FePt_3$ but did not investigate the higher temperature region of alloys of a concentration above 63 at.% Pt. The tentative maximum ordering point of FePt has been drawn according to the related Fe-Pd phase diagram; the FePt + $FePt_3$ boundary has been constructed to agree with findings of Crangle and Shaw [7] who determined the temperature and compositions limits of $FePt_3$ by high temperature X-ray diffraction photometry. The two ordered phases FePt and $FePt_3$ are seperated by a two phase field as confirmed by [4] and [7]. The eutectoidal temperature is tentative and constructed in analogy with the Fe-Pd system.

The magnetic structure of ordered Fe-Pt alloys has been measured by [4, 12, 14, 17].

References

1 Buckley, R.A.; Hume-Rothery, W.: J. Iron and Steel Inst. 193 (1959) 61
2 Isaac, E.; Tammann, G.: Z. Anorg. Chem. 55 (1907) 63
3 Dunne, D.P.; Wayman, C.M.: Metall. Trans. 4 (1973) 137
4 Kussman, A.; von Rittberg, G.: Z. Metallkd. 42 (1950) 470
5 Chang, H.; Sastri, S.: Metall. Trans. A 11 (1980) 905
6 Berkowitz, A.E.; Donahoe, F.J.; Franklin, A.D.; Steijn, R.P.: Acta Metall. 5 (1957) 1
7 Crangle, J.; Shaw, J.A.: Philos. Mag. 7 (1962) 207
8 Sundaresen, M.; Gerasimov, Y.I.; Geiderikh, V.A.; Vasileva, I.A.: Zh. Fiz. Khim. 37 (1963) 2,462
9 Kajiwara, S.; Owen, W.S.: Metall. Trans. 5 (1974) 2,047
10 Skinner, D.; Miodownik, A.P.: Platinum Met. Rev. 22 (1978) 21
11 Tadaki, T.; Shimizu, K.: Scr. Metall. 9 (1975) 771
12 Sumiyama, K.; Shiga, M.; Morioka, M.; Nakamura, Y.: J. Phys. F: Met. Phys. 9 (1979) 1,665
13 Foos, M.; Frante, C.; Gantois, M.: In: Shape Memory Effects in Alloys (Ed. Perkins, J.) AIME Int. Symp., Toronto, 1975. New York: Plenum Press 1975, p. 407
14 Men'shikov, A.Z.; Dorofeev, Yu.A.; Kazantsev, V.A.; Sidorov, S.K.: Fiz. Met. Metalloved. 38 (1974) 505
15 Martelly, J.: Ann. Phys. 9 (1938) 318
16 Fallot, M.: Ann. Phys. 10 (1938) 291
17 Bacon, G.E.; Wilson, S.A.: Proc. Phys. Soc. 82 (1963) 620

Fe–Pr

Iron-Praseodymium, see Fe-R: Iron-Rare Earth Metals, Fig. 58

Fe–Pu   Iron–Plutonium

Figs. 54, 55

The phase diagram in Figure 54 is a combination of summary publications by Feschotte and Livey [1] and Ellinger et al. [2] with some changes pertaining to the transformation temperatures of Pu [3].

Alloyed Fe and Pu form two intermetallic compounds, $FePu_6$ (tetragonal: MnU_6 structure) with $a = 1.404$ nm, $c = 0.5355$ nm, and two modifications of Fe_2Pu. The cubic (high temperature form) has a Cu_2Mg type structure. The information concerning the cell dimensions vary from $a = 0.715$ nm to $a = 0.719$ nm [2]. The hexagonal modification of Fe_2Pu (low temperature form) with $a = 0.564$, $c = 1.837$ nm is stable beween 760 and 1,020°C according to Avivi [4].

The solubility of Pu in α(Fe) between 847–877°C and γ(Fe) between 977–1,007°C is very low, < 0.02 at.% and ~ 0.05 at.% respectively, but increases rapidly to about 1 at.% at 1,297°C.

Solubilities and phase relationships in the δ'-Pu region have been studied by Elliott and Larson [5] and the δ-Pu and ϵ-Pu region by Mardon et al. [6], the latter employing differential thermal and chemical analyses as well as dilatometric, X-ray and microscopical examinations (Fig. 55).

The transition temperatures of plutonium suggested by [3] are as follows: α (monoclinic)/β (bc monoclinic) 122°C; β/γ (fc orthorhombic) 207°C; γ/δ (fc cubic) 315°C; δ/δ' (bc tetragonal) 457°C; δ'/ϵ (bc cubic) 479°C.

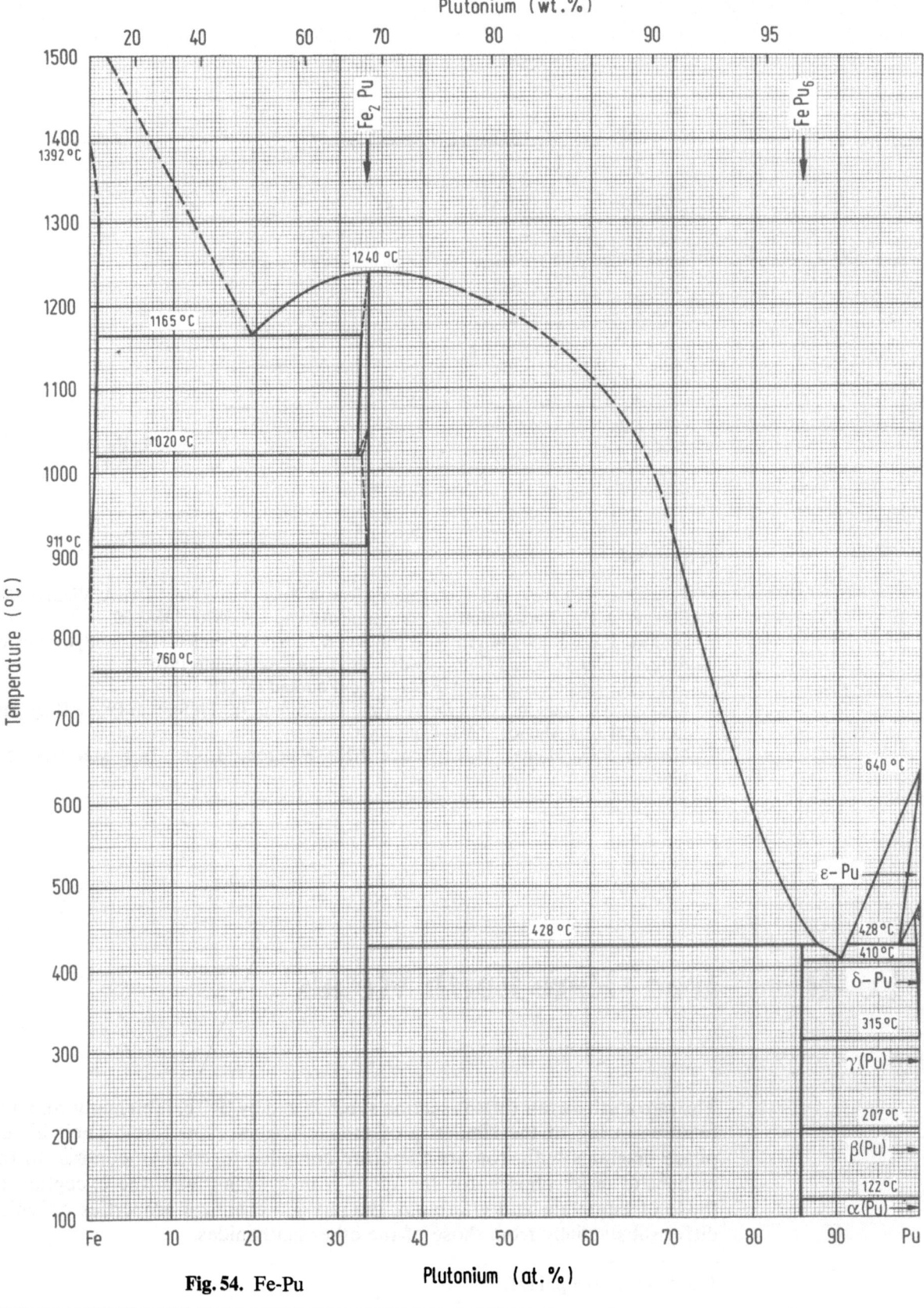

Fig. 54. Fe-Pu

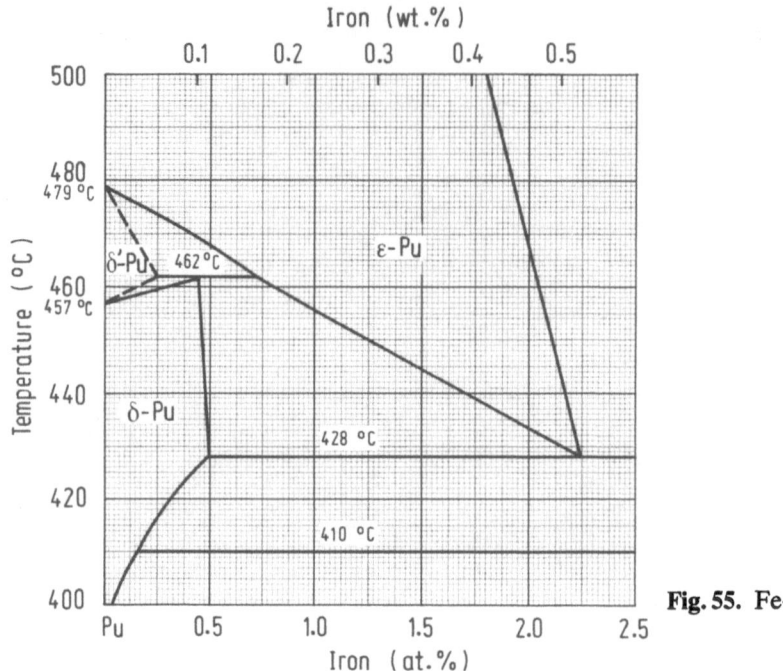

Fig. 55. Fe-Pu

References

1 Feschotte, P.; Livey, D.T.: At. Energy Rev. 4. Spec. Issue No. 1 IAEA, Vienna 1966
2 Ellinger, F.H.: Constitution of Plutonium Alloys. LA-3870 (1968) 39.
3 Oetting, F.L.; Rand, M.H.; Ackermann, R.J.: The Chemical Thermodynamics of Actinide Elements and Compounds, Part 1. IAEA, Vienna 1976
4 Avivi, E.: French Rep. CEA-R2444 (1964) quoted by [2]
5 Elliott, R.O.; Larson, A.C.: Delta-Prime Plutonium. In: Plutonium Phase Diagrams studied at Los Alamos. 1961, Chap. XXIV
6 Mardon, P.G.; Haines, H.R.; Pearce, J.H.; Waldron, M.B.: J. Inst. Met. 86 (1957) 166

Fe–R   Iron–Rare Earth Metals[1]

The rare earth metals (in this section symbolyzed by 'R') exhibit many common features owing to the similarity of their electronic structures. Their binary phase diagrams with iron are therefore compiled together in contrast to the alphabetic order chosen for the other iron systems, with the exception of those of europium and ytterbium, the physico-chemical properties of which differ substantially from those of the other lanthanides.

---

1 References see p. 117f.

In his monograph entitled "Rare Earth Alloys", K.A. Gschneidner summarized and evaluated the literature on this subject which appeared till 1961. The present assessments are based on the work of Gschneidner continuing it till 1981.

The rare earth metals considered in the present section all have a transformation at high temperatures [1–4]. The melting points, entropies of fusion and the eutectic temperatures of the binary systems with iron increase with the atomic weight of R. All these systems have a eutectic on the R-rich side in equilibrium with Fe_2R. They form up to four stable compounds, namely $Fe_{17}R_2$ with either the Th_2Zn_{17} or the Th_2Ni_{17} structure, $Fe_{23}R_6$ with the Th_6Mn_{23} structure, Fe_3R with the $PuNi_3$ structure and Fe_2R with the $MgCu_2$ (Laves) structure. The number of compounds per system increases with decreasing radius of R. Thus, the lighter metals (Ce, Pr, Nd and possibly Pm) form only two compounds with Fe, Sm forms three and the heavier metals form four. The Laves phase exhibits magnetic ordering in the systems with Sm, Tb, Dy, Ho, Er, Tm and Lu [5].

The appearance of a phase of composition Fe_5R mentioned occasionally in the literature (e.g. [6–8]) has not been confirmed by thorough investigations by others [3, 9–13]. It may also be assumed that Fe, in contrast to Co and Ni, does not form compounds of the $CaCu_5$ type with the R metals. However, compounds with ca. 11 at.% R may exist with two structures, hexagonal Th_2Ni_{17} or rhombohedral Th_2Zn_{17}, which are related to the $CaCu_5$ type. Buschow [14] and Givord et al. [15] describe and discuss the appearance of these two structures.

Further, according to O'Keefe et al. [16], a phase with the $CaCu_5$ structure does not exist in Fe-R systems because the distance between the Fe-atoms in two consecutive planes would be too short.

As far as can be concluded from the few experimentally determined solubilities in the pure metals in the solid state, the maximum solubility is small and does not exceed 1 wt.% R in iron [11].

Where, in the following, the respective investigators have made no measurements of the melting and transformation temperatures of the pure lanthanides, these have been taken from recent literature in IPTS-68, or have been estimated.

The R-Fe systems are discussed in the following sequence: Ce, Pr, Nd, Pm, Sm, Gd, Tb, Dy, Ho, Er, Tm, Lu.

Iron–Cerium

The investigations of the system Fe-Ce have been critically discussed by Gschneidner and Verkade [17]. The diagrams in Figs. 56, 57 are from this source. Figure 56 in turn stems from a redetermination by Buschow and van Wieringen [18] who carried out thermal, metallographic, magnetic and X-ray investigations using Fe of 99.94% purity, that of Ce not being reported. The two phases found, $Fe_{17}Ce_2$ and Fe_2Ce, are both formed peritectically and show no significant homogeneity ranges. In contrast, Gebhart et al. [19] deduce

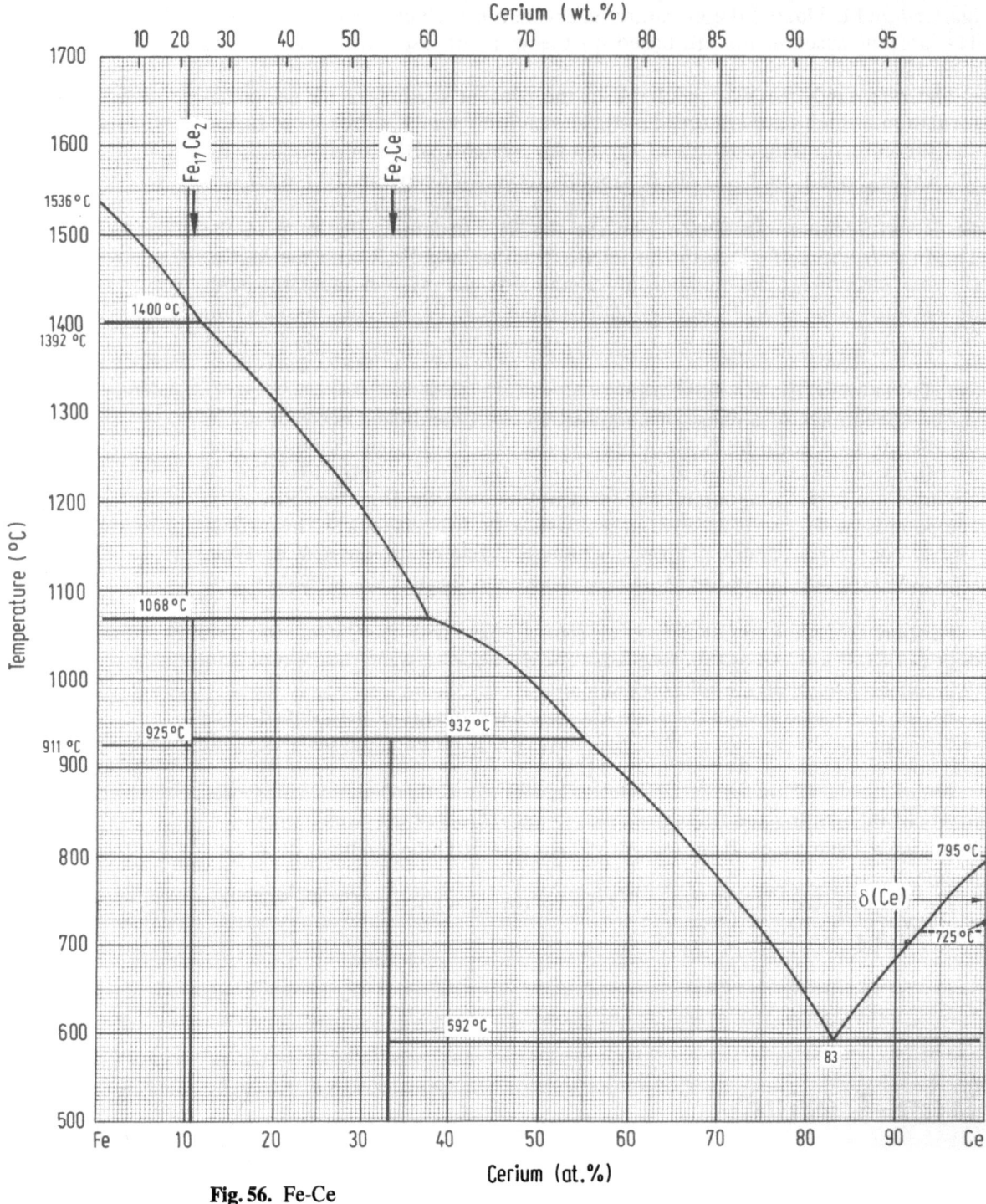

Fig. 56. Fe-Ce

from their measurements small ranges of solid solubility. $Fe_{17}Ce_2$ appears in two modifications: hexagonal and rhombohedral. The transition temperature has not been established, and it would probably be difficult to do so. An alloy with 89.5 at.% Fe quenched from 1,025°C showed only the rhombohedral β structure.

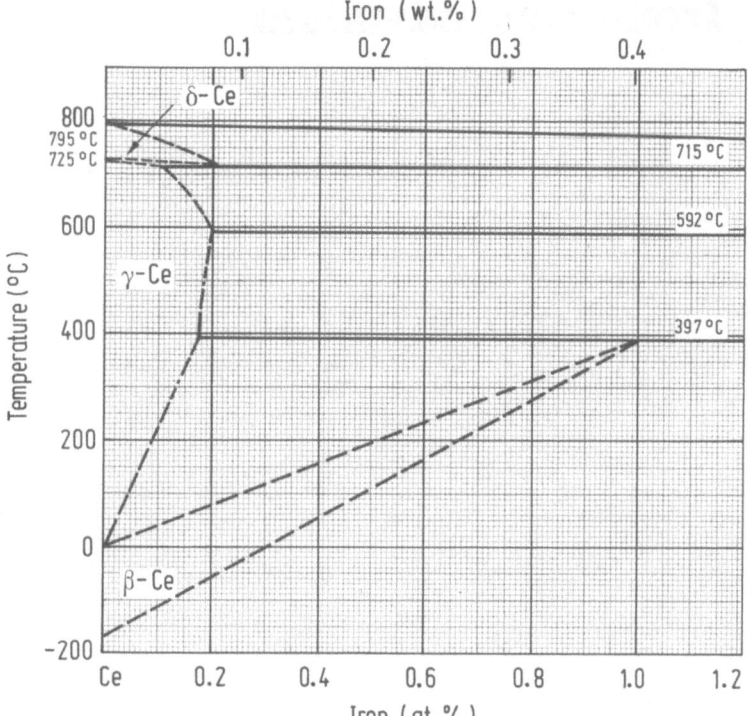

Fig. 57. Fe-Ce

The melting point of cerium (795 °C) has been taken from [17] and agrees within the limits of error with a value suggested by Hultgren et al. [D]. The transformation temperature γ/δ(Ce) was found at 725 °C [17] and the eutectic temperature at 592 °C (83 at.% Ce).

Investigations of the mutual solid solubilities are incomplete. Solely some data for the Ce-rich range are available deduced by Tucker et al. [20] from observations in the ternary system Ce-Fe-Pu (Fig. 57).

The thermodynamic properties for the dilute solution of cerium in liquid iron have been estimated by Gatellier [56].

Fe_2Ce is ferromagnetic with $T_C = 230$ K. Investigations of the Mössbauer effect and bulk magnetization on $\beta(Fe_{17}Ce_2)$ indicate an antiferromagnetic/ferromagnetic transition near 70 K and an antiferromagnetic/paramagnetic transition at about 270 K.

The relevant structural data are as follows:

Compound	Structure type	Lattice parameters, nm	Refs.
Fe_2Ce	$MgCu_2$	$a = 0.7302$	6, 21, 22, 18
$\alpha(Fe_{17}Ce_2)$	Th_2Ni_{17}	$a = 0.8490$, $c = 0.8281$	18
$\beta(Fe_{17}Ce_2)$	Th_2Zn_{17}	$a = 0.8493$[a], $c = 1.241$[a]	23, 18, 14, 24, 25

[a] Lattice parameters are for the hexagonal unit cell

Fe–Pr Iron–Praseodymium

Fig. 58

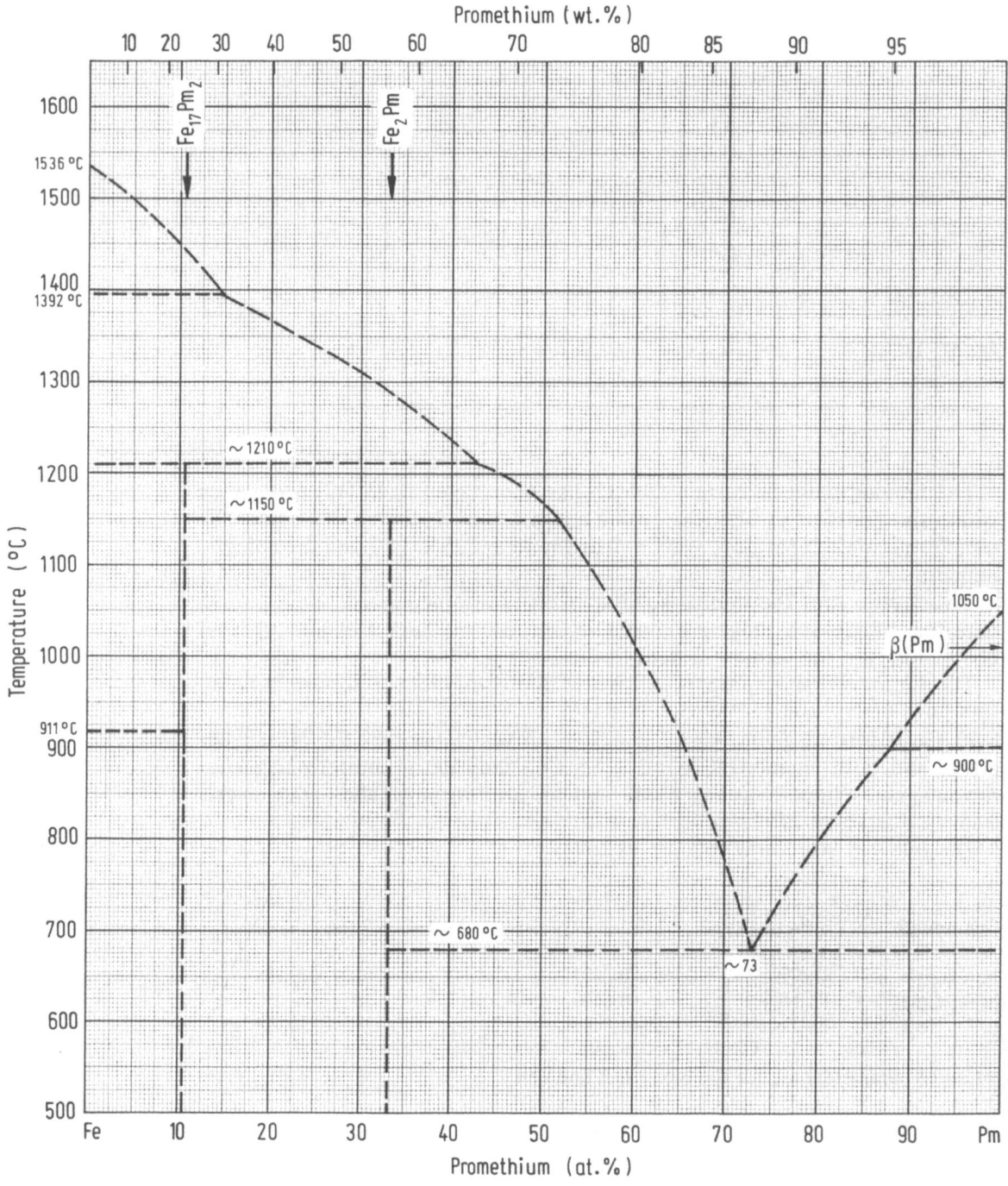

Fig. 58. Fe-Pr

Phase relationships in the system Fe-Pr have been published by Moffat [4]. Two compounds, $Fe_{17}Pr_2$ and Fe_2Pr, are stable. In addition, metallographic and X-ray observations in the range 11.1–16.7 at.% Pr of Ray et al. [26] are at hand. They found a hexagonal compound to which they assigned the stoichoimetry Fe_7Pr. Neutron diffraction carried out by Weik et al. [27] produced a rhombohedral Th_2Zn_{17} structure. Most probably, this compound corresponds to the $Fe_{17}Pr_2$ reported by Kripyakevich and Frankevich [23] and by Moffat [4] (see also [24]). The peritectic temperature of formation of $Fe_{17}Pr_2$ is 1,108°C [4], the one for 'Fe_7Pr' mentioned by [26] 1,165°C. In view of the introductory remarks (Fe-R), the temperature for the peritectic formation of Fe_2Pr, 670°C, suggested by [4] may need reconsideration. Skrabek [28] could not confirm the expected $MgCu_2$ structure.

Values for the α/β transformation temperature of praseodymium, 792°C [6], 796°C [D] and 790°C [4], agree rather well, whereas those for the melting point, 931°C [D] and 917°C [4] 918°C (IPTS-68), do less so. The latter value has been adopted in Fig. 58 which is an attempt to present the phase relationships after [4] under consideration of the introductory (Fe-R) remarks.

Iron–Neodymium

The system Fe-Nd (Fig. 59) has been investigated up to 80 at.% Nd by Terekhova et al. [29] using thermal and X-ray analyses, metallography, microhardness tests and measurements of the electrical resistance. Carbonyle iron (99.9%) and neodymium of 99.0% purity served to melt the alloys in an arc-furnace under a helium pressure of 399–532 mbar. Then the samples were sealed in evacuated silica capsules and annealed for 130 hours at 900°C (Fe-rich alloys) and at 600°C (Nd-rich). It emerged that the time for equilibration of samples with 40–65 wt.% Nd did not suffice, and a further 400 hours annealing time was applied.

There are two stable compounds: $Fe_{17}Nd_2$ and Fe_2Nd. The eutectic temperature is 640°C. The α/β transformation temperature of neodymium of 860°C agrees well with those given by [D] (856°C) and [6] (862°C). The two compounds form peritectically and are to all intents and purposes 'line compounds'. For Fe_2Nd, a Laves structure is suggested. Metallographic studies of alloys with 83.3–88.9 at.% Fe by Ray [30] pointed to a composition Fe_7Nd formed peritectically at 1,230°C with a hexagonal structure. This is likely to be the accepted $Fe_{17}Nd_2$ phase [24].

Hardness measurements showed that Nd additions to Fe increase the hardness, the specimens becoming so brittle that in the 15–60 wt.% range microhardness tests had to be employed. Determinations of the electrical resistivity agreed with the results of the hardness tests [29].

The maximum solid solubility of Fe in $\alpha(Nd)$ is stated by [29] to be ca. 4 at.% which appears to be improbably high compared with results of other Fe-R systems. Finally, the lattice constants for $Fe_{17}Nd_2$ may be recorded:

$Fe_{17}Nd_2$, Th_2Ni_{17} structure, hexagonal, $a = 0.859$ nm, $c = 1.247$ nm [29,25].

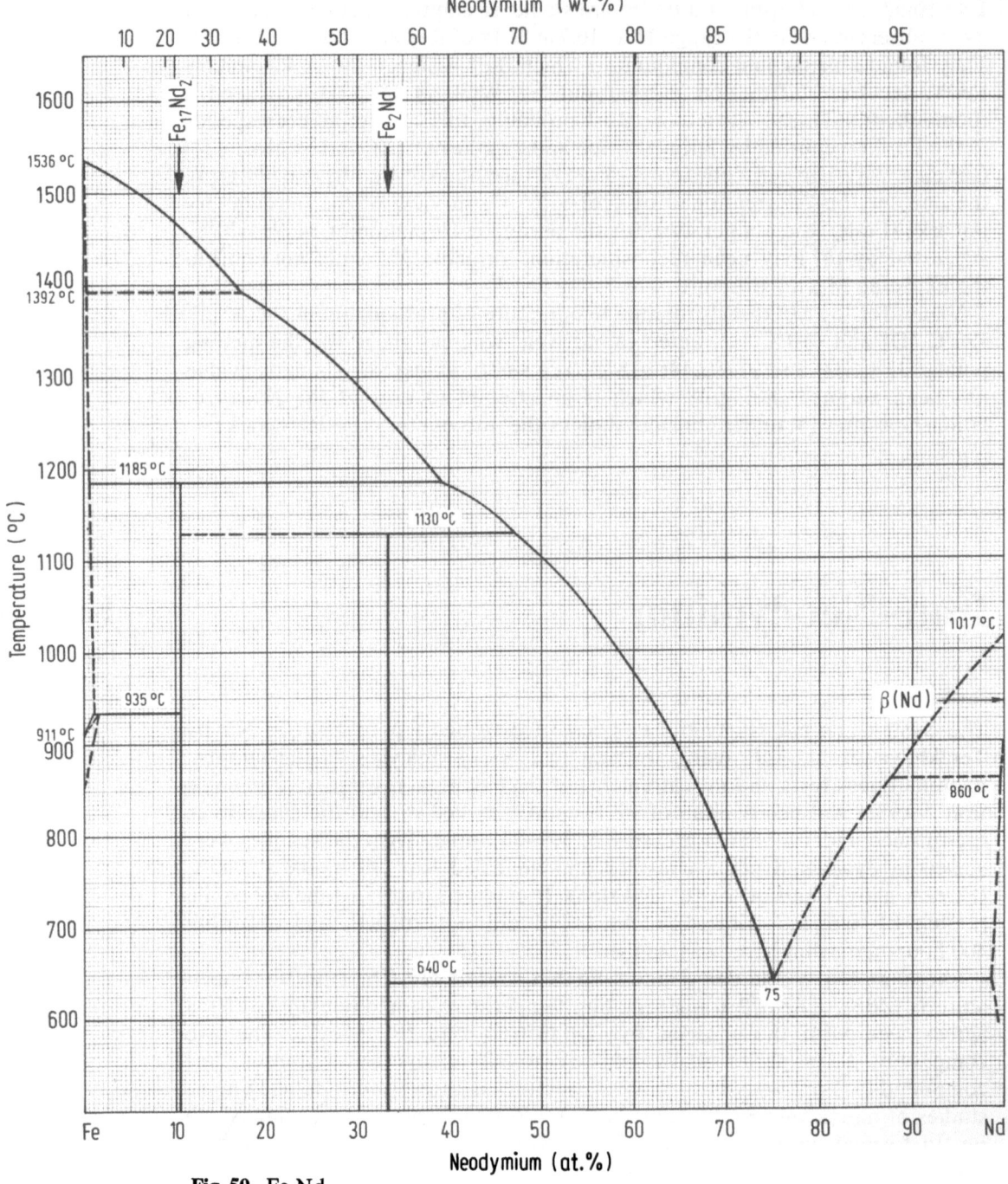

Fig. 59. Fe-Nd

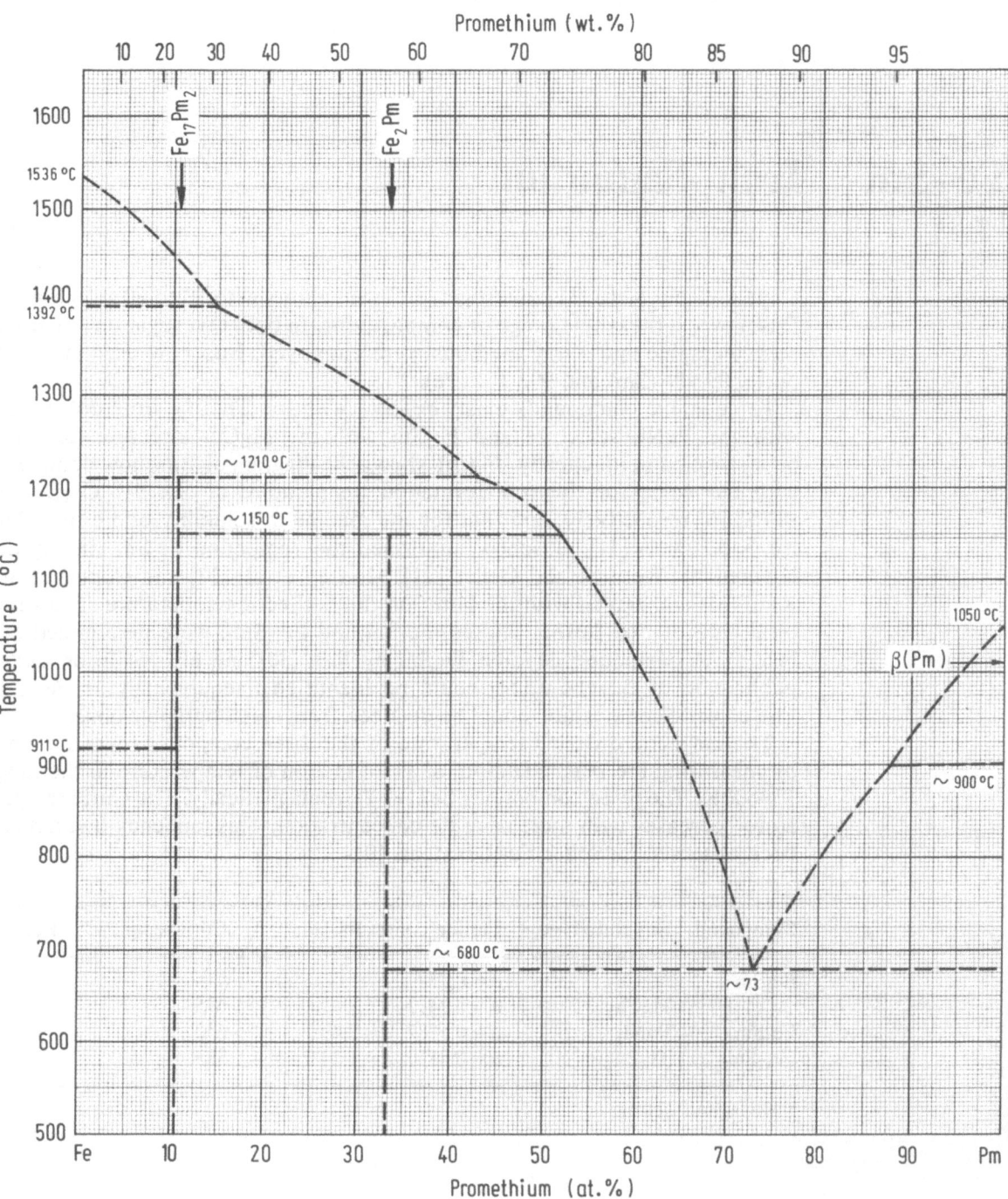

Fig. 60. Fe-Pm

The phase diagram of the Fe-Pm system in Fig. 60 has been estimated by the present compiler based on the considerations set out in the introduction (Fe-R). According to these, two 'line compounds' are expected to be stable, namely $Fe_{17}Pm_2$ and Fe_2Pm. The estimated data are as follows. Melting point of Pm: 1,050 °C (which agrees with the value listed by Hultgren et al. [D]), its transformation temperature: 900 °C, the eutectic temperature: 680 °C, and the eutectic composition: 73 at.% Pm. The mutual solubilities in the component metals are likely to be very small.

Fe–Sm Iron–Samarium

Fig. 61

Copeland and Kato [31] investigated Fe-Sm alloys with 30 to 80 at.% Sm and Buschow [9] the whole range of compositions. In both investigations, thermal, metallographic and X-ray methods were employed. Buschow used metals of purity 99.9 % (Sm) and 99.99 % (Fe).

Three stable compounds exist: $Fe_{17}Sm_2$ with the Th_2Zn_{17} structure, Fe_3Sm with the $PuNi_3$ structure and Fe_2Sm with the $MgCu_2$ structure, all three being formed peritectically. Fig. 61 shows the diagram of Buschow completed with the value for the transformation temperature of Sm 918 °C [D] and 917 °C [6] respectively, and that for the maximum solubility of Fe in α(Sm), i.e. <0.54 at.% [31]. Buschow confirmed the compound Fe_2Sm already discovered by Wernick and Geller [21], Mansey et al. [22] and Copeland and Kato [31] and agreed with the latter on the peritectic temperature. The structural data of two further phases, Fe_3Sm and $Fe_{17}Sm_2$, were determined (see below). A compound denoted Fe_7Sm by Ray [30] and also so in an earlier publication of Buschow [14] could be identified unequivocally as $Fe_{17}Sm_2$. The eutectic composition could be deduced from metallographic observations.

The magnetic ordering temperature of the three compounds was obtained by measuring the magnetization, σ, as a function of temperature in the range 77–800 K. The three compounds behave ferromagnetically. The Curie temperature decreases with increasing Fe concentration. This tendency had already been observed for other Fe-R systems [9].

Compound	Structure type	Lattice constants, nm	Curie temp., °C	Ref.
Fe_2Sm	$MgCu_2$	$a = 0.738$		28
	$MgCu_2$	$a = 0.7417$	675	9
Fe_3Sm	$PuNi_3$	$a = 0.5187$, $c = 2.491$[a]	650	9
$Fe_{17}Sm_2$	Th_2Zn_{17}	$a = 0.8570$, $c = 1.244$[a]	385	9

[a] Hexagonal setting

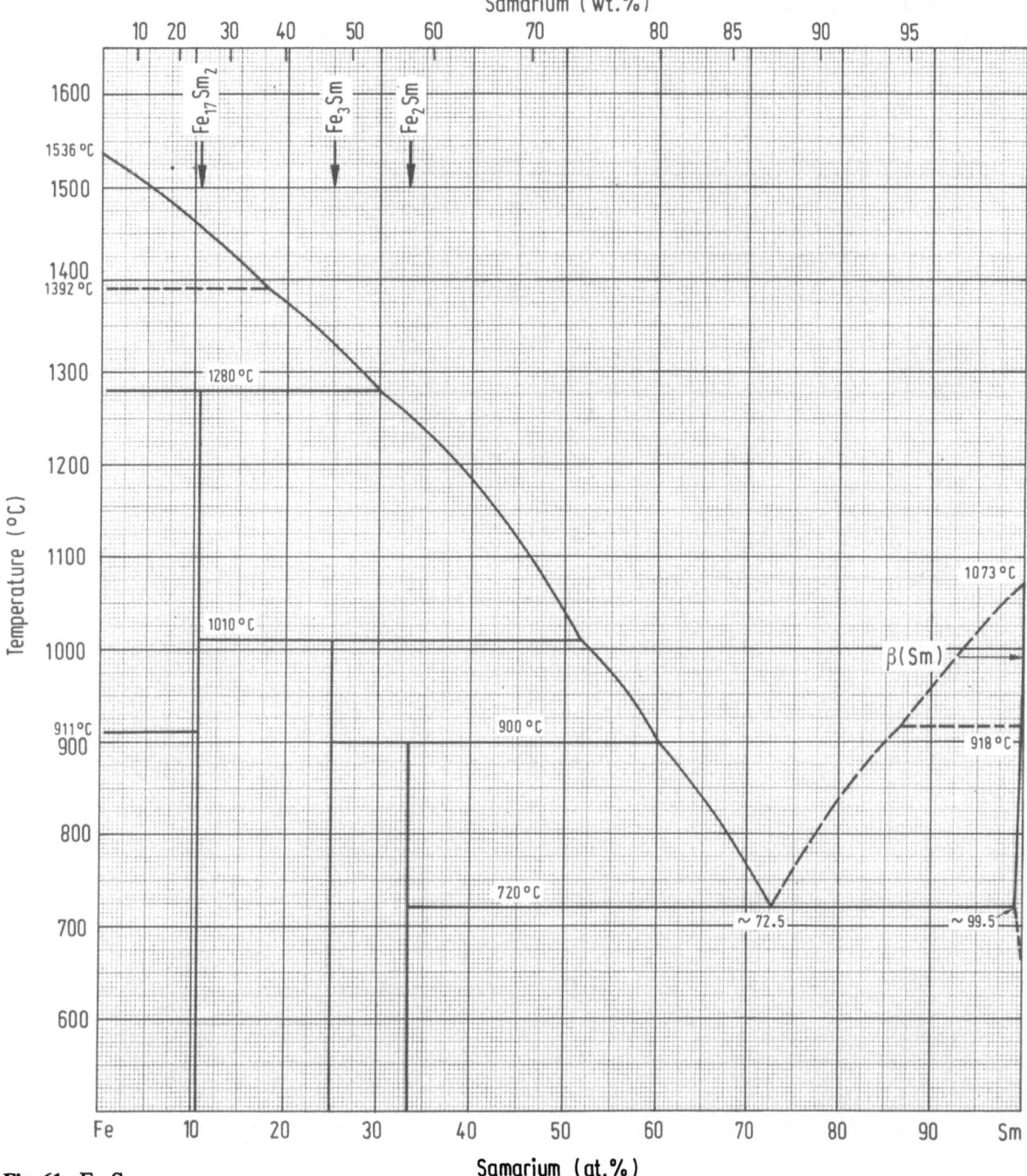

Fig. 61. Fe-Sm

Iron-Europium, see p. 38

Iron-Europium, see p. 38

Fe–Eu

Fe–Gd Iron–Gadolinium

Fig. 62

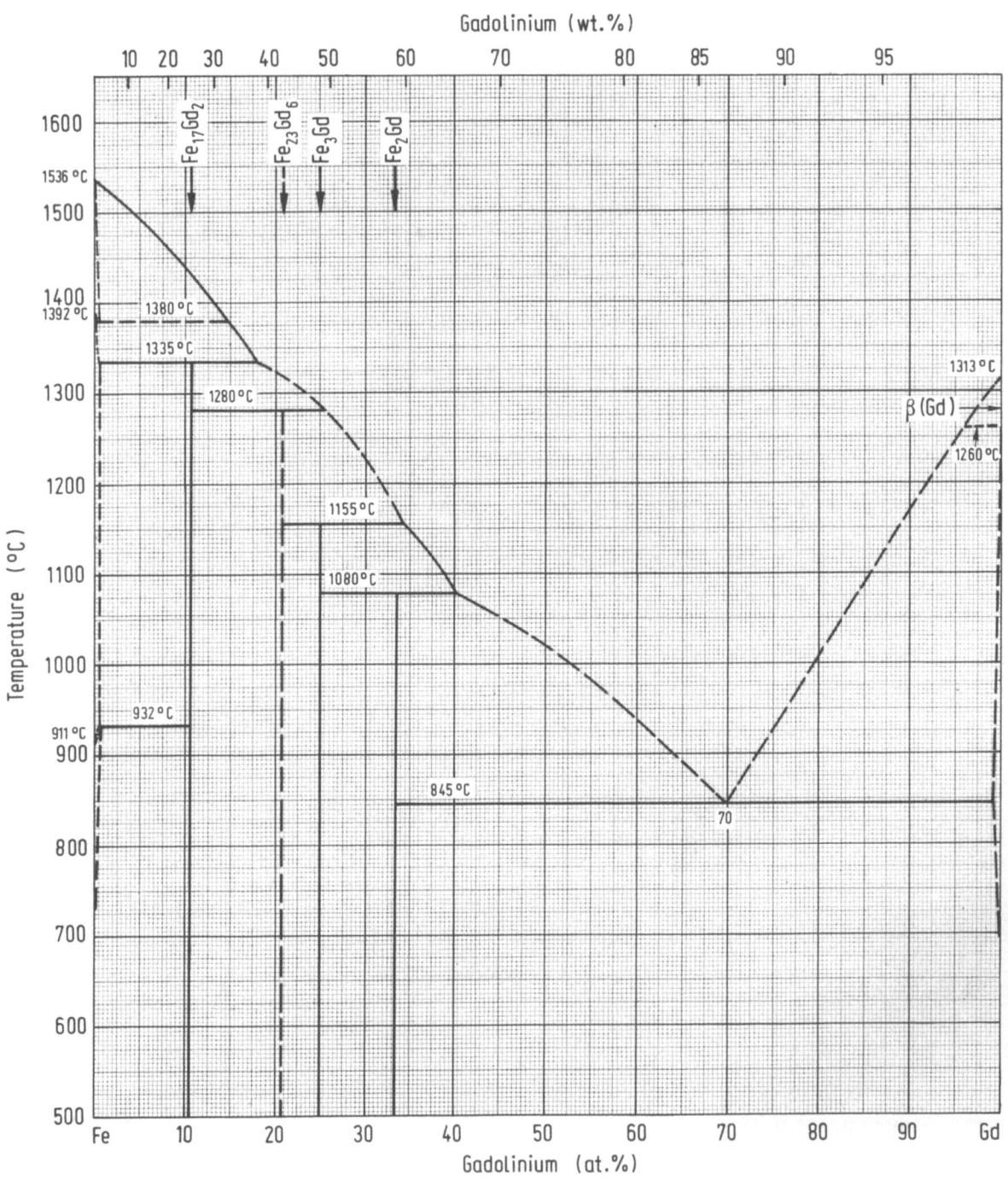

Fig. 62. Fe-Gd

Equilibria and structures in the system Fe-Gd have been investigated by quite a number of teams [31–35, 37–41]. The phase diagram presented by Elliott [B] and based in principle on the work of Copeland and Kato [31] has been confirmed by Burov et al. [40], Spedding et al. [41] and Moffat [4].

The system consists essentially of three compounds and one eutectic. $Fe_{17}Gd_2$ appears in two structural modifications [25, 32–35] (see below). The two other established compounds are Fe_3Gd and Fe_2Gd. Recent investigations [4] report a fourth compound, $Fe_{23}Gd_6$, which in view of the introductory considerations might be stable.

Figure 62 represents the modified phase diagram after [31] who investigated the system by means of thermal, metallographic and X-ray methods using a gadolinium of >99.7% purity. The melting point of gadolinium was accurately determined by Beaudry and Spedding [43]. Their result (1,313°C) agrees well with the value (1,315°C) adopted by Hultgren et al. [D]. (Bruzzone et al. [42] give 1,300°C.) For the α/β transformation temperature of Gd, the following values have been suggested: 1,270°C [4], 1,261°C [D], 1,264°C [6], 1,238°C [42] and 1,233°C [43]. A temperature of 1,260°C has here been accepted.

The eutectic temperature of 845°C [31] agrees reasonably well with the 849°C of [41]. For the eutectic composition, 66.9 [31] and 72.0 at.% Gd [41] were found. The only information regarding the solid solubilities is a statement by [31] that α-Gd dissolves less than 0.6 at.% Fe.

Klemm [55] reported that the three compounds are ferromagnetic with saturation values smaller than those of pure nickel. Nesbitt et al. [36] found that Fe_2Gd is ferrimagnetic which is more probable in view of the low saturation value.

There is general agreement regarding the structure of the intermetallics but there is some scatter in the values of the lattice dimensions. Here only selections or mean values are reproduced:

Compound	Structure type	Lattice constants, nm	Refs.
Fe_2Gd	$MgCu_2$	$a = 0.740$	7, 28, 37–39
Fe_3Gd	$PuNi_3$	$a = 0.873$, $\alpha = 34°19'$	31
or	rhombohed.	$a = 0.472$, $\alpha = 39°42'$	34
$Fe_{17}Gd_2$	Th_2Zn_{17}	$a = 0.854$, $c = 1.242$	25, 33, 35
do.	Th_2Ni_{17}	$a = 0.845$, $c = 0.844$	25, 34

Fe–Tb   Iron–Terbium

Fig. 63

The phase diagram of the Fe-Tb system in Fig. 63 was determined by Dariel et al. [13] by means of metallographic and X-ray methods, DTA and electronic microprobe technique. The purity of iron and terbium was 99.99% and 99.9% respectively. Samples were arc-melted under Zr-gettered argon and then heat-treated.

The system entails four compounds with the stoichiometries: Fe_2Tb, Fe_3Tb, $Fe_{23}Tb_6$ and $Fe_{17}Tb_2$, the last mentioned appearing in two modifications, the rhombohedral Th_2Zn_{17} structure on the Tb-rich side and the hexagonal Th_2Ni_{17} structure on the Fe-rich side of the stoichiometric composition. The existence of Fe_2Tb, Fe_3Tb, $Fe_{23}Tb_6$ and $Fe_{17}Tb_2$ had already been noticed earlier [44], but not all investigators could confirm the stability of $Fe_{23}Tb_6$. There is also some disagreement about $Fe_{17}Tb_2$.

The temperature of the α/β transformation (1,318°C) in Fig. 63 stems from [13]. The eutectic was located at 847°C and 72 at.% Tb. The confidence limit of the temperatures of the peritectic and eutectic temperatures, obtained by thermal analysis, was stated as ± 5 K. The solid solubilities are so small that they turned out to be outside the experimental possibilities of [13]: no influence on the α/β transformation could be observed on adding Fe to Tb. However, the DTA results showed a small increase of the α/γ transformation of Fe by Tb and a small decrease of the γ/δ transformation.

According to Koon et al. [45], Clark and Belson [46] and Klimker et al. [47], the Fe-Tb compounds, in particular the Laves phase Fe_2Tb, exhibit interesting magnetic properties: giant magnetostriction and huge ΔE effects.

The structural data somewhat condensed are as follows:

Compound	Structure type	Lattice constants, nm		Refs.
Fe_2Tb	$MgCu_2$	$a = 0.730$		28
do.	$MgCu_2$, distorted	$a = 0.5190$,	$c = 1.282^a$	13
Fe_3Tb	$PuNi_3$	$a = 0.514$,	$c = 2.461^a$	13
$Fe_{23}Tb_6$	Th_6Mn_{23}	$a = 1.208$		13,23
$Fe_{17}Tb_2$ (Tb-rich)	Th_2Zn_{17}	$a = 0.852$,	$c = 1.243^a$	13,23
$Fe_{17}Tb_2$ (Fe-rich)	Th_2Ni_{17}	$a = 0.847$,	$c = 0.832$	13

[a] Hexagonal setting

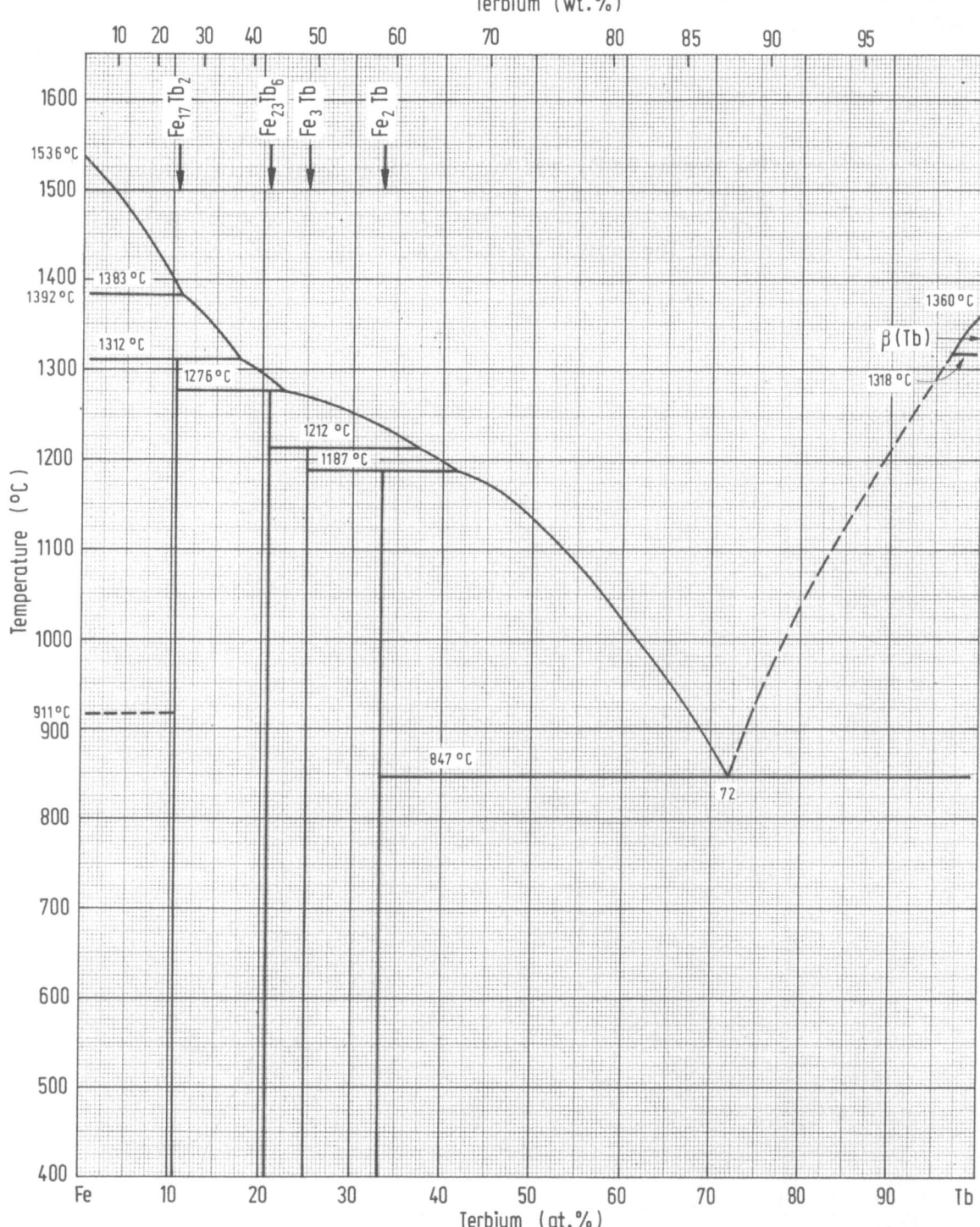

Fig. 63. Fe-Tb

Fe–Dy   Iron–Dysprosium

Fig. 64

The system Fe-Dy was investigated by X-ray, metallographic and thermal analyses in the range 25–100 at.% Fe by van der Goot and Buschow [10]. Magnetic measurements in the temperature range 4.2–1,200 K were also carried out. The purity of the component metals was 99.9% Dy and 99.99% Fe.

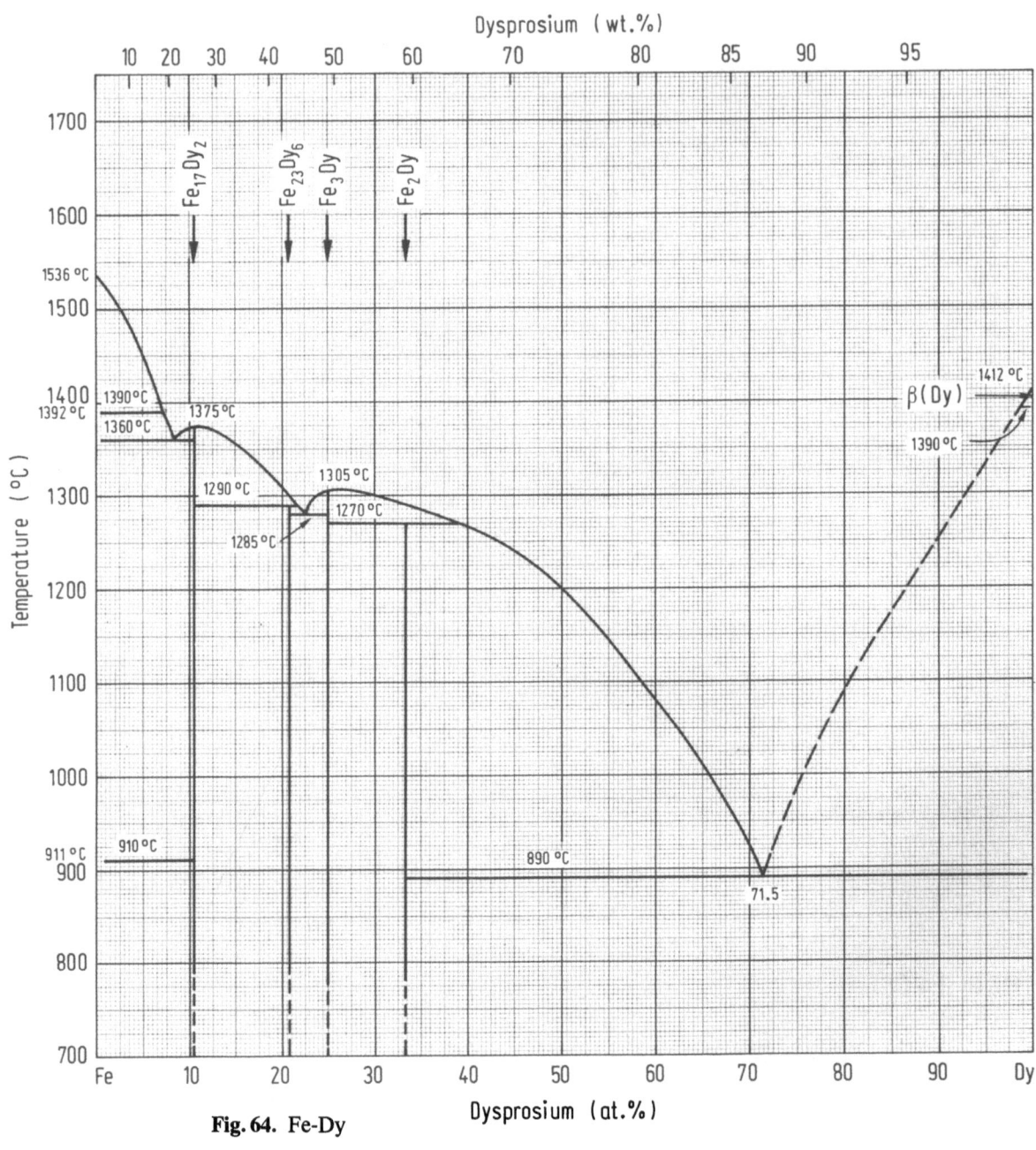

Fig. 64. Fe-Dy

Figure 64 reproduces the suggested phase diagram [10]. There are four compounds, Fe_2Dy, Fe_3Dy, $Fe_{23}Dy_6$ and $Fe_{17}Dy_2$, and a eutectic. In contrast to the Fe-R systems so far discussed, not all the compounds are formed peritectically. $Fe_{17}Dy_2$ and Fe_3Dy melt congruently. As for the structures, see below. X-ray investigations by Nassau et al. [7] had indicated that Fe_2Dy and 'Fe_5Dy' form peritectically. Results of [10] show that $Fe_{23}Dy_6$ originates from a eutectic reaction at 1,285°C. The other eutectic was located at 71.5 at.% Dy and 890°C [10], the temperature corresponding to the 855°C observed by Kato and Copeland [31, 48].

Information regarding the solid solubilities is confined to a note by Elliott [B] according to which the solubility of Fe in Dy is less than 0.3 at.%.

The magnetic studies showed that the Curie temperature of the compounds decreases with increasing Fe content. Fe_3Dy and $Fe_{23}Dy_6$ exhibit compensation points [10].

The structural data are as follows:

Compound	Structure type	Lattice constants, nm	Refs.
Fe_2Dy	$MgCu_2$	$a = 0.732$ (mean)	10, 28, 49
Fe_3Dy	$PuNi_3$	$a = 0.5116^a$, $c = 2.455^a$	10
$Fe_{23}Dy_6$	Th_6Mn_{23}	$a = 1.206$	10, 23
$Fe_{17}Dy_2$	Th_2Ni_{17}	$a = 0.8453$, $c = 0.8287$	10

[a] Hexagonal setting

Iron–Holmium

The only publication on the phase equilibria in the system Fe-Ho seems to be that by Roe and O'Keefe [11]. Before this, solely the structures of $Fe_{17}Ho_2$ [23] and Fe_2Ho [28, 7, 21] had been investigated. [11] employed X-ray, metallographic and DTA methods on alloys made of 99.9% pure Fe and 99.0% pure Ho by arc-melting.

Figure 65 shows the phase diagram of [11]. Since neither the Fe-rich nor the Ho-rich concentration ranges were investigated, the transformation and melting temperatures of holmium have been taken from Hultgren et al. [D]. Solid solubilities have so far not been determined. As mentioned earlier, in Fe-R systems they are likely to be much less than 1%.

There are four compounds, the structural data of which are as follows:

Compound	Structure type	Lattice constants, nm	Refs.
$Fe_{17}Ho_2$	Th_2Ni_{17}	$a = 0.844$, $c = 0.832$	11, 23
$Fe_{23}Ho_6$	Th_6Mn_{23}	$a = 1.204$	11, 23
Fe_3Ho	$PuNi_3$	$a = 0.5084^a$, $c = 2.545^a$	11
Fe_2Ho	$MgCu_2$	$a = 0.728$ (mean)	7, 11, 21, 28

[a] Hexagonal setting

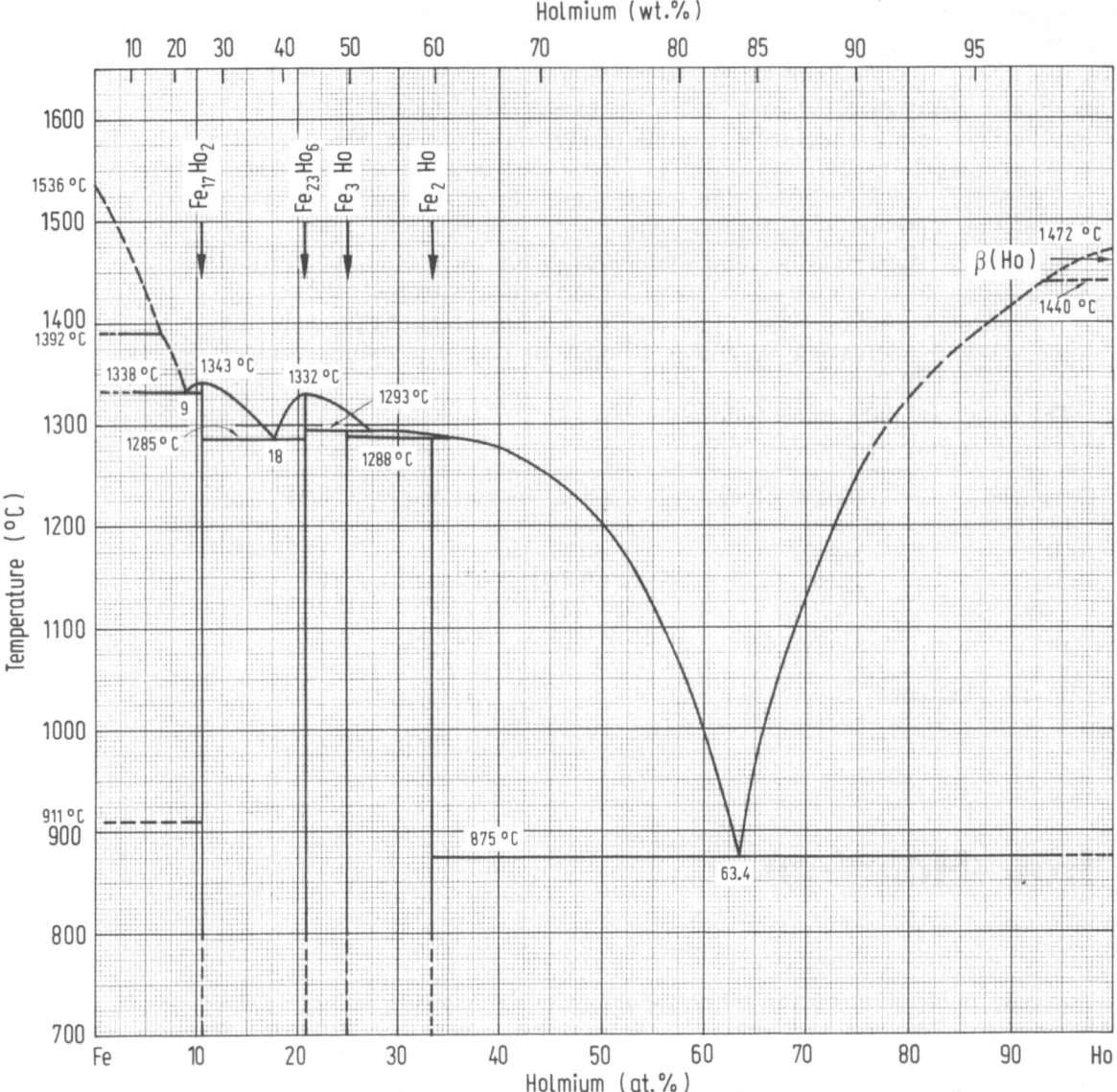

Fig. 65. Fe-Ho

The phase relationships in the system Fe-Er were investigated at about the same time independently by Buschow and van der Goot [3] and by Meyer [12] both using 99.9% pure erbium and 99.99% pure iron and employing X-ray, metallographic and thermal analyses. Meyer produced alloys of 0.5–1 g weight

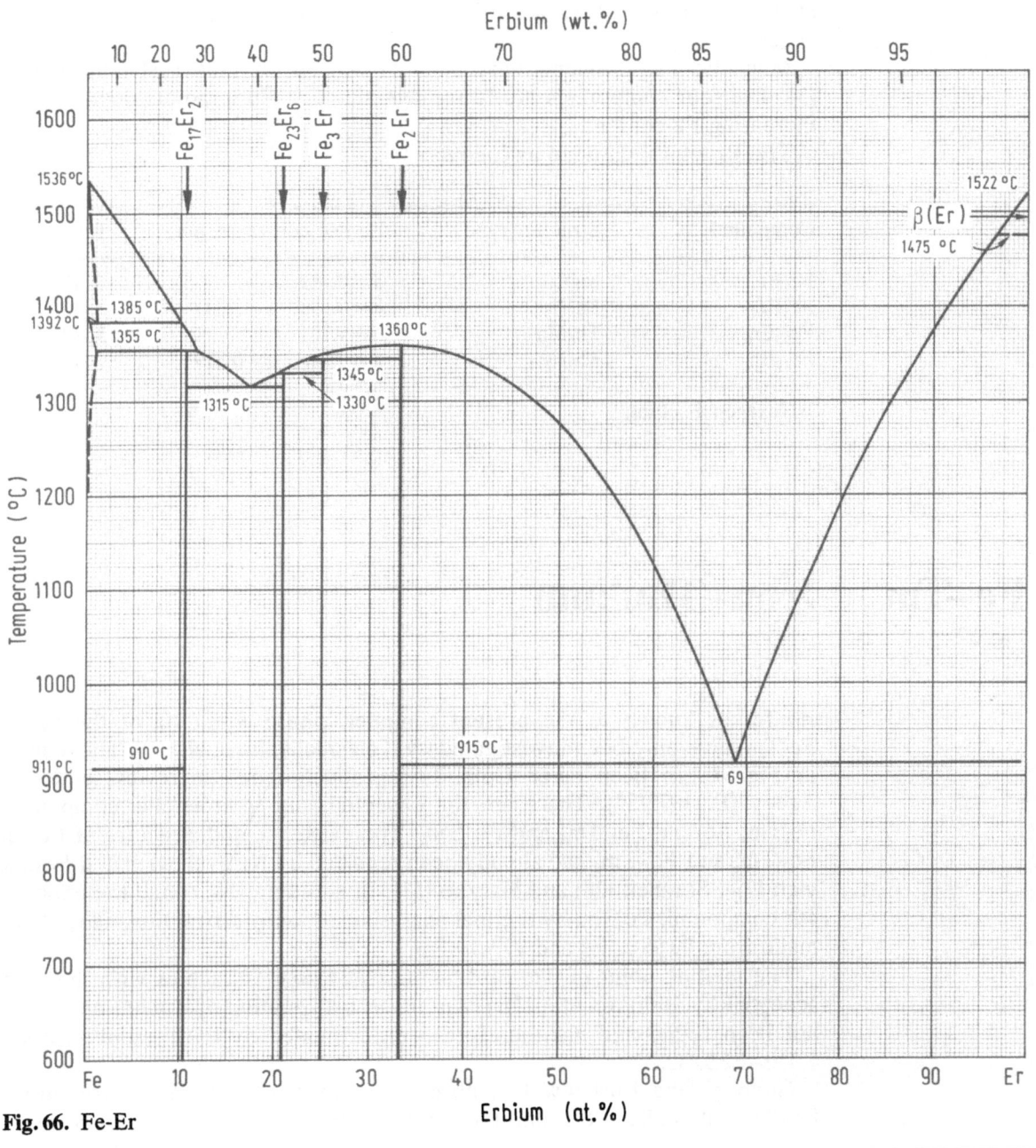

Fig. 66. Fe-Er

in an induction furnace, Buschow and van der Goot specimens weighing ca. 10 g in an arc-furnace, both working under purified argon. The investigators agree on the number of intermetallic compounds, their structures and the phase relationships. Only the reaction horizontals, found by thermal analysis, are somewhat higher in [3] than in [12].

Figure 66 represents the phase diagram submitted by [3]. There are four compounds, Fe_2Er, Fe_3Er, $Fe_{23}Er_6$ and $Fe_{17}Er_2$, the structural data of which are listed below. Fe_2Er melts congruently at 1,360 °C. The maximum solubility of Er in δ-Fe is stated to be 1 at.% [12]. Its solubility in α-Fe could be traced microscopically up to 0.015 at.%. The solubility of Fe in Er is less than 0.6 at.% [48]. The transformation temperature α/β(Er) (ca. 1,475 °C) is an estimate by the present compiler. The Fe-rich eutectic lies at 890 °C and 75 at.% Er according to [48], at 905 °C and 70 at.% Er according to [12] and at 915 °C and 68 at.% Er according to [3]. The last mentioned value has been accepted.

Magnetic measurements have been carried out on polycrystalline material [3]. The four compounds are ferromagnetic. The Curie points decrease with increasing Fe-content.

The structural data are as follows:

Compound	Structure type	Lattice constants, nm	Refs.
Fe_2Er	$MgCu_2$	$a = 0.729$	3, 12
Fe_3Er	$PuNi_3$	$a = 0.509^a$, $c = 2.447^a$	3, 12
$Fe_{23}Er_6$	Th_6Mn_{23}	$a = 1.201$	3, 12, 23
$Fe_{17}Er_2$	Th_2Ni_{17}	$a = 0.843$, $c = 0.829$	3, 12

[a] Hexagonal setting

Fe–Tm   Iron–Thulium
Fig. 67

As far as known to the present compiler, phase equilibria of the system Fe-Tm have only been investigated by Kolesnichenko et al. [50]. The diagram in Fig. 67 has been taken from its reproduction in the review edited by Ageev [E: 1972/52]. There are the expected four intermetallic compounds, Fe_2Tm, Fe_3Tm, $Fe_{23}Tm_6$ and $Fe_{17}Tm_2$. The existence and structure of Fe_2Tm, $Fe_{23}Tm_6$ and $Fe_{17}Tm_2$ had already been reported by Kripyakevich and Frankevich [23], Skrabek [28] and Haszko [51]. In addition, [50] found the expected 4th compound, Fe_3Tm, by thermal analysis and X-ray diffraction, alloys being made from 99.98% Fe and 99.9% Tm.

The melting point of Tm of 1,547 °C (IPTS-68) [50] agrees with the value accepted by Hultgren et al. [D]. An α/β transformation of Tm has apparently not been found and has therefore been estimated by the present assessor: 1,510–1,520 °C.

Gubbens and Buschow [52] observed a change in magnetic structure of $Fe_{17}Tm_2$ at 72 K. At room temperature, the compound is paramagnetic.

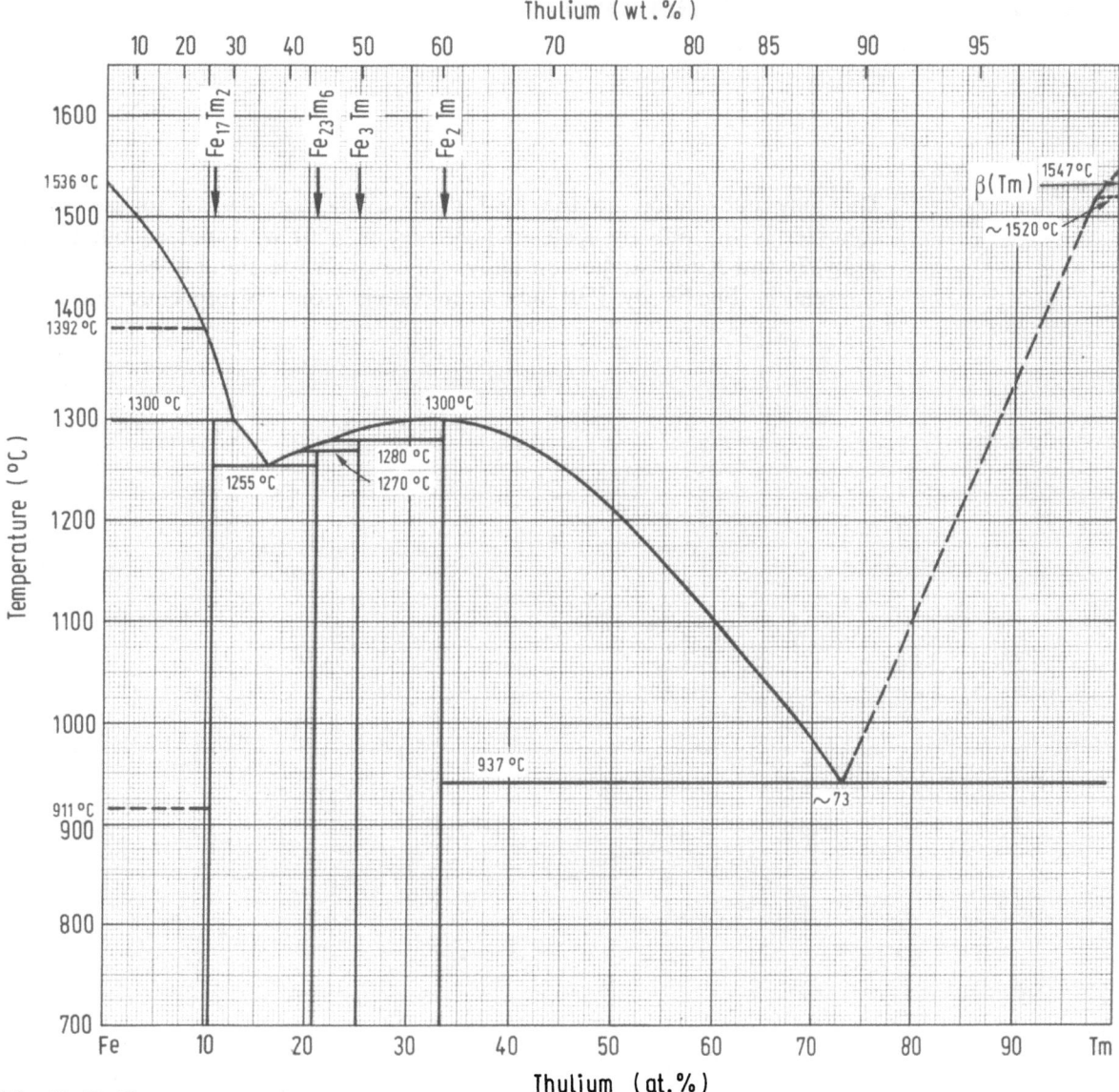

Fig. 67. Fe-Tm

The crystallographic structures are as follows:

Compound	Structure type	Lattice constants, nm	Refs.
Fe$_2$Tm	MgCu$_2$	$a = 0.724$ (mean)	28, 50, 51
Fe$_3$Tm	PuNi$_3$	–	50
Fe$_{23}$Tm$_6$	Th$_6$Mn$_{23}$	$a = 1.198$	23, 50
Fe$_{17}$Tm$_2$	Th$_2$Ni$_{17}$	$a = 0.8301$, $c = 0.8298$	50

Iron-Ytterbium, see p. 170

Fe–Yb

115

Fe–Lu   Iron–Lutetium

Fig. 68

The only publication on the Fe-Lu phase diagram known to the present compiler is the one by Kolesnichenko et al. [50] who carried out thermal analyses and X-ray studies on alloys made of 99.9% Fe and 99.2% Lu. The diagram in Fig. 68 by [50] has been taken from its reproduction in the

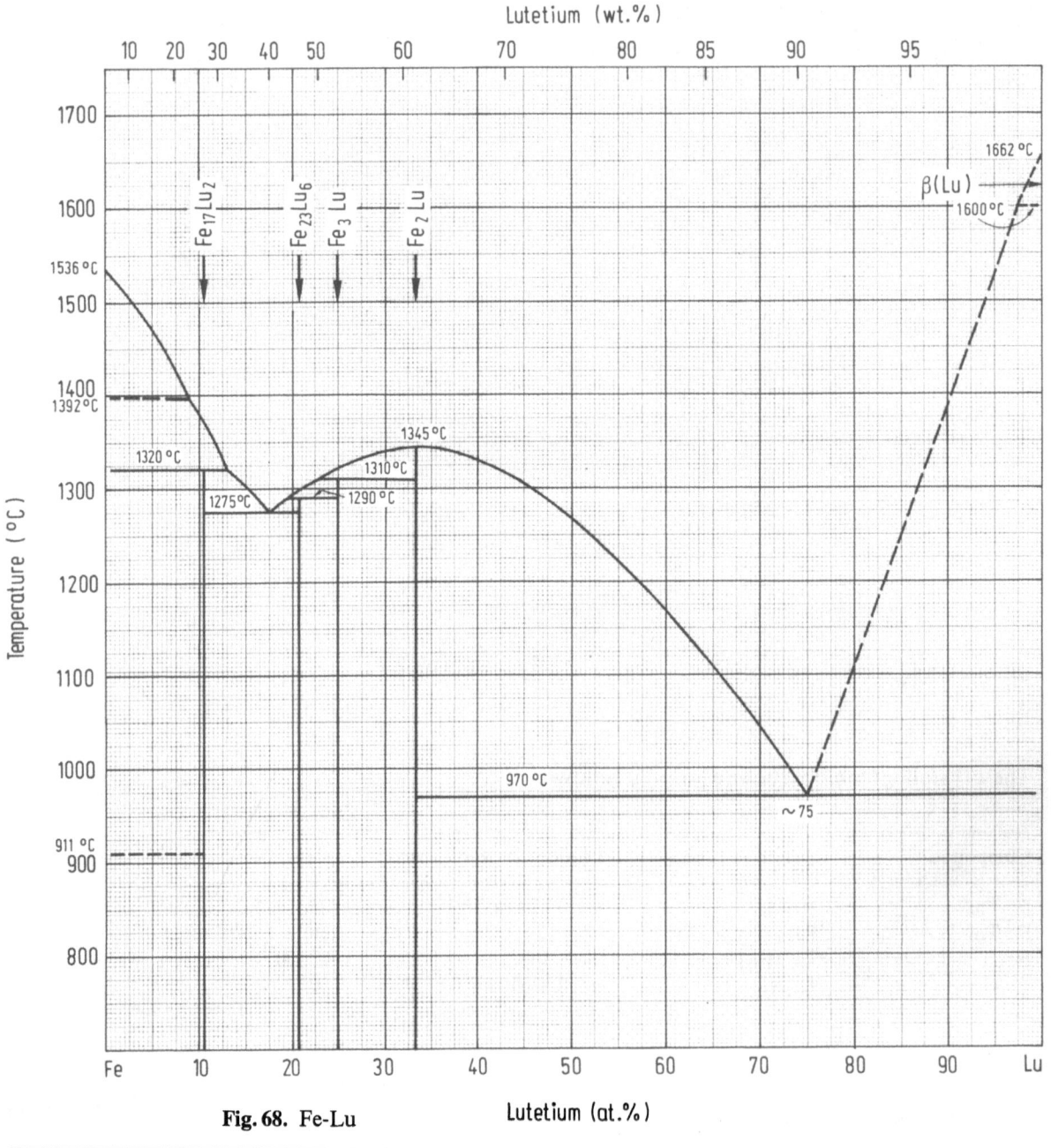

Fig. 68. Fe-Lu

review edited by Ageev [E: 1972/48]. There are four intermetallic compounds: Fe_2Lu, Fe_3Lu, $Fe_{23}Lu_6$ and $Fe_{17}Lu_2$. The existence and the structures of Fe_2Lu, $Fe_{23}Lu_6$ and $Fe_{17}Lu_2$ had already been reported by Kripyakevich and Frankevich [23], Skrabek [28], Dwight [53] and Givord et al. [15]. The expected fourth compound was discovered by [50] and assigned the $PuNi_3$ structure. [15], who investigated the Th_2Ni_{17} structure in the Fe-Lu system by X-ray and neutron diffraction, found that the composition extends from $Fe_{8.5}Lu$ to $Fe_{9.5}Lu$. Lattice parameters are given below.

The melting point of lutetium observed by [50], i.e. 1,650°C, is lower than the value suggested by Hultgren et al. [D], 1,662°C. An α/β transformation was not observed by [50], but a high-temperature β phase was identified by Spedding and Daane [2]. A transformation temperature α/β has therefore here been estimated (ca. 1,600°C).

Compound	Structure type	Lattice constants, nm	Refs.
Fe_2Lu	$MgCu_2$	$a = 0.7215$	28, 50, 53
Fe_3Lu	$PuNi_3$	–	50
$Fe_{23}Lu_6$	Th_6Mn_{23}	$a = 1.20$	23, 50
$Fe_{17}Lu_2$	Th_2Ni_{17}	$a = 0.837,\ c = 0.826$	50

References for Iron-Rare Earth Metals

1 Spedding, F.H.; Hanak, J.J.; Daane, A.H.: J. Less Common Met. 3 (1961) 110
2 Spedding, F.H.; Daane, A.H.: Trans. Metall. Soc. AIME 230 (1964) 568
3 Buschow, K.H.J.; van der Goot, A.S.: Phys. Status Solidi 35 (1969) 515
4 Moffatt, W.G.: The Handbook of Binary Phase Diagrams. General Electric Comp. 1976–1981
5 Ilarraz, J.; del Moral, A.: Phys. Status Solidi A 51 (1979) K 41
6 Gschneidner, K.A.: Rare Earth Alloys. New York: van Nostrand 1961
7 Nassau, K.; Cherry, L.V.; Wallace, W.E.: Phys. Chem. Solids 16 (1960) 123
8 Copeland, M.; Kato, H.: Physics and Material Problems of Reactor Control Rods. IAEA, Vienna 1964, p. 295
9 Buschow, K.H.J.: J. Less Common Met. 25 (1971) 131
10 van der Goot, A.S.; Buschow, K.H.J.; J. Less Common Met. 21 (1970) 151
11 Roe, G.J.; O'Keefe, T.J.: Metall. Trans. 1 (1970) 2565
12 Meyer, A.: J. Less Common Met. 18 (1969) 41
13 Dariel, M.P.; Holthuis, J.T.; Pickus, M.R.; J. Less Common Met. 45 (1976) 91
14 Buschow, K.H.J.: J. Less Common Met. 11 (1966) 204
15 Givord, D.; Lemaire, R.; Moreau, J.M.; Roudant, E.: J. Less Common Met. 29 (1972) 361
16 O'Keefe, T.J.; Roe, G.J.; James, W.J.: J. Less Common Met. 15 (1968) 357
17 Gschneidner, K.A.; Verkade, M.E.: Selected Cerium Phase Diagrams. Rare Earth Inf. Center, Energy & Minerals Resources Res. Inst., Iowa State Univ., Sept. 1974
18 Buschow, K.H.J.; van Wieringen, J.S.: Phys. Status Solidi 42 (1970) 231
19 Gebhart, J.M.; Etter, D.E.; Tucker, P.A.: Conf. Preprints, 6th Rare Earth Res. Conf., Gatlinburg, Tenn. 1967, p. 452
20 Tucker, P.A.; Etter, D.E.; Gebhart, J.M.: 'Plutonium 1965' (Eds. Kay, A.E.; Waldron, M.B.) London: Chapman & Hall 1967, p. 392
21 Wernick, J.H.; Geller, S.: Trans. Metall. Soc. AIME 218 (1960) 866
22 Mansey, R.C.; Raynor, G.V.; Harris, I.R.: J. Less Common Met. 14 (1968) 329
23 Kripyakevich, P.I.; Frankevich, D.P.: Sov. Phys. Crystallogr. 10 (1966) 468
24 Johnson, G.; Wood, D.H.; Smith, G.S.; Ray, A.E.: Acta Crystallogr. (B) 24 (1968) 274

Fe–R

25 Kripyakevich, P.I.; Terekhova, V.F.; Zarechnyuk, O.S.; Burov, I.V.: Sov. Phys. Crystallogr. 8 (1963) 203
26 Ray, A.E.; Sternat, K.; Feldmann, D.: Proc. 3rd Conf. Rare Earth Res. 1963 (Ed. Vorres, K.S.). New York: Gordon & Breach 1964, p. 443
27 Weik, H.; Fisher, P.; Hälg, W.; Stoll, E.; Eyring, L.: Proc. 4th Conf. Rare Earth Res. 1964 (Ed. Vorres, K.S.). New York: Gordon & Breach 1965, p. 19
28 Skrabek, E.A.: Thesis, Univ. of Pittsburgh 1962
29 Terekhova, V.F.; Maslova, E.V.; Savitsky, Ye.M.: Russ. Metall. 6 (1965) 50
30 Ray, A.E.: Proc. 5th Conf. Rare Earth Res. 1965. New York: Gordon & Breach
31 Copeland, M.I.; Kato, H.: Proc. 2nd Conf. Rare Earth Res. (Eds. Nachman, J.F.; Lundin, C.E.). New York: Gordon & Breach 1962, p. 133
32 Copeland, M.I.; Krug, M.; Armantrout, C.E.; Kato, H.: U.S. Bur. Mines, Rep. Invest. No. 5925 (1962)
33 Savitski, E.M.; Terekhova, V.F.; Burov, I.V.; Chistiakov, O.D.: Russ. J. Inorg. Chem. 6 (1961) 883; Tsvetn. Metal. 33, 11 (1960) 59
34 Novy, V.F.; Vickery, R.C.; Kleber, E.V.: Trans. Metall. Soc. AIME 221 (1961) 580; Vickery, R.C.; Sexton, W.C.; Novy, V.; Kleber, E.V.: J. Appl. Phys. 31 (1960) 366s
35 Kripyakevich, P.I.; Gladyshevski, E.I.: Sov. Phys., Crystallogr. 6 (1961) 118
36 Nesbitt, E.A.; Wernick, J.H.; Corenzwit, E.: J. Appl. Phys. 30 (1959) 365
37 Baenziger, N.C.; Moriarty, J.L.: Acta Crystallogr. 14 (1961) 948
38 Hubbard, W.M.; Adams, E.: J. Phys. Soc. Jpn. 17, Suppl. B–I (1962) 143
39 Mansmann, M.; Wallace, W.E.: J. Chem. Phys. 40 (1964) 1167
40 Burov, I.V.; Terekhova, V.F.; Savitskij, E.M.: Voprosy Teorii i Primeneniya Redkozemelnykh Metallov. Akad. Nauk SSSR, Moscow (1964) 116 translated as U.S. Dept. Comm., JPRS-28849 (1964) 148
41 Spedding, F.H., et al.: U.S. At. Energy Comm. IS-700 (1963) C15
42 Bruzzone, G.; Fornosini, M.L.; Merio, F.: J. Less Common Met. 25 (1971) 295
43 Beaudry, B.J.; Spedding, F.H.: Metall. Trans. 5 (1974) 1631
44 Ray, A.E.: Proc. 7th Conf. Rare Earth Res. Caronado, Calif., New York: Gordon & Breach 1968, p. 473
45 Koon, N.; Schindler, A.; Carter, F.: Phys. Lett. (A) 37 (1971) 413
46 Clark, A.E.; Belson, H.S.: Phys. Rev. (B) 5 (1972) 3642
47 Klimker, H.; Rosen, M.; Dariel, M.P.; Atzmony, V.: Phys. Rev., (B) 10 (1974) 2968
48 Kato, H.; Copeland, M.I.: U.S. At. Energy Comm., USBM-U-887 (QPR13) (1961) 4
49 Baenziger, N.C.; Moriarty, J.L.: Acta Crystallogr. 14 (1961) 948
50 Kolesnichenko, V.F.; Terekhova, V.F.; Savitskij, E.M.: Metalloved. Tsvet. Met. Splavov Nauka (1972) 31
51 Haszko, S.E.: Trans. Metall. Soc. AIME 218 (1960) 958
52 Gubbens, P.C.M.; Buschow, K.H.J.: J. Appl. Phys. 44 (1973) 3739
53 Dwight, A.E.: U.S. At. Energy Comm., ANL-6516 (1961) 259
54 Dwight, A.E.; Kimball, C.W.: Acta Crystallogr. (B) 30 (1974) 2791
55 Klemm, W.: FIAT Rep. German Sci. 1939–46, Inorg. Chem. III (1948) 255
56 Gatellier, C.; Commission des Communautés Européennes CECA No. 7210-CA/3/303, Nov. 1981

Fe–Re   Iron–Rhenium

Fig. 69

A tentative phase diagram for the region 0–40 at.% Re is presented in Fig. 69. It is based on results obtained by Eggers [1] who carried out thermal analysis and microscopical investigations supported by X-ray analysis in the Fe − Fe_3Re_2 composition range. He established the existence of a tetragonal

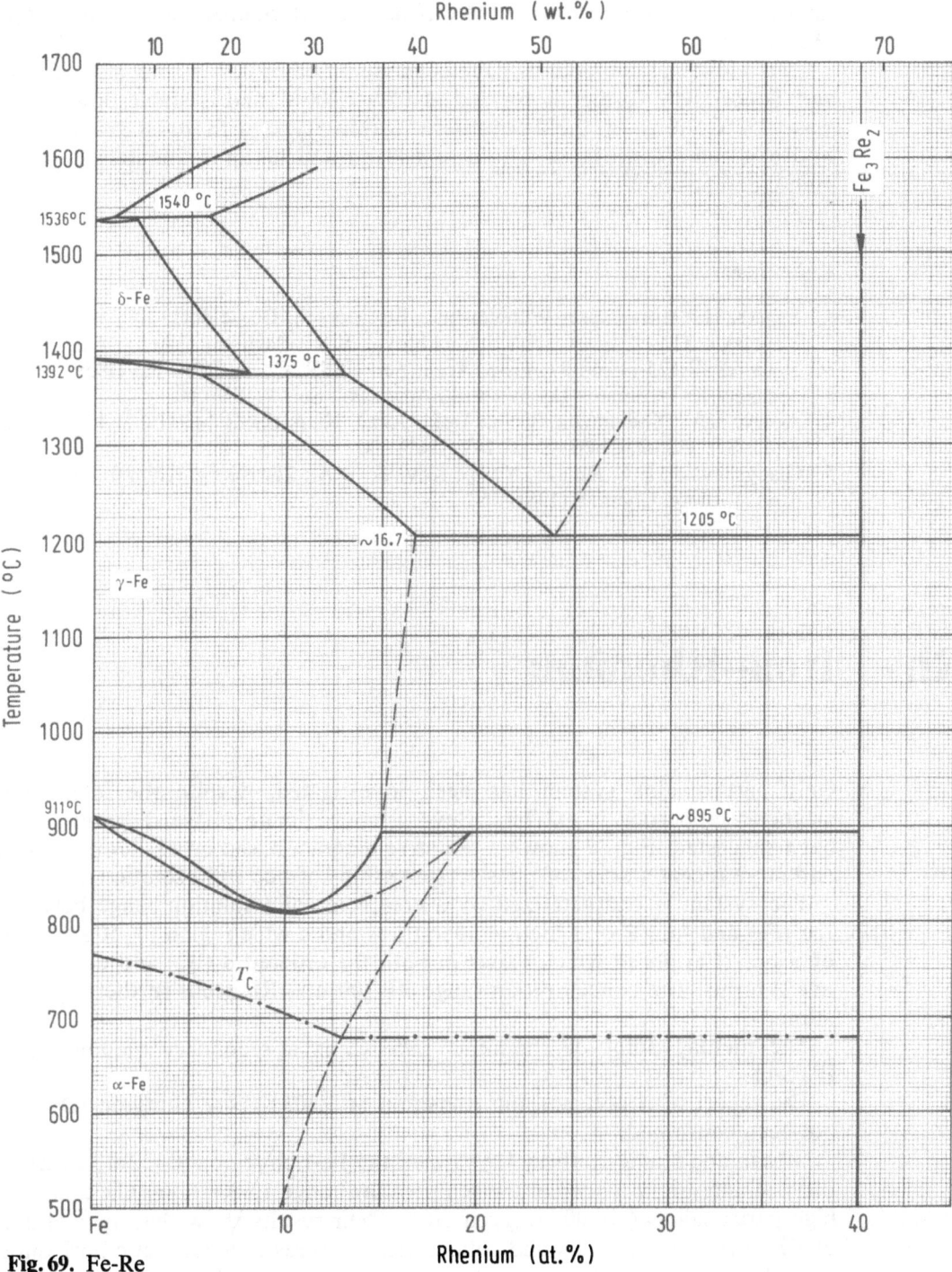

Fig. 69. Fe-Re

compound Fe$_3$Re$_2$ with the parameters $a = 0.904$ nm and $c = 0.472$ nm confirmed by Kopetski et al. [2] who quote slightly smaller parameters. According to Hansen and Anderko [A] this diagram does not represent true equilibrium conditions, but only the phase relationships.

According to Niemiec and Trzebiatowski [3] Fe$_3$Re$_2$ is of the σ type structure which is supported by findings of [2] on an alloy containing 45.4 at.% Re (annealed at 1,500 °C) showing the diffraction pattern characteristic of σ type phases.

Two additional compounds have been identified by X-ray analysis, ~FeRe$_2$ and Fe$_2$Re$_3$. The former is formed at 950–1,200 °C and is stable down to at least 750 °C: it is of the bcc ordered α(Mn) type with $a = 0.8978$ nm [4]. Fe$_2$Re$_3$ is cubic, β(Mn) type with $a = 0.643$ nm, and has approximately stoichiometric composition [5].

References

1 Eggers, H.: Mitt. Kaiser-Wilhelm-Inst. Eisenforsch. Düsseldorf, 20 (1938) 147
2 Kopetskii, C.V.; Ageev, N.V.; Shekhtova, V.S.; Savitskij E.M.: Dokl. Akad. Nauk SSSR 125 (1959) 87; Proc. Acad. Sci. USSR, Chem. Sect. 125 (1959) 175
3 Niemiec, J.; Trzebiatowski, W.: Bull. Acad. Pol. Sci. 4 (1956) 601
4 Ageev, N.V.; Shekhtman, V.Sh.: Dokl. Akad. Nauk SSSR 143 (1962) 1091; Proc. Acad. Sci. USSR, Chem. Sect. 143 (1962) 300
5 Gladyshevskii, E.I.; Kuzma, Yu.B.; Kovalik, D.A.: Zh. Neorg. Khim 9 (1964) 665; Russ. J. Inorg. Chem. 9 (1964) 368

Fe–Rh   Iron–Rhodium

Fig. 70

Only part of this system has been investigated, despite the interesting magnetic properties it exhibits. Several approximate phase diagrams have been suggested. Fig. 70 shows a tentative diagram based on earlier proposals and more recent studies of phase relationships of bcc solid solutions.

The system is characterized by a complete solid solution range between fcc (Fe) and fcc (Rh) at high temperatures and a wide region (0–50 at.% Rh) of a bcc (Fe) phase which exists in both the ordered α$_1$(B2) and disordered α(A2) forms. The concentration range 60–100 at.% Rh needs further investigation.

The melting point of rhodium, 1,963 °C, is a secondary melting point IPTS-68.

The report of a polymorphic transition at 1,303 K (see Hultgren [D]) has not been confirmed.

Gibson and Hume-Rothery [2] studied the Fe-rich region (0–22 at.% Rh) by thermal analysis *in vacuo* in the temperature range 1,380–1,560 °C employing high purity alloys (<0.007 oxygen etc.). Their results show that the addition of Rh lowers the liquidus and solidus temperatures to a peritectic horizontal at 1,515 °C and a minimum. Both curves remain more or less constant and are within a few degrees of each other between 4.3 and 20 at.% Rh, so that the actual location of the minimum is difficult to place.

The solid state equilibrium diagram belongs to the type of the expanded γ field where the temperature of the A$_4$ transformation is raised by the addition of solute. The γ-Fe-rich solution is formed peritectically from liquid and δ-Fe solid solutions at 1,515 °C, δ-Fe (2.9 at.%) + liquid (4.3 at.%) ⇌ γ-Fe (3.4 at.%).

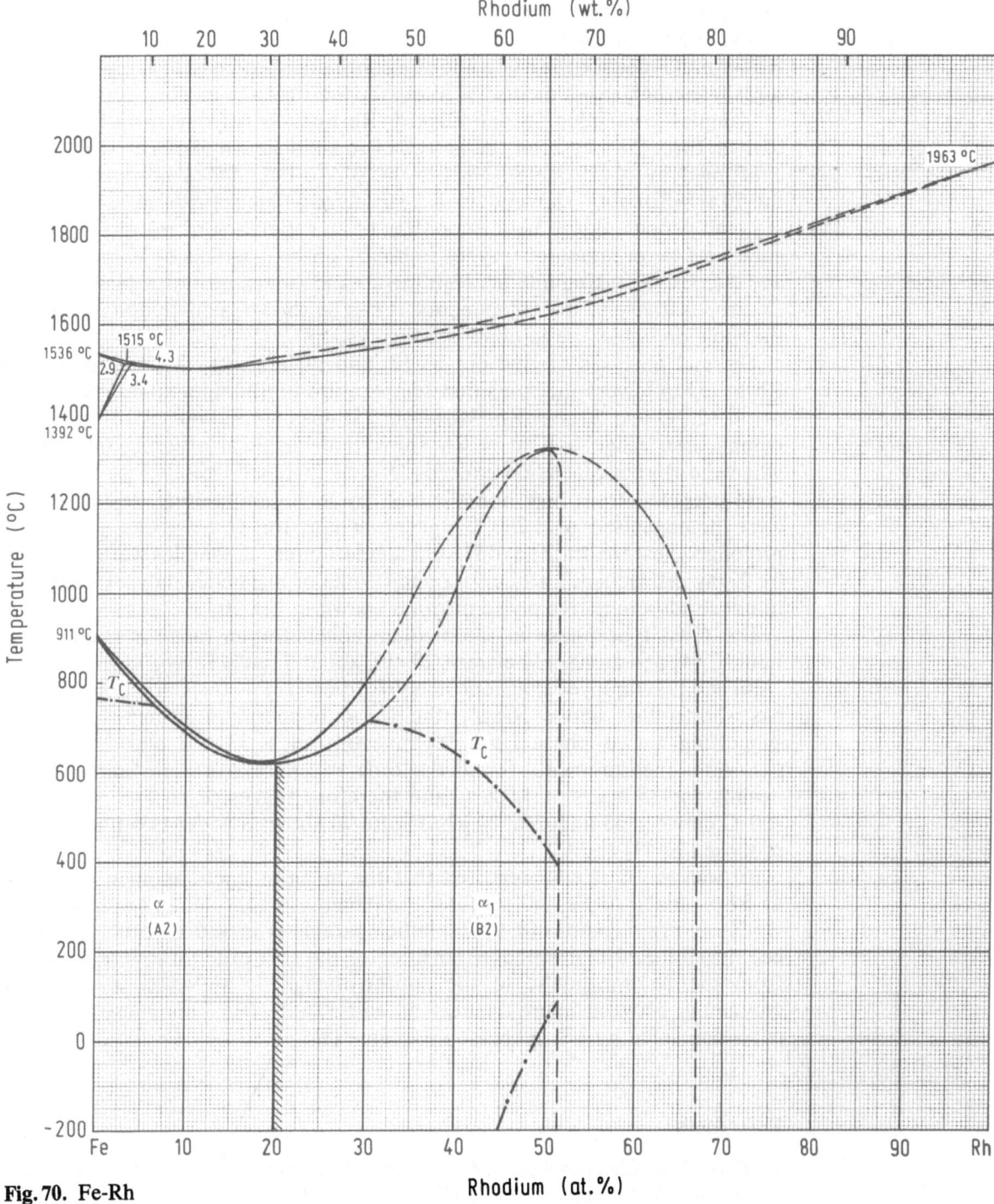

Fig. 70. Fe-Rh

Figure 70 is based on the tentative phase diagram of Chao et al. [3], Shirane et al. [4] supplemented by observations of Schwerdtfeger and Zwell [5]. [3] surveyed and critically assessed earlier results by Fallot [6] and Fallot and Hocart [7, 4], de Bergevin and Muldawer [8] and Muldawer and Bergevin [9]. They also carried out experiments (i. e. lattice spacings and Mössbauer spectra) on alloys in the concentration range 25–90 at.% Rh to study the metastable fcc/Fe-Rh alloys. Alloys were prepared by induction melting of Fe (99.99%) and Rh (99.9%) in an argon atmosphere.

The γ-Fe solid solution range has not been established experimentally. However, [3] deduce the existence of a continuous fcc Fe field at high temperatures from their experiments. [9] found the fcc structure in a brief investigation of samples containing 25 and 80 at.% Rh quenched from 1,000°C to room temperature. [5] measured the activities in solid Fe-Rh alloys at 1,200°C and confirmed by X-ray measurements the phase boundaries proposed by [3]: the 70 at.% Rh alloy showing the fcc structure whereas the 60 at.% Rh sample is heterogenous: fcc + bcc. According to an estimation of regions of preference for the formation of γ types in binary phase diagrams by Kubaschewski and Alcock [F] taking into account the atomic radius, heat of sublimation and electronegativity of the components, it can be shown, a continuous γ phase is the most probable one to dominate the system.

Thermal analyses carried out by [8] and [9] indicate, that an addition of Rh to α(Fe) lowers the transition temperature until a minimum is reached at about 625°C and 20 at.% Rh. Mössbauer studies carried out by [4] in the bcc composition range 0–52 at.% Rh show that chemical ordering takes place when the Rh concentration exceeds 20 at.% (hatched lines in Fig. 70). Alloys were prepared from high purity metals by levitation melting, and the ingots were homogenised at 1,200°C under vacuum for at least one day. After prolonged heat treatment at different temperatures, all samples were examined by X-ray diffraction. Lattice data and results of magnetic analyses were used to construct a tentative phase diagram shown in Fig. 70.

The CsCl type phase in the medium concentration range is stable up to 1,300°C [4] and at 400°C between 20 and about 52 at.% Rh [6, 7, 8].

[6] and [7] employed magnetic analysis to determine the Curie points and [7] has shown that above 50 at.% the Curie temperature becomes constant, suggesting a two phase field. [8] investigating a 53 at.% alloy by lattice-parameter measurements found the ordered phase still dominant after annealing for 35 hours at 1,000°C and therefore suggested an (α + γ) range at 400°C between 55 and 69 at.% Rh. Figure 70 shows a wider two phase region proposed by [3] and [4].

[6] was the first who reported that the mean magnetic moment increases as Rh is added to Fe and reaches a maximum at approximately 25 at.% Rh. By X-ray investigations [8, 9] and lattice parameter measurements [4, 7] it has been established that a phase transition of the first order at low temperature from the ferro-magnetic to the antiferro-magnetic state occurs in alloys containing between approximately 48 and 52 at.% Rh. This phase transition is accompanied by 1% decrease in volume [8].

References

1 Kats, S.A.; Chekhovskoi, V.Ya.; Gorina, N.B.; Polyakova, V.P.; Savitskii, E.M.: Teplofiz. Vys. Temp. 15 (1977) 1309
2 Gibson, W.S.; Hume-Rothery, W.: J. Iron Steel Inst. 189 (1958) 243
3 Chao, C.C.; Duwez, P.; Tsuei, C.C.: J. Appl. Phys. 42 (1971) 4282
4 Shirane, G.; Chen, C.W.; Flinn, P.A.: Phys. Rev. 131 (1963) 183
5 Schwerdtfeger, K.; Zwell, L.: Trans. AIME 242 (1968) 631
6 Fallot, M.: Ann. Phys. 10 (1938) 291
7 Fallot, M.; Hocart, R.: Rev. Sci. 77 (1939) 498
8 de Bergevin, F.; Muldawer, L.: Compt. Rend. 252 (1961) 1347
9 Muldawer, L.; de Bergevin, F.: J. Chem. Phys. 35 (1961) 1904

Iron–Ruthenium

The phase boundaries in Fig. 71 have been constructed according to the work of several teams of investigators [1–4]. The liquidus-solidus gap and the δ-Fe ⇌ γ-Fe equilibria are based on thorough thermal measurements carried out by Gibson and Hume-Rothery [3]. The peritectic reaction Liq + Ru (s.s.) ⇌ γ-Fe was found by Obrowski [4] to occur at 1,590 °C. The range

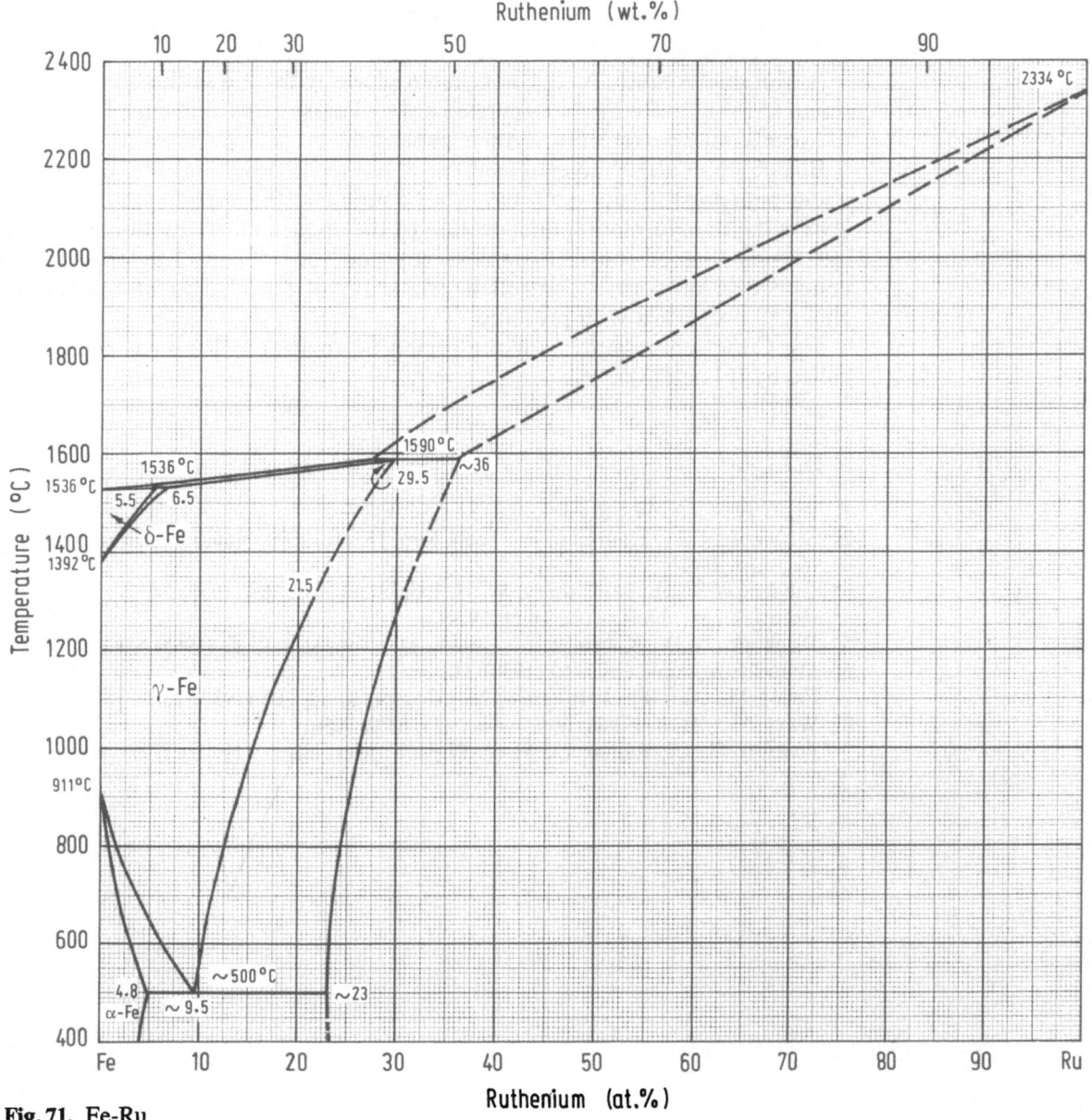

Fig. 71. Fe-Ru

Fe–Ru   450–1,300°C was investigated by Raub and Plate [1] who employed X-ray techniques, dilatometry and microscopy. A re-investigation of the whole system by Stepakoff and Kaufman [2], who employed specific heat and vapour pressure measurements and X-ray, chemical and microscopical analyses, confirmed the results obtained and summarized by [1].

The system is dominated by a large solid solubility of Fe in Ru. The hcp Ru phase is separated from the solutions in α-and γ-Fe by a fairly narrow two-phase field, γ being formed peritectically at 1,590°C and decomposed eutectoidally at ca. 500°C. [3] investigated the Fe-rich region between 1,380 and 1,560°C with alloys made of high-purity metals (99.96% Fe and 99.8% Ru). The depression of the δ liquidus and δ solidus curves is too small to be measurable, and the curves are drawn as nearly horizontal lines.

Using 99.97% Fe and 99.95% Ru, [1] confirmed the earlier observation by Wever [8] and Fallot [7] that addition of Ru to Fe widens the γ-Fe area at the expense of the α-Fe solution. The results of the X-ray and microscopical studies of [1] are incorporated in Fig. 71. The vapour pressure data of [2] at 1,327°C are in excellent agreement with those found by [1].

A Mössbauer study of the hexagonal phase with ^{57}Fe in the concentration range 15–30 at.% Ru has been carried out by Pearson and Williams. For details, the reader may be referred to their paper [9].

Blackburn et al. [5] investigated the phase transformations of Fe-Ru alloys under high pressure up to 50 kbar. They found that the application of pressure lowers the transformation temperature of the diffusionless reaction bcc $\rightleftharpoons$ fcc but raises that of the diffusionless reaction fcc $\rightleftharpoons$ hcp. The pressure dependence is of course controlled primarily by the change of the thermo-chemical properties with pressure.

The Curie temperatures for the region up to 10 at.% Ru have been established by [6] and [7].

References

1   Raub, E.; Plate, W.: Z. Metallkd. 51 (1960) 477
2   Stepakoff, G.L.; Kaufman, L.: Tech. Rep. 13, No. 2600, Naval Research, Washington, Apr. 1967
3   Gibson, W.S.; Hume-Rothery, W.: J. Iron Steel Inst. 189 (1958) 243
4   Obrowski, W.: Naturwiss. 46 (1959) 624
5   Blackburn, L.D.; Kaufman, L.; Cohen, M.: Acta Metall. 13 (1965) 533
6   Martelly, J.: Ann. Phys. 10 (1938) 318
7   Fallot, M.: Ann. Phys. 10 (1938) 291; Compt. Rend. 205 (1937) 227
8   Wever, F.: Arch. Eisenhüttenwes. 2 (1928/29) 739
9   Pearson, D.I.C.; Williams, J.M.: J. Phys. F: Met. Phys. 9 (1979) 1797

Thermodynamic analyses of the phase relationships in the Fe-S system have been carried out by Schürmann and Henke (0–50 at.% S) [1] and by Charma and Chang (0–100%) [2] up to a pressure of 1 bar S_2. The predicted phase equilibria are in close agreement with the experimental temperature vs. composition diagram.

Figure 72 represents the experimentally established phase relationships

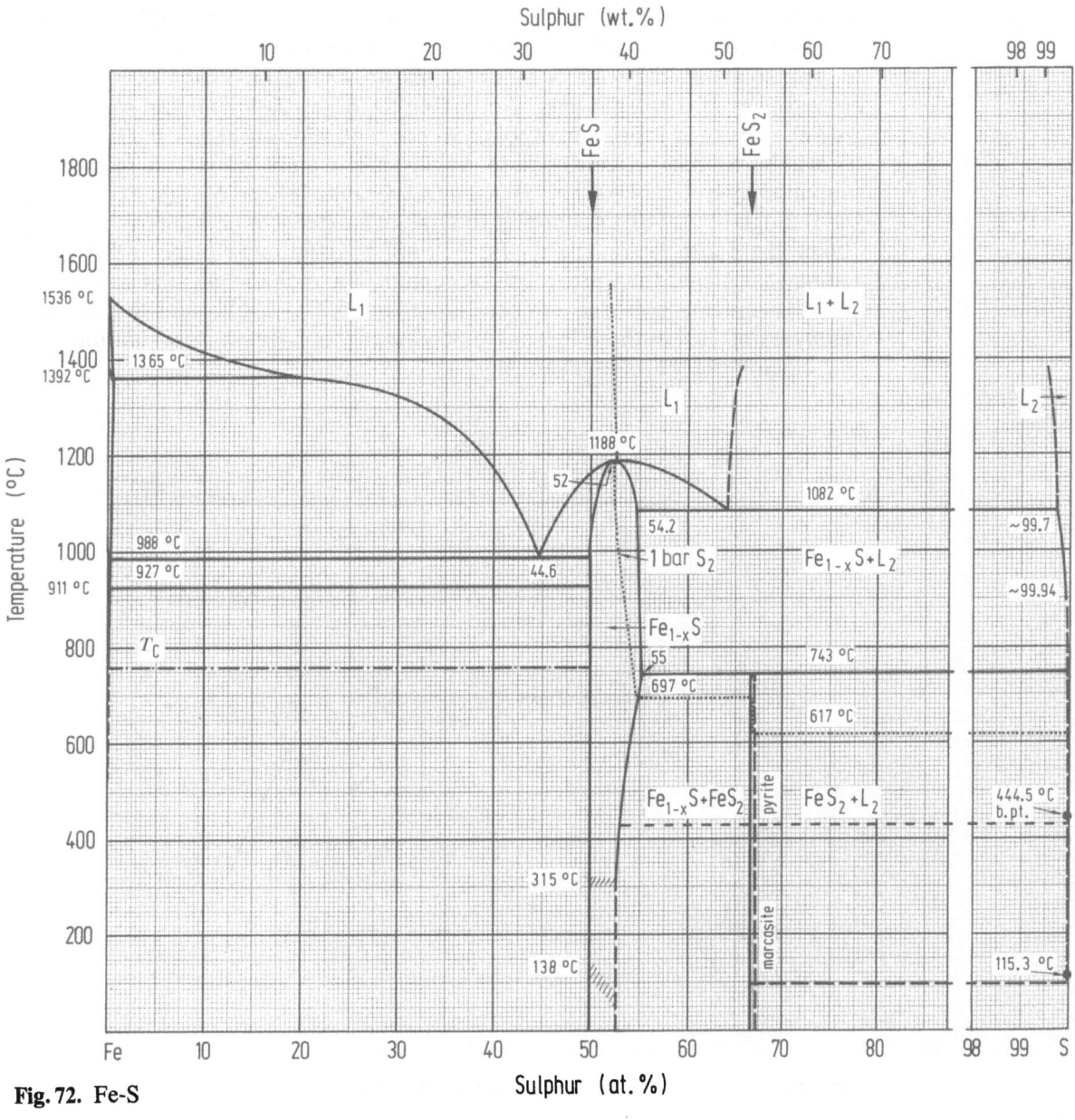

Fig. 72. Fe-S

of the system Fe-S. The part diagram Fe-FeS has been constructed mainly from data of Hansen [A] supplemented by results of Nagamori et al. [3], Burgmann et al. [4] and Rau [5] (Fe$_{1-x}$S) as well as the solubility data of [6–10], the part diagram FeS-S from Hansen [A], Elliott [B] and Shunk [C] as well as Arnold [11] and Rau [5]. The system is characterized by a congruently melting phase (1,188°C, 52 at.% S), Fe$_{1-x}$S, a eutectic between Fe and FeS at 988°C and 44.6 at.% S and a monotectic at 1,082°C and ~64.5 at.% S between FeS and S.

Fe-FeS region

The course of the liquidus line in the Fe-FeS concentration range has been reviewed and summarized several times (e.g. [1, 2]). Data published by Burgmann et al. [4] and Nagamori et al. [3] in addition to those perused by Hansen [A] and Shunk [C] have been accepted in the construction of Fig. 72. The saturation curve of S$_2$ (vapour) at 1 bar in the temperature range 1,200–1,600°C is based on the studies of Schürmann and Henke [12].

The solid solubility of S in Fe is shown in Fig. 73, which represents an assessment of available data. According to these, the maximum solubility in α(Fe), γ(Fe) and δ(Fe) corresponds to concentrations of 0.033, 0.09 and about 0.24 at.% S, respectively. These values are mainly based on information supplied by α: Grabke [8], γ: Hansen [A], Grabke [8] and Margot et al. [9] and δ: Barloga et al. [10]. According to Grabke's extensive gravimetric investigations, the peritectic reaction γ-Fe + 'FeS' ⇋ α-Fe takes place at 927°C, i.e. somewhat higher than previously accepted. The radiochemical measurements of [9] were made in the temperature range 950–1,250°C and the chemical analyses of [10] in the δ-Fe range. Schürmann and Henke assessed the γ-Fe/δ-Fe equilibria thermodynamically.

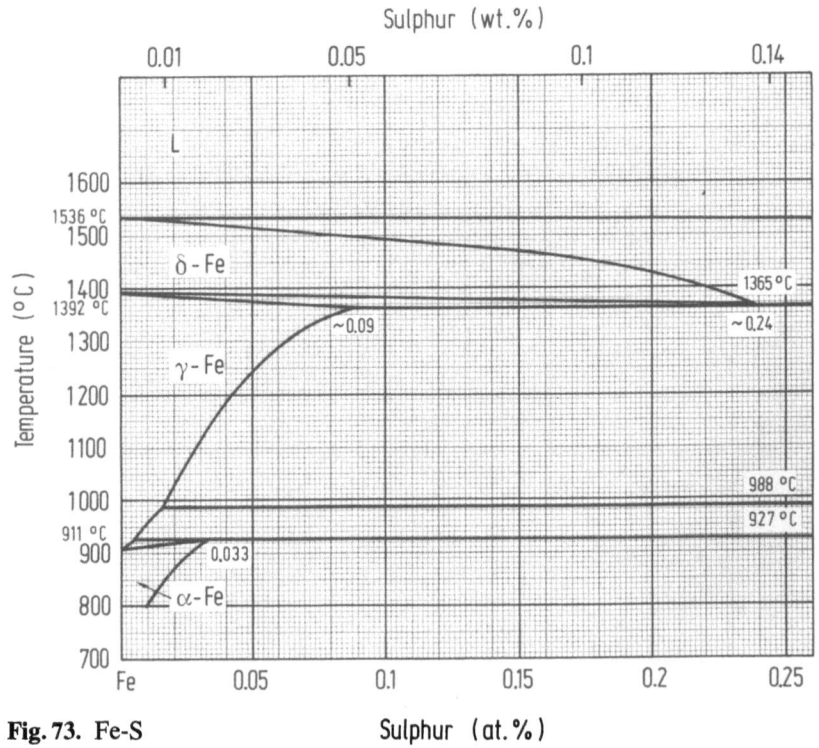

Fig. 73. Fe-S

Fe$_{1-x}$S melts congruently at 1,188 °C, exhibits a broad homogeneity range, crystallizes in an NiAs structure and has two low-temperature modifications. It has been investigated extensively because of its interesting electrical, magnetic and thermodynamic properties. In the present assessment, the compiler has concentrated on the experimentally established phase boundaries but less so on the low-temperature phase changes. Thus, the transformation temperature of Fe$_{1-x}$S at 315 °C (from specific heat measurements of Hirone et al. [14]) and at 138 °C [A] are only shown as hatched lines. As pointed out in the 'Introduction', accurate phase boundaries at such relatively low temperatures are very difficult to obtain.

The phase boundaries in the range 500–1,188 °C are well established by recent investigations of Rau (547–1,100 °C) [5] and Burgmann et al. (700–1,300 °C) [4], Rau applying pressures up to more than 100 bar. He interpreted his results by the formation and interaction of Fe vacancies and by additional incorporation of Fe on sulphur sites. Calculated phase boundaries [5] agree well with findings of [4] and [11]. Their compositions are given by: Fe$_7$S$_8$, Fe$_9$S$_{10}$, Fe$_{10}$S$_{11}$, Fe$_{11}$S$_{12}$ and FeS. All these except the last one show superstructures of the NiAs type (B8) owing to Fe point defects. (For details, see Koto et al. [15].)

Iron disulphide, FeS$_2$, melts incongruently at (742 ± 1) °C [16, 11] and decomposes at a pressure of 1 bar at 697 °C [17]. It occurs in two modifications, the orthorhombic marcasite type (C18) and the cubic pyrite type (C2). Heat capacity measurements by Grönvold and Westrum [18] point to a transformation temperature of about 425 °C. A slight compositional difference between the two modifications may occur but has not conclusively been established (Kullerud et al. [16] and Kullerud [19]). The S-rich phase boundary is taken to be FeS$_{2.0}$.

FeS-S region

Heating experiments by Arnold [11] carried out in evacuated, sealed silica tubes indicate that in the system FeS-S two immiscible liquids (L$_1$ and L$_2$) can coexist in equilibrium with crystalline Fe$_{1-x}$S and vapour at a temperature of 1,092 °C [11] or 1,082 °C [20], confirming results reported by Kullerud [20]. The phase relationships at this temperature are of the monotectic type with the monotectic point at about 64.5 at.% S (extrapolated).

Pyrrhotite, Fe$_{1-x}$S, in the temperature interval 742–1,082 °C can coexist in equilibrium with L$_2$ and vapour, and between 425 and 742 °C with pyrite and vapour. As mentioned above, pyrite melts incongruently at 742 °C, and the invariant "assemblage" [11], Fe$_{1-x}$S, FeS$_2$, L$_2$ and vapour can coexist in equilibrium.

In Fig. 72 the solidus curve of Fe in S-rich liquid (pyrrhotite liquidus) in the temperature range 1,082–742 °C stems from [11]. The solubility of Fe decreases from 99.7 at.% S (1,082 °C) to 99.9 at.% S (895 °C).

References

1 Schürmann, E.; Henke, H.J.: Gießereiforsch. 24 (1972) 1
2 Charma, R.C.; Chang, Y.A.: Metall. Trans. (B) 10 (1979) 103
3 Nagamori, M.; Hatakeyama, T.; Kameda, M.: Trans. Jpn. Inst. Met. 11 (1970) 190
4 Burgmann, W.; Urbain, G.; Frohberg, M.G.: Mem. Sci. Rev. Metall. 65 (1968) 568
5 Rau, H.: J. Phys. Chem. Solids 37 (1976) 425

6   Rosenquist, T.; Dunicz, L.: Trans. Metall. Soc. AIME 194 (1952) 604
7   Turkdogan, E.T.; Ignatowicz, S.; Pearson, J.: J. Iron Steel Inst. 180 (1955) 349
8   Grabke, H.J.: Private Communication 1980
9   Margot, E.; Venard, B.; Barbouth, N.; Oudar, J.: R. Acad. Sci. Paris, Ser. C 272 (1971) 373
10   Barloga, A.M.; Bock, K.R.; Parlee, N.: Trans. Metall. Soc. AIME 221 (1961) 173
11   Arnold, R.G.: Econ. Geol. 66 (1971) 1121
12   Schürmann, E.; Henke, H.J.: Giessereiforsch. 23 (1971) 165
13   Schürmann, E.; Henke, H.J.: Giessereiforsch. 24 (1972) 12
14   Hirone, T.; Maeda, S.; Chiba, S.: J. Phys. Soc. Jpn. 9 (1954) 500
15   Koto, K.; Morimoto, N.; Gyobu, A.: Acta Crystallogr. B31 (1975) 2759
16   Kullerud, G.; Yoder, H.S.: Econ. Geol. 54 (1959) 533
17   Rosenquist, T.: J. Iron Steel Inst. 176 (1954) 37
18   Grønvold, F.; Westrum, E.F.: J. Chem. Thermodyn. 8 (1976) 1039
19   Kullerud, G.: In: Researches in Geochemistry (Ed., Abelson). New York: Wiley 2 (1967) 286
20   Kullerud, G.; Yoder, H.S.: Ann. Rep. Dir. Geophys. Lab., 1960/61, Carnegie Inst., Washington, Year Book 60, p. 174

Fe–Sb   Iron–Antimony

Figs. 74, 75

The phase relationships in the system Fe-Sb have been re-investigated by Maier and Wachtel [1] with the aid of magneto-thermal analyses supported by X-ray and DTA measurements in the temperature range 220–1,600°C with alloys prepared from pure iron (99.99%, main impurities: 0.0036% O, 0.005% C) and antimony also of 99.99% purity. The general pattern of the phase diagram previously published by Hansen [A] was confirmed, except for the Fe-rich region. The revised diagram is shown in Fig. 74. The melting point of antimony, 630.755°C, is a secondary reference point.

The system is characterized by two intermediary phases, β (NiAs type) [2] and $FeSb_2$ (marcasite type, C18). The β phase melts congruently at 1,019°C with a composition of 42 at.% Sb. A transformation at 620°C quoted in the literature [4] was not confirmed by [1]. However, a magnetic transformation at about 220°C was found. The paramagnetic Curie temperature is positive on the Fe-rich and negative on the Sb-rich side. $FeSb_2$ is formed peritectically at 738°C and is stable at the stoichiometric composition with no significant width [1]. Its magnetic susceptibility depends on the heat treatment, the compound being ferromagnetic in the annealed state below 565°C and paramagnetic in the quenched state.

The solubility of Sb in bcc (Fe) has been newly determined by Nageswararao et al. [5] who measured lattice parameters, by Predel and Frebel [6] who employed X-ray and DTA analyses, and most recently by Takayama et al. [9] who based their results on X-ray diffraction and electron-probe micro-analyses of long-time (150–7,000 hrs.) annealed samples of high purity. The observations of all three teams are in substantial agreement. The data of [9] have been selected for the construction of Fig. 74. Solubilities of 4.89, 4.19, 3.92, 3.70, 3.37, 3.08, 2.77 and 2.58 at.% Sb correspond to the temperatures 950, 900, 850, 800, 750, 700, 650 and 600°C. The magnetic transition

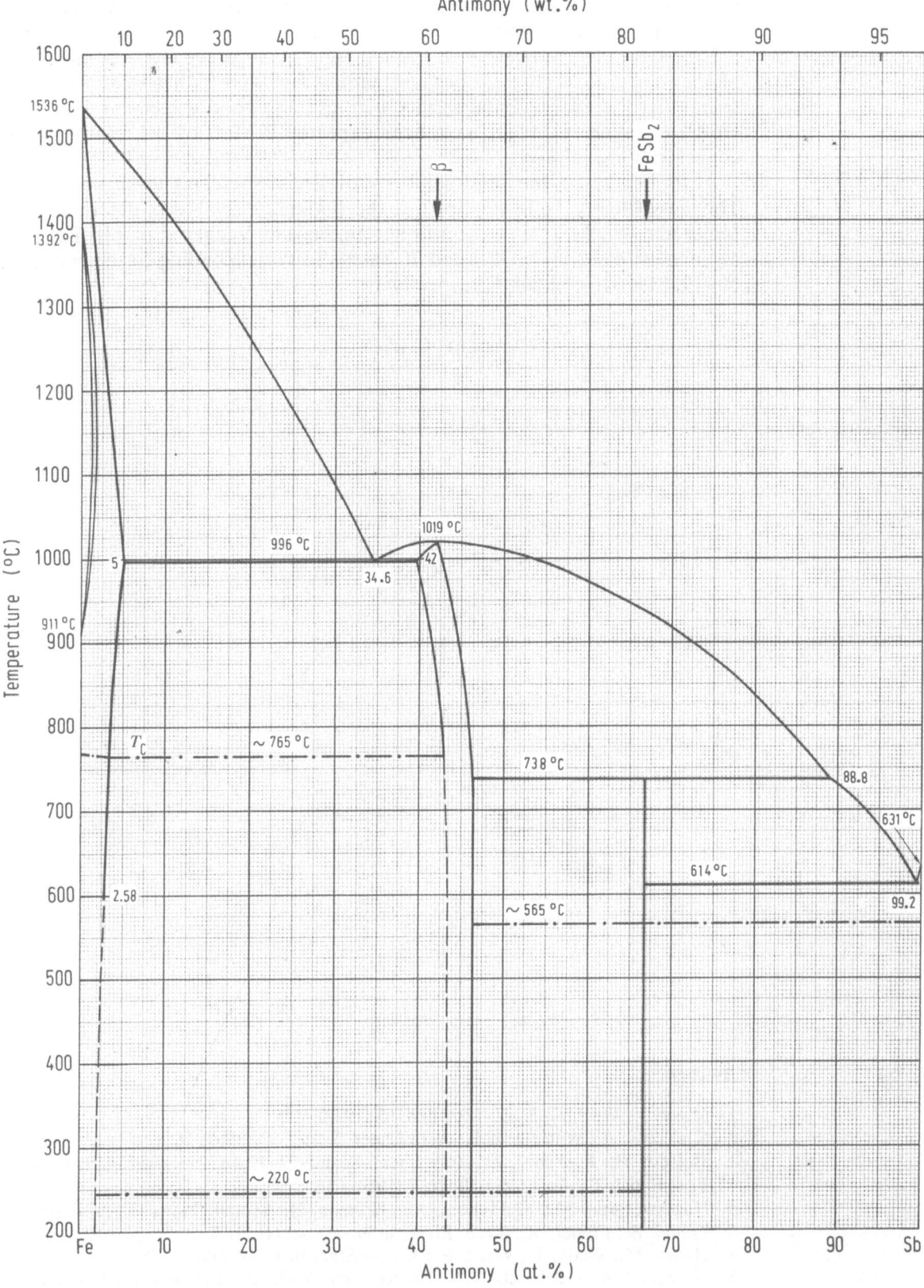

Fig. 74. Fe-Sb

does not appear to influence the course of the phase boundary. The eutectic temperature is quoted as being 1,002 °C by [5] and as 996 °C by [1], the latter value being adopted in Fig. 74.

Hillert et al. [7] redetermined the solubility of Sb in fcc (Fe) and the width of the (α + γ) field (Fig. 75) because earlier data were rather scattered. They found that the experimental results were in fairly good agreement with van't Hoff's equation, as might have been expected in view of the relatively low concentrations involved.

According to [1] the solid solubility of Fe in Sb seems to be very small. The Fe-solubility in liquid Sb (i.e. the liquidus curve) was determined by Maier and Wachtel [1] who confirmed findings of Vecher et al. [8]. They are as follows: 85.82, 81.0, 78.7 and 76.7 at.% Sb corresponding to temperatures of 779, 819, 848 and 868 °C respectively.

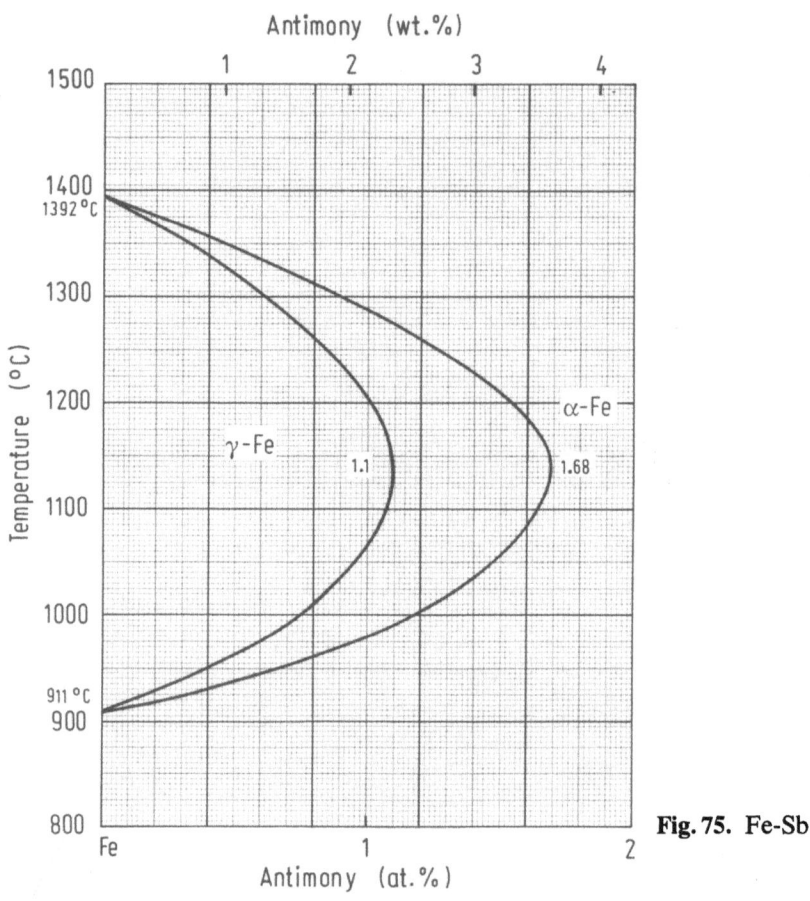

Fig. 75. Fe-Sb

References

1 Maier, J.; Wachtel, E.: Z. Metallkd. 63 (1972) 411
2 Kjekshus, A.; Walseth, K.P.: Acta Chem. Scand. 23 (1969) 2621
3 Holseth, H.; Kjekshus, A.: Acta Chem. Scand. 22 (1968) 3273 and 23 (1969) 3043
4 Rosenquist, T.: Acta Crystallogr. 7 (1954) 636
5 Nageswararao, M.; McMahon, C.J.; Herman, H.: Metall. Trans. 5 (1974) 1061
6 Predel, B.; Frebel, M.: Arch. Eisenhüttenwes. 42 (1971) 42
7 Hillert, M.; Wada, T.; Wada, H.: J. Iron Steel Inst. 205 (1967) 539
8 Vecher, A.A.; Geiderich, V.A.; Gerassimov, Ya.I.: Zh. Fiz. Khim. 35 (1961) 1578
9 Takayama, T.; Wey, M.Y.; Nishizawa, T.: Trans. Jpn. Inst. Met. 22 (1981) 315

The phase diagram of the system Fe-Sc is shown in Fig. 76. It is essentially that of Naumkin et al. [1], but incorporates data reported by [1], Ikeda et al. [2, 3], Hellawell [4], Dwight [5], Protasov et al. [6] and Gladyshevskii et al. [7].

The system exhibits two compounds, Fe_2Sc, i.e. a Laves phase with a small homogeneity range and appearing in three modifications, and $FeSc_3$ formed peritectoidally at about 800°C from $\beta(Sc)$ solid solution and Fe_2Sc.

The melting and transformation points of scandium are 1,541°C and 1,336°C (IPTS-68) respectively.

The addition of scandium to iron lowers the melting point of Fe [1, 4] as well as the γ/δ transition temperature and raises the α/γ transformation temperature [4]. X-ray, thermal analysis, dilatometric measurements and metallographic investigations were used by [1] to determine the entire system. 39 alloys were prepared from 99.91% carbonyl Fe and vacuum distilled Sc, 99.8% (impurities: Si: 0.001, Ti: 0.05, Al: 0.01, Ca: 0.005, Fe: 0.06, Cu: 0.1, N: 0.03 and O: 0.1 wt.%)

The samples were arc-melted under helium, homogenised at 900°C for 150 hours (for alloys up to 31 at.% Sc) and at 725°C for 200 hours and quenched.

The Laves phase was found to melt congruently at about 1,600°C. The composition and temperature of formation of the Fe-rich and Sc-rich eutectic were found to be at about 9 at.% Sc (1,200°C) and about 80 at.% Sc (910°C). The solubility of scandium in α- and $\gamma(Fe)$ does not exceed 0.4 at.% Sc [4] and the solid solubility of Fe in $\alpha(Sc)$ is 0.8 at.% Fe at 700°C [1].

In a re-assessment of the Fe-Sc phase diagram Bodak et al. [8] using X-ray, microscopy and DTA analyses, confirm the feature of the already existing diagram. However, the results concerning the composition and temperature of the eutectica are at variance, the temperatures being about 50°C higher than those reported by [1], and the Sc-rich compound was found to correspond to a composition $FeSc_7$ to be compared with $FeSc_3$ proposed by [1]. (A translation of the Ukranian publication [8] was not available to the compiler).

A number of investigators have studied the crystal stucture and the magnetic and electrical resistivity of Fe_2Sc. In a more recent investigation, [8] was able to identify all three Laves modifications, namely: λ_1 ($MgZn_2$-type), λ_2 ($MgCu_2$-type) and λ_3 ($MgNi_2$-type). They suggested a temperature range of 1,295–1,525°C for the cubic modification (λ_2) and a low temperature range for the (λ_1) type, thus confirming findings by [6, 7] who observed a cubic structure, and of [3] and [5] who reported a $MgZn_2$- and $MgNi_2$-type structure respectively, both hexagonal. [3] investigated the magnetism of the (λ_1) modification on 33.3 at.% Sc arc-melted samples, prepared from 99.99% and 99.9% Sc which were homogenized and quenched. Their observations establish conclusively that the hexagonal $MgZn_2$-type phase is ferromagnetic, the Curie temperature T_C being (542 ± 9) K. According to Nevitt et al. [10] and [2], λ_3 and λ_2 are also ferromagnetic. Electrical resistivity measurements of the Laves phases have been carried out by Ikeda and Nakamachi [9]. For a discussion of the mutual correlation between electrical resistivity and magnetic properties, the reader may be referred to their publication [9].

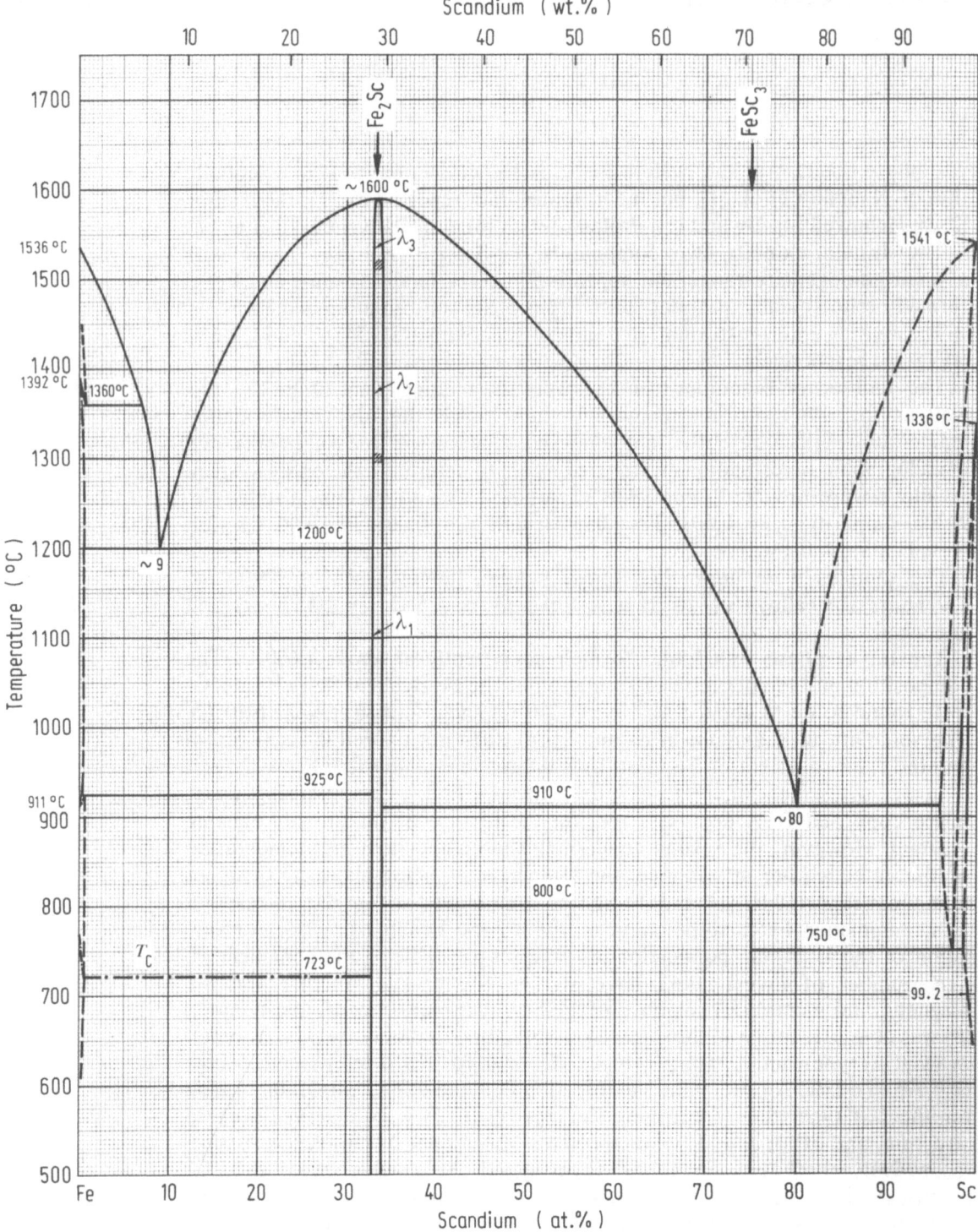

Fig. 76. Fe-Sc

1 Naumkin, O.P.; Terekhova, V.F.; Savitskij, E.M.: Izv. Akad. Nauk SSSR, Met. (1969) 3, 125
2 Ikeda, K.: Z. Metallkd. 68 (1977) 195
3 Ikeda, K.; Nakamichi, T.; Yamada, T.; Yamamoto, M.: J. Phys. Soc. Jpn. 36 (1974) 611
4 Hellawell, A.: J. Less Common Met. 4 (1962) 101
5 Dwight, A.E.: Trans. ASM 53 (1961) 479
6 Protasov, V.S.; Kripyakevich, P.I.; Cherkashin, E.E.: Kristallografiya 11 (1966) 689
7 Gladyshevskii, E.I.; Kripyakevich, P.I.; Kuzma, Yu.B.; Protasov, V.S.: In: Savitskij, E.M.; Terekhova, V.F.; (Eds.): Voprosy Teorii i Primeneniya Redkozemelnykh Metallov. Akad. Nauk SSSR, Moscow 1964, pp. 153–154
8 Bodak, O.I.; Kotur, B.Ya.; Gavrilenko, I.S.; Markiv, V.Ya.; Ivanchenko, V.G.: Dopov. Akad. Nauk Ukr. RSR, A Apr. (1978) 365
9 Ikeda, K., Nakamachi, T.: J. Phys. Soc. Jpn. 39 (1975) 963
10 Nevitt, M.V. et al.: Proc. Int. Conf. Magnetism. Nottingham, 1964, p. 153

Iron–Selenium

Fe–Se
Fig. 77

The phase diagram (Fig. 77) has been adopted in its main features from a publication by Schuster et al. [1], who constructed the phase boundaries from their own investigations in the region 20–66 at.% Se and data published by other investigators. Two liquid miscibility gaps, two compounds, namely tetragonal $\beta(Fe_{1.04}Se)$ and orthorhombic $\varepsilon(FeSe_2)$, and several $Fe_{1-x}Se$ (NiAs-related structures δ, δ', γ, γ') were observed. The Fe-rich hexagonal δphase transforms to a high temperature modification δ' of unknown structure, and undergoes a λtransformation to the monoclinic γ' phase.

Various results relating to the phase relationships in the Fe-Se phase diagram have been published, e.g. by Kullerud [2], Svendsen [3], Dutrizac et al. [4], Haraldsen and Grønvold [5, 6]. However, there are major differences in the Fe-rich and NiAs-phase composition range. For clarification, [1] re-investigated the region 20–66 at.% Se. They used samples prepared from thin Fe-sheets (purity 99.9%), thin iron-wire (99.9%) and Se-shots (99.999%), sealed under vacuum in quartz capsules, which were re-capsuled and heated at 1,000–1,100°C for about 20 days. X-ray and thermal measurements supported by isopiestic methods were employed to establish the phase relationships represented in Fig. 77.

Phase diagram below 750°C

The monoselenide $\beta(Fe_{1.04}Se)$ exists between 49.0–49.4 at.% Se. It decomposes peritectoidally at 457°C [6] and crystallizes with the tetragonal PbO-structure. Values for the lattice constants reported by [1] ($a = 0.377_5$ nm and $c = 0.552_7$ nm) are in good agreement with results quoted by Grønvold [6], Hägg and Kindström [7] and Reddy and Chetty [8].

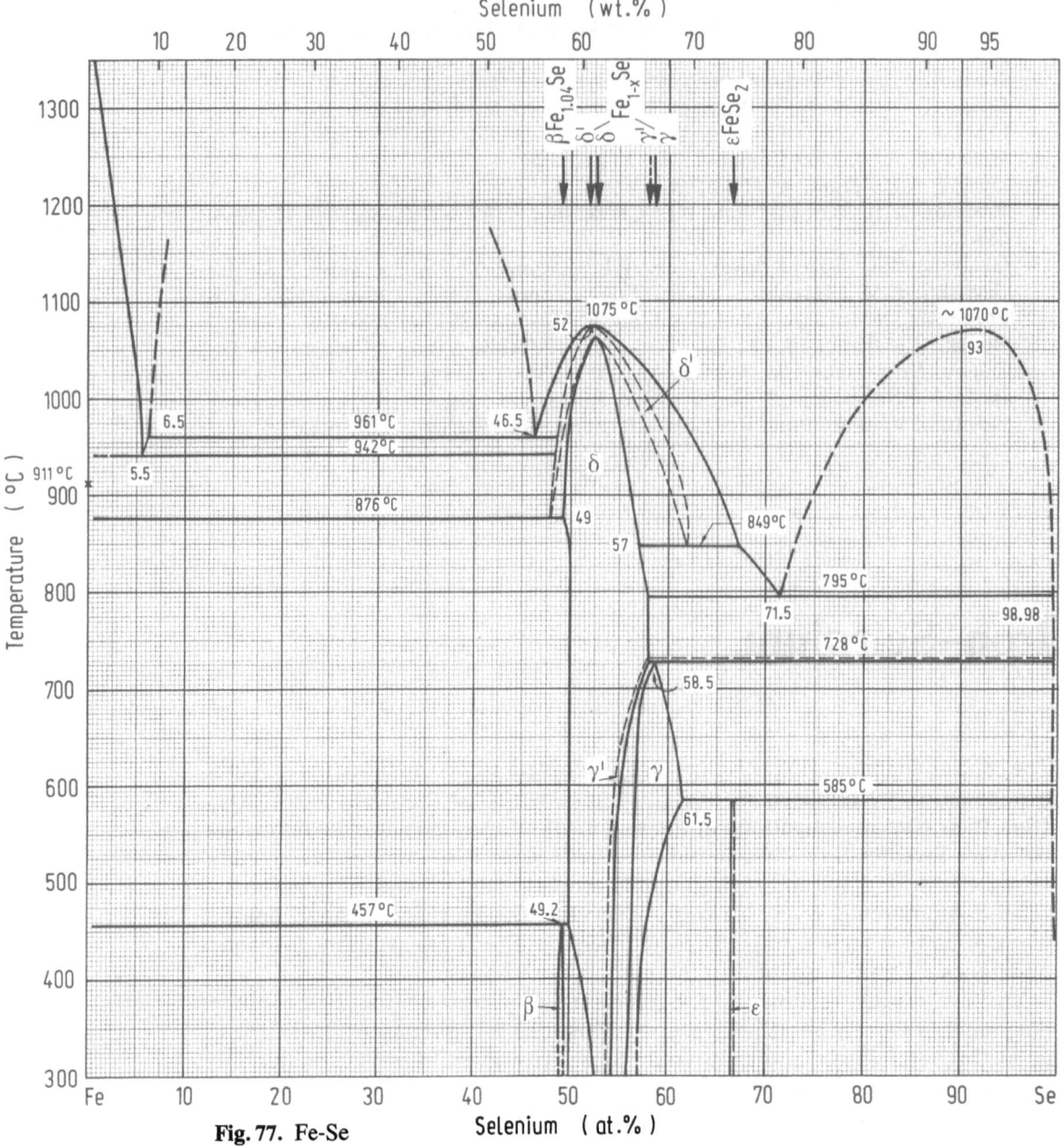

Fig. 77. Fe-Se

The $Fe_{1-x}Se$ phases having NiAs-related structures have an extended range of homogeneity. Samples quenched from 380°C and 550°C, respectively, and investigated by X-ray analysis are hexagonal δphase in the range 51.5–53.5 at.% Se and 51.5–54.3 at.% Se respectively. The lattice constant (350°C) is in good agreement with results quoted by Okazaki and Hirakawa [9]. At about 54 at.% Se the δphase undergoes a λtransformation to the mono-clinic γ'phase which is stable at 550°C between 54–54.6 at.% Se and exhibits a monoclinically distorted structure with a simple c'axis. In the adjoining two phase field (54.6–56.4 at.% Se) two monoclinic phases with very similar structure coexist, namely γ' and γ with similar a- and b-axes but simple c'- and a double c-axis, respectively; γ is formed pertectically at 728°C and

58.5 at.% and exists between 56.4 and 57.5 at.% Se (380°C) with a maximum extension at 585°C (56.7–61.5 at.% Se).

The paramagnetic, antiferromagnetic and ferrimagnetic properties of iron-selenides with NiAs-type structure has been investigated and described by [10] and [13].

The orthorhombic marcasite structure of the $\varepsilon(FeSe_2)$ phase with a narrow range of homogeneity was confirmed by [1]. The peritectic decomposition temperature of 585°C found by [1] is a confirmation of results obtained by [3] and [4].

Thermal and X-ray analyses by [1], magnetic measurements by Terzieff and Komarek [10] and isopiestic studies by Schuster et al. [11] have here been used to construct the region 40–70 at.% Se in the temperature range 300–750°C.

Phase diagram above 750°C

According to [1] and [2] the Fe-rich region is characterized by three invariant equilibria at higher temperatures. A monotectic at 961°C and 46.5 at.% Se, a eutectic at 942°C and 5.5 at.% Se and a eutectoid at 876°C. The Se-rich region is marked by a eutectoid at 849°C, a monotectic at 795°C and 71.5 at.% Se [1, 2] with a miscibility gap extending from 71.5–~98 at.% Se [1], 71.5–~99 at.% Se [2] and 73.9–99.98 at.% Se respectively [4]. According to [4] the miscibility gap is asymmetrical. Their results have been incorporated in Fig. 77 and are shown as dotted line. The experimental data were obtained by visual and quenching methods. The pressure over the melts was not 1 bar but rather the equilibrium Se pressure at a particular temperature-composition.

The hexagonal δphase changes to a high-temperature modification δ' of undetermined structure at 52.8 at.% Se and a transformation temperature of 1,065°C. The congruent melting point of δ' has been observed at 52.0 at.% Se and 1,075°C.

The solid solubility of Se in Fe is negligibly small according to [2, 4] and [12]. Information relating to the solubility of iron in liquid selenium are inconsistent. The value of 0.06 wt.% at 790°C submitted by [4] seems to be the most probable one.

References

1 Schuster, W.; Mikler, H.; Komarek, K.L.: Monatsh. Chem. 110 (1979) 1153
2 Kullerud, G.: Carnegie Inst. Wash. Year Book 67 (Publ. 1969) 175 (1967–1968)
3 Svendsen, S.R.: Acta Chem. Scand. 26 (1972) 3757
4 Dutrizac, J.E.; Janjua, M.B.I.; Toguri, J.M.: Can. J. Chem. 46 (1968) 1171
5 Haraldsen, H.; Grønvold, F.: Struct. Rep. 9 (1955) 97
6 Grønvold, F.: Acta Chem. Scand. 22 (1968) 1219
7 Hägg, G.; Kindström, A.L.: Z. Phys. Chem. B22 (1933) 453
8 Reddy, K.V.; Chetty, S.C.: Phys. Stat. Sol. (A) 32 (1975) 585
9 Okazaki, A.; Hirakawa, K.: J. Ohys. Soc. Jpn. 11 (1956) 930
10 Terzieff, P.; Komarek, K.L.: Monatsh. Chem. 109 (1978) 651
11 Schuster, W.; Ipser, H.; Komarek, K.L.: Monatsh. Chem. 110 (1979) 1171
12 Jain, B.K.; Singh, A.K.; Chandra, K.: J. Phys. F: Met. Phys. 8 (1978) 2625
13 Terzieff, P.; Komarek, K.L.: Monatsh. Chem. 109 (1978) 1037

Fe–Si   Iron–Silicon

Figs. 78, 79

A set of interdependent thermodynamic values, consistent with the Fe-Si phase diagram, based on a critical assessment of all the published phase diagram and thermodynamic data has recently been submitted by Chart [1]. Simultaneously, Schürmann and Hensgen [2] re-determined the melting and solid state equilibria in the region Fe-FeSi of samples prepared from "Rein-Eisen" or electrolyte iron and Si of 99.6% purity. The results are based on thermal and differential thermal analyses as well as isothermal holding tests. The re-investigation confirmed the experimental results reported in the literature.

The melting point of pure silicon has been accepted as $(1,412 \pm 2)$ °C (IPTS-68) [1, 10]. Figure 78 shows with minor modifications the phase diagram selected by [1]. This diagram incorporates experimental data recently reported by [2]. It is based primarily on the information presented by Hansen [A], Elliott [B], Shunk [C], Köster and Gödecke [3, 4] and the theoretical and experimental work of Inden, Schlatte and Pitsch [5–9]. In addition the work of 32 investigators has been taken into account. The main qualitative difference between the phase diagram proposed by [1] and that accepted by earlier critical assessments of the thermodynamical properties by Hultgren [D] and Chart [10] is the nature of the ordering reactions in the bcc Fe-rich solutions.

The liquidus and solidus curves are essentially those accepted by [A] combined with data reported by Übelacker [11–13] and confirmed by [2] for the Fe-rich region, the findings of [3, 4] and [2] for the area adjacent to Fe_2Si and investigations by Piton and Fay [14] and Wachtel and Mager [15] of the region adjacent to $FeSi_2$.

The boundaries of the γ-loop have been adopted from the work of Fischer et al. [16] and of [11] whose observations are in excellent agreement. The $(\alpha + \gamma)$ region at the vertex extends from 3.19–3.8 at.% Si, a lower concentration than that reported in earlier publications, owing probably to the higher purity of the alloys employed by [11] and [16] (Fig. 79).

In the Fe-rich region, the bcc alloys exist in both ordered (B2 and $D0_3$) and disordered (A2) modifications and are referred to as α_2, α_1 and α respectively. A considerable amount of experimental and theoretical work has been undertaken to determine the composition and temperature ranges over which these phases are stable and to establish the nature of the order/disorder transformations. The current status is represented in Fig. 78. The second-order (continuous) reactions are shown by hatched lines, implying that they do not represent compositions at which abrupt physical changes take place but indicate temperatures at which the rate of decrease of the degree of order with increasing temperature is a maximum.

The phase $\alpha_1(D0_3)$ exists in a high and low-temperature modification: [3] and Inden [17]. The boundary drawn in Fig. 78 is based on work by [4, 12, 13] and Lecocq [18] whose results are in very good agreement. The temperature of the A2/B2 transitions decrease more rapidly below the Curie temperature (T_C) owing to the magnetic interaction energy. A two phase field $(B2 + D0_3)$ is stable below T_C which has been proven by theoretical calculations and experimental investigations [5–9].

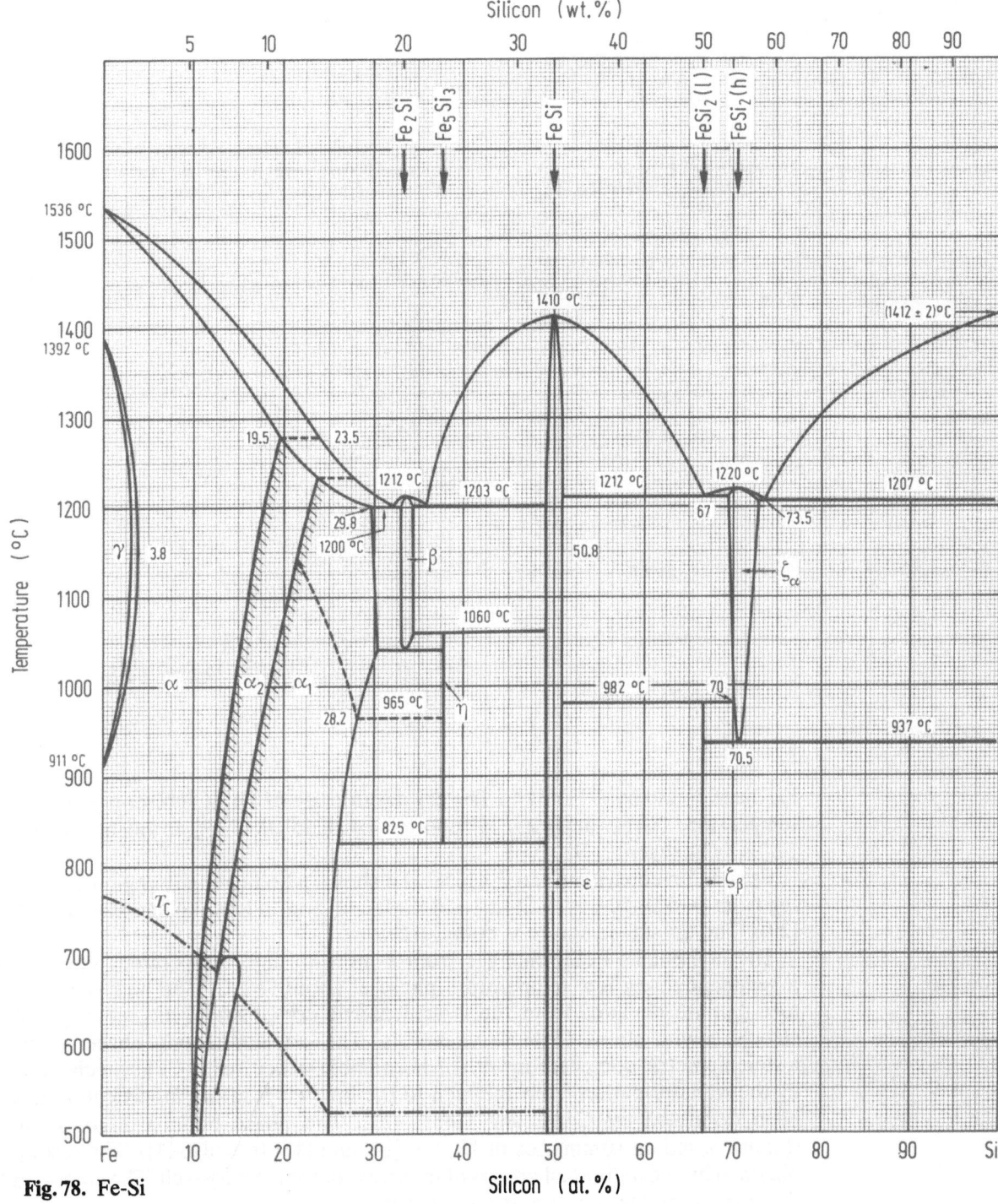

Fig. 78. Fe-Si

Compounds

The following compounds are stable: Fe$_2$Si, (β), Fe$_5$Si$_3$, (η), FeSi, (ϵ), and FeSi$_2$ in a high and low temperature modification, ζ_α and ζ_β respectively.

The stability and homogeneity ranges of Fe$_2$Si and Fe$_5$Si$_3$ have been taken from work by [3, 4] and [10]. A homogeneity range for Fe$_2$Si of about 1 at.%

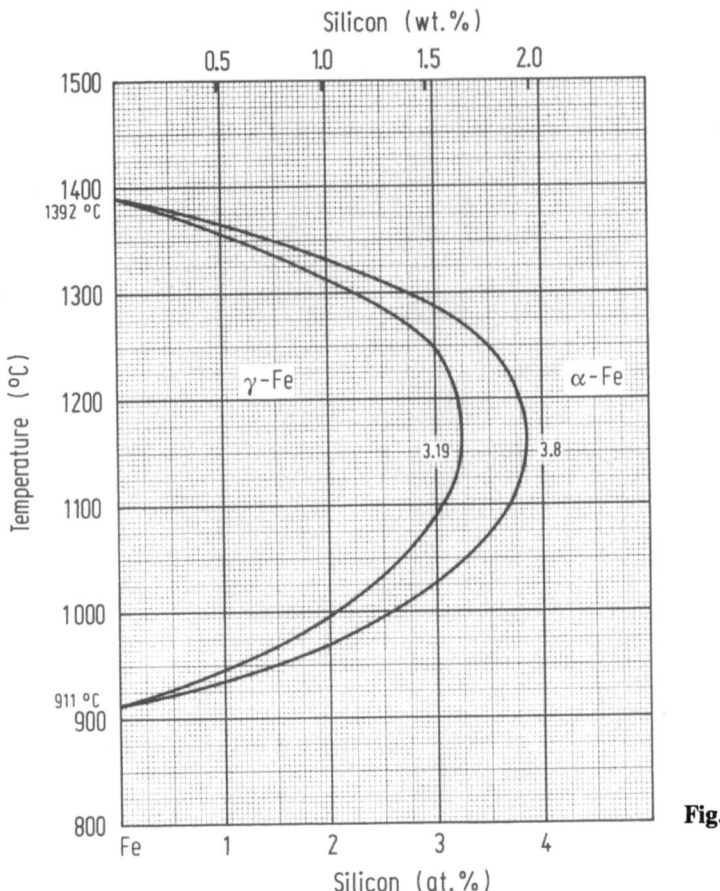

Fig. 79. Fe-Si

has been found by [3] and [4]. The crystal stucture of Fe_2Si determined by high temperature X-ray analysis, is reported by Kudielka [19] to be hexagonal. Fe_5Si_3 is hexagonal, $D8_8$ type.

FeSi melts congruently at 1,410°C and shows a homogeneity region extending from ~49.0–50.8 at.% Si at 1,150°C ([A, C] and Sidorenko and Rabinovich [20]). The crystal structure is cubic, B20.

The compound $FeSi_2$ changes from a metallic high temperature phase to a semiconducting low temperature phase at about 982°C. The high temperature $\zeta_\alpha$phase deviates from the stoichiometric composition, has a tetragonal structure with about 13% Fe vacancies. The semiconducting phase ζ_β has a stoichiometric composition with orthorhombic structure [21]. The mechanism of the transformation from the high to the low temperature modification has been discussed by Corre and Genin [22]. The phase boundaries of $FeSi_2$ (ζ_α) are based on summaries of [B, C], [14] and [15] (see also [1]). The results obtained by the individual groups of investigators agree very well. The eutectoid decomposition temperature is a compromise.

The extremily low solubility of iron in silicon has been measured by Weber and Riotte [23] using neutron activation analysis (NAA) and ERP of quenched samples in the temperature range 900–1,200°C. The method employed allowed the determination of both, the total and the interstitial Fe concentration. The total content is given, e.g., as 4.6×10^{14} atoms/cm³ at 1,000°C and 1.5×10^{16} atoms/cm³ at 1,200°C, corresponding to 9.2×10^{-7} and 3.0×10^{-5} at.% Fe respectively.

References

1 Chart, T.G.: Commission des Communautés Européennes CECA No. Research Project 7210 – CA/3/303, Nov. 1981
2 Schürmann, E.; Hensgen, U.: Arch. Eisenhüttenweg. 51 (1980) 1
3 Köster, W.; Gödecke, T.: Z. Metallkd. 59 (1968) 602
4 Köster, W.: Trans. Iron Steel Inst. Jpn. 14 (1974) 387
5 Inden, G.; Pitsch, W.: Z. Metallkd. 62 (1971) 627
6 Inden, G.; Pitsch, W.: Z. Metallkd. 63 (1972) 253
7 Schlatte, G.; Inden, G.; Pitsch, W.: Z. Metallkd. 65 (1974) 94
8 Schlatte, G.; Pitsch, W.: Z. Metallkd. 66 (1975) 660
9 Schlatte, G.; Pitsch, W.: Z. Metallkd. 67 (1976) 462
10 Chart, T.G.: High Temp. High Press. 2 (1970) 461
11 Übelacker, E.: C. R. Acad. Sci. 261 (1965) 976
12 Übelacker, E.: Colloq. Int. CNRS 1976 (157) 171
13 Übelacker, E.: Mem. Sci. Rev. Metall. 64 (1967) 183
14 Piton, J.P.; Fay, M.F.: C. R. Acad. Sci. Ser. C 266 (1968) 514
15 Wachtel, E.; Mager, T.: Z. Metallkd. 61 (1970) 762
16 Fischer, W.A.; Lorenz, K.; Fabritius, H.; Hoffmann, A.; Kalwa, G.: Arch. Eisenhüttenwes. 37 (1966) 79
17 Inden, G.: Private communication to [1] (1978)
18 Lecocq, Y.; Lecocq, P.: Colloq. Int. CNRS 1967 (157) 165
19 Kudielka, H.: Z. Kristallogr. 145 (1977) 177
20 Sidorenko, F.A.; Rabinovich, B.S.: Tr. Ural, Politekh. Inst., 1965 (144) 71
21 Nishida, I.: Phys. Rev. B7 (1973) 2710
22 Le Corre, C.; Genin, J.M.: Phys. Status Solidi B, 51 (1972) 85
23 Weber, E.; Riotte, H.G.: J. Appl. Phys. 51 (1980) 1484

Iron-Samarium, see Fe-R: Iron-Rare Earth Metals, Fig. 61

Fe–Sm

Iron–Tin

Fe–Sn

Figs. 80, 81

The system Fe-Sn has been studied by numerous authors [1–14]. The diagram shown in Fig. 80 is essentially that suggested by Hansen [A]. Its main features are a broad miscibility gap in the liquid state above a monotectic temperature of 1,130°C [2, 7], a solution in γ(Fe) in the shape of a closed loop [3, 5, 11, 12], and five intermediary phases represented by the formulae Fe_3Sn, Fe_3Sn_2, $FeSn$, $FeSn_2$ and γ (NiAs structure).

The Fe-rich liquidus was determined by Campbell et al. [8] and Wever and Reinecken [10]. Their results are in good agreement and are adopted in Fig. 80. The Sn-rich liquidus was taken from solubility measurements carried out by Davey [6]. According to this author, they are accurate for the temperature range 232°C up to 1,000°C, but uncertain at higher temperatures. They are as follows: The temperature in °C 232, 300, 400, 500, 600, 700, 800, 900, 1,000, 1,100, 1,130 (monotectic) correspond to 0.0022, 0.0098, 0.052, 0.17, 0.46, 1.3, 3.2, 5.8, 8.2, 17.5, 34.7 at.% Fe.

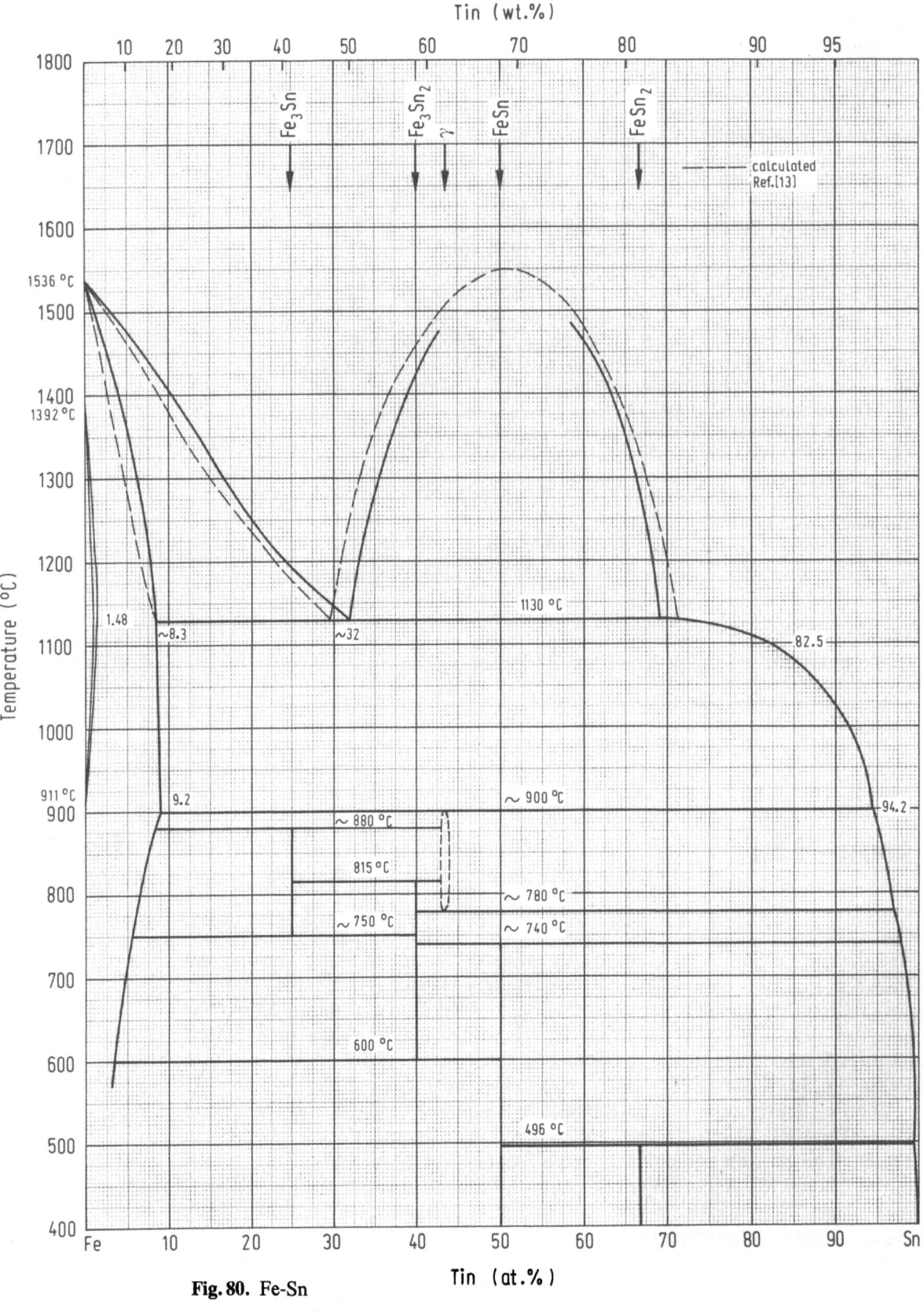

Fig. 80. Fe-Sn

The Sn-rich eutectic in the Fe-Sn system was very accurately determined by Connolly and McAllan [4] who used resistance thermometry with a resolution of 0.0001°C. It was found at a concentration of 0.001 at.% Fe and 231.96°C (IPTS 1968).

In view of conflicting data on the miscibility gap, Shiraishi and Bell [7] re-investigated this region. Their results agree with those given previously by Mills and Turkdogan [9] and they are as follows. The temperatures in °C 1,145, 1,217, 1,305, 1,392 and 1,490 correspond to 68.1 and 31.0 at.% Fe, 63.2 and 35.0 at.% Fe, 64.1 and 33.0 at.% Fe, 58.9 and 36.5 at.% Fe and 59.2 and 38.5 at.% Fe. The phase boundary drawn in Fig. 80 is that calculated by [13] and recently (1981) confirmed by experimental results (mass spectrometry) [16].

Speight [3], Hillert et al. [5], Tréheux et al. [11] and Thwaites and Chatterjeee [12] published data on the solubility limits in γ(Fe). Their results are in good agreement. However, it should be mentioned that the boundaries reported are based on measurements made on alloys containing about 0.02 at.% C. According to [11] and [12], a decrease in the carbon content moves the boundaries to lower but not well established Sn concentrations. Figure 81 represents the calculated γ-loop [13] based on data reported by [3, 5, 11].

The solid solubilities of Sn in α(Fe) were determined by lattice-parameter measurements [1]. Accordingly, the solubility ranges from a maximum of 9.2 at.% at 900°C to 3.2 at.% at 600°C. This, however, differs from data

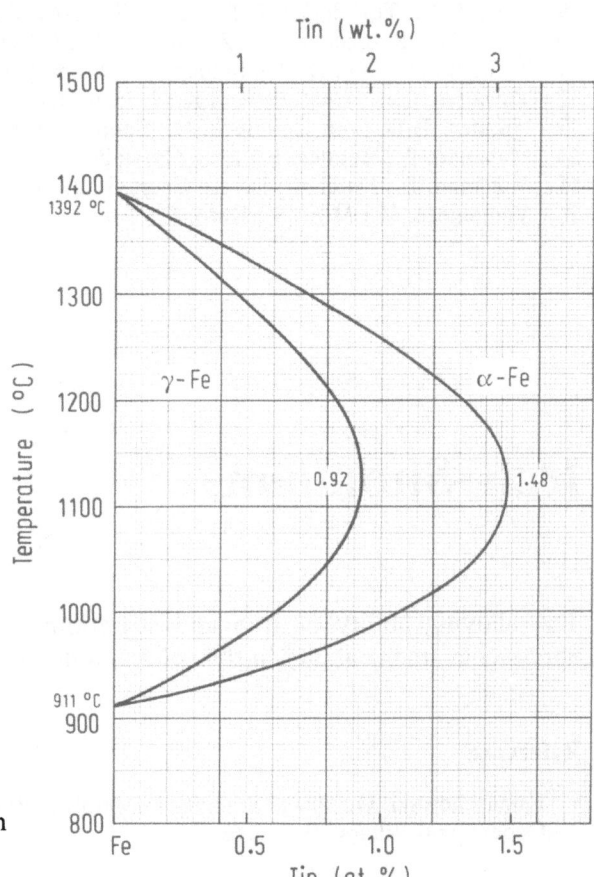

Fig. 81. Fe-Sn

suggested by Hansen [A]. In view of purer materials and experimental care taken by [1], these results have been adopted in Fig. 80.

The controversy regarding the compositions and dissociation temperatures of the compounds in the Fe-Sn system [11] is an indication of the sluggish diffusion rates in this system and also point to the small Gibbs energy differences between stable phases [13]. In the present assessment the compounds and their stability ranges are basically those suggested by Hansen [A], as mentioned earlier, excepting those for the Fe_3Sn_2 region.

Malaman et al. [14, 15] determined the structure, magnetic properties and stability range of Fe_3Sn_2 and found it belongs to the rhombohedral system, with $z = 6$ in a hexagonal unit cell of the following dimensions: $a = b = 0.5344$ nm and $c = 1.9845$ nm; it is ferromagnetic and stable between 600 and 815 °C.

A thermodynamic evaluation of the Fe-Sn system has been undertaken by Nüssler et al. [13], to whose paper the reader may be referred.

References

1 Nageswararao, M.; McMahon, C.J.; Herman, H.: Metall. Trans. 5 (1974) 1061
2 Wagner, S.; St. Pierre, G.: Metall. Trans. 3 (1972) 2873
3 Speight, E.A.: Met. Sci. J. 6 (1972) 57
4 Connolly, J.J.; McAllan, J.V.: Acta Metall. 23 (1975) 1209
5 Hillert, M.; Wada, H.; Wada, T.: J. Iron Steel Inst. 205 (1967) 539
6 Davey, T.R.A.: Trans. Inst. Min. Metall. 76 (1976) C 66
7 Shiraishi, S.Y.; Bell, H.B.: Min. Metall. 77 (1968) C 104
8 Campbell, A.N.; Wood, J.H.; Skinner, G.B.: J. Am. Chem. Soc. 71 (1949) 1729
9 Mills, D.C.; Turkdogan, E.T.: Trans. TMS-AIME 230 (1964) 1202
10 Wever, F.; Reinecken, W.: Z. Anorg. Allg. Chem. 151 (1926) 349
11 Tréheux, D.; Duc, D.; Guirakdeno, P.: Mem. Sci. Rev. Metall. 71 (1974) 289
12 Thwaites, C.H.; Chatterjee, S.K.: J. Iron Steel Inst. 210 (1972) 581
13 Nüssler, H.D.; von Goldbeck, O.; Spencer, P.J.: CALPHAD 3 (1979) (1) 19
14 Malaman, B.; Roques, B.: Acta Crystallogr. B32 (1976) 1348
15 Malaman, B.; Fruchart, D.; Le Caer, G.: J. Phys. F: Met. Phys. 8 (1978) 2389
16 Yamamoto, M.; Mori, S.; Kato, E.: Tetsu To Hagane 67 (1981) 1952

Fe–Sr Iron–Strontium

The alloying ability of iron and strontium is likely to be so small that it approaches immiscibility in the solid and liquid states.

Reference

1 Kubaschewski, O.: Phase Stability in Metals and Alloys. Proc. Batelle Colloquium, Geneva and Villars 1967, p. 63

The phase relationships in the Fe-Ta system in the range 0–15 at.% Ta [1, 2] are well established. An investigation of the complete system, using 11 alloy compositions, has been carried out by Raman [7] (Fig. 82). Previous work has been reviewed by Hansen [A] and Elliott [B].

The phase diagram is of the contracted γfield type and entails two intermediary phases, Fe_2Ta, C14, a $MgZn_2$-type Laves phase and $\mu(FeTa)$, D8$_5$, with Fe_7W_6 structure, both showing a significant range of homogeneity.

Fe-rich region (0–12 at.% Ta) (Fig. 83)

An accurate and detailed re-determination in the temperature range 900–1,550°C was carried out by Sinha and Hume-Rothery [1]. Alloys were prepared from 99.96/99.97% Fe (melting point 1,535.5°C) and 99.99% Ta, arc-melted, quenched and homogenized. Thermal analysis was used to determine the liquidus and eutectic horizontal, while metallographic and dilatometric techniques were employed to establish the solid state relationships. These results were used to calculate the Gibbs energy changes for the equilibria liquid $\leftrightarrows$ δ, liquid $\leftrightarrows$ γ and δ $\leftrightarrows$ γ. Fischer et al. [2] investigated the Fe-rich area in the temperature interval 900–1,400°C employing susceptibility measurements on 9 high purity alloys and also calculated the Gibbs energy changes. [2] found the temperature of the eutectoidal as well as the one of the peritectoidal to be somewhat lower than previously reported by Genders and Harrison [9] and [1]. The width of the $(\gamma + \delta)$ two phase field (0.86–1.38 at.% Ta) at 1,215°C according to [2] lies in between earlier findings. The solubility in $\gamma(Fe)$ (0.4 at.% Ta) at 965°C agrees with results reported in the literature. The maximum solid solubility in $\alpha(Fe)$ (0.71 at.% Ta) at 965°C [2] confirms the results by [1] and is in good agreement with an extrapolation of the solubility data, determined by lattice parameter measurements (600–900°C) by Abrahamson and Lopata [12].

Raman [7] investigated the 20–90 at.% Ta region studying specimens prepared from high purity material (99.98% Fe and 99.9% Ta) which were arc-melted under argon, annealed at 1,300°C and quenched. His findings are based on X-ray analysis supplemented by metallographic examinations and are represented in a tentative phase diagram (Fig. 82).

He confirmed the phase Fe_2Ta reported by several investigators, e.g. Burlakov and Kogan [3], Kuo [4], Elliott and Rostocker [5]. It was found to melt congruently at about 1,775°C [6] and to have a homogeneity range of about 28–36 at.% Ta at 1,300°C [7]. It is a Laves phase and apparently crystallizes only in the hexagonal $MgZn_2$-type structure ($a = 0.481$–0.483 nm, $c = 0.787$–0.784 nm); see also [10]. The axial ratio ($c/a = 1.633$) is nearly ideal at the Fe-rich side and decreases with increasing Ta content [7]. The magnetic behaviour and the electrical resistivity have been studied by Kai et al. [13] and Ikeda and Nakamachi [14] respectively. A then unknown phase stable at a composition between 49 and 54 at.% Ta (FeTa) has been identified by [7]. It has a rhombohedral (D8$_5$) structure isotypic with Fe_7W_6 and is denoted μ. That the ideal composition is not covered by its homogeneity region is not unusual with intermetallic compounds the deviation depending on its Gibbs

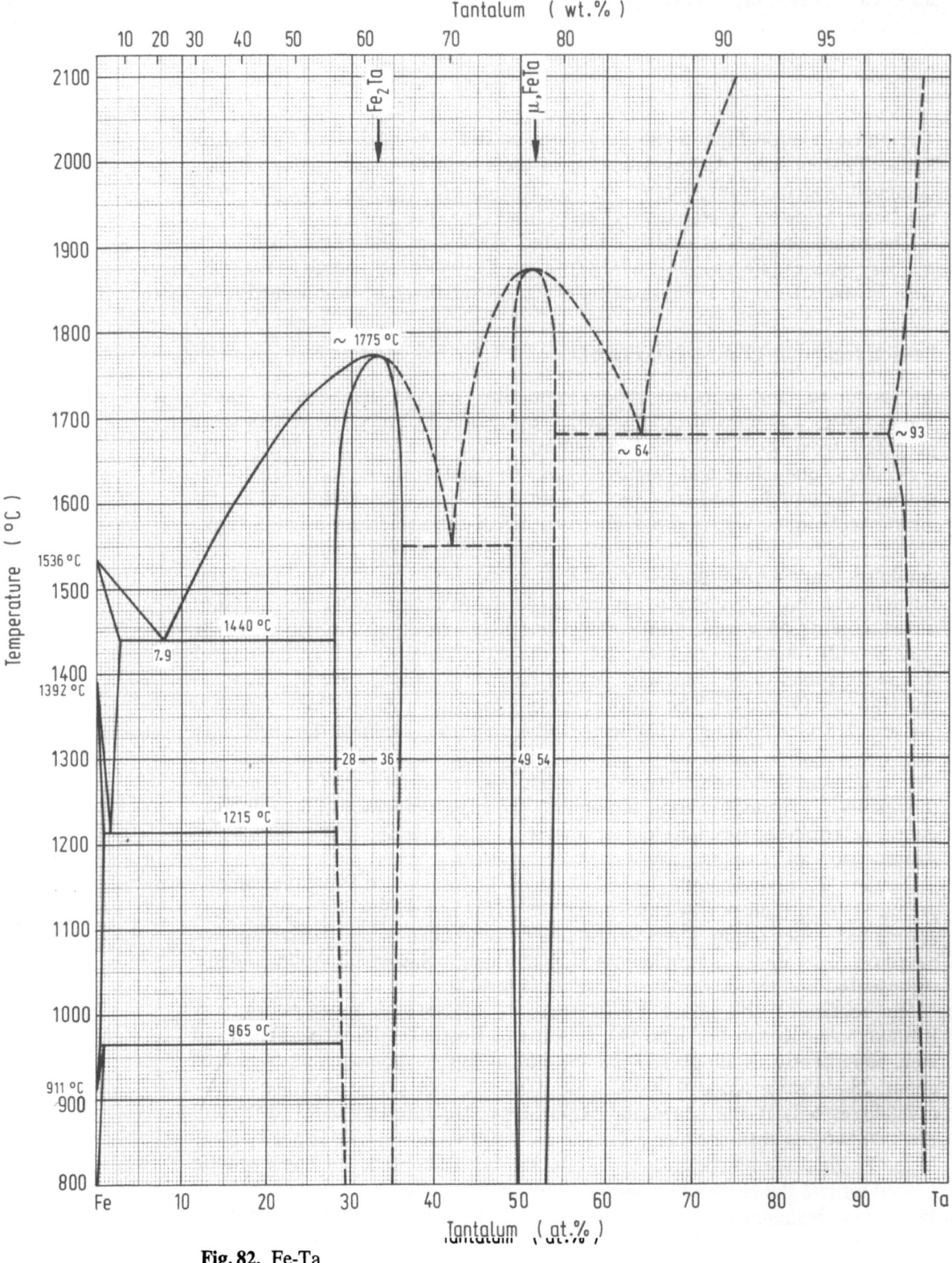

Fig. 82. Fe-Ta

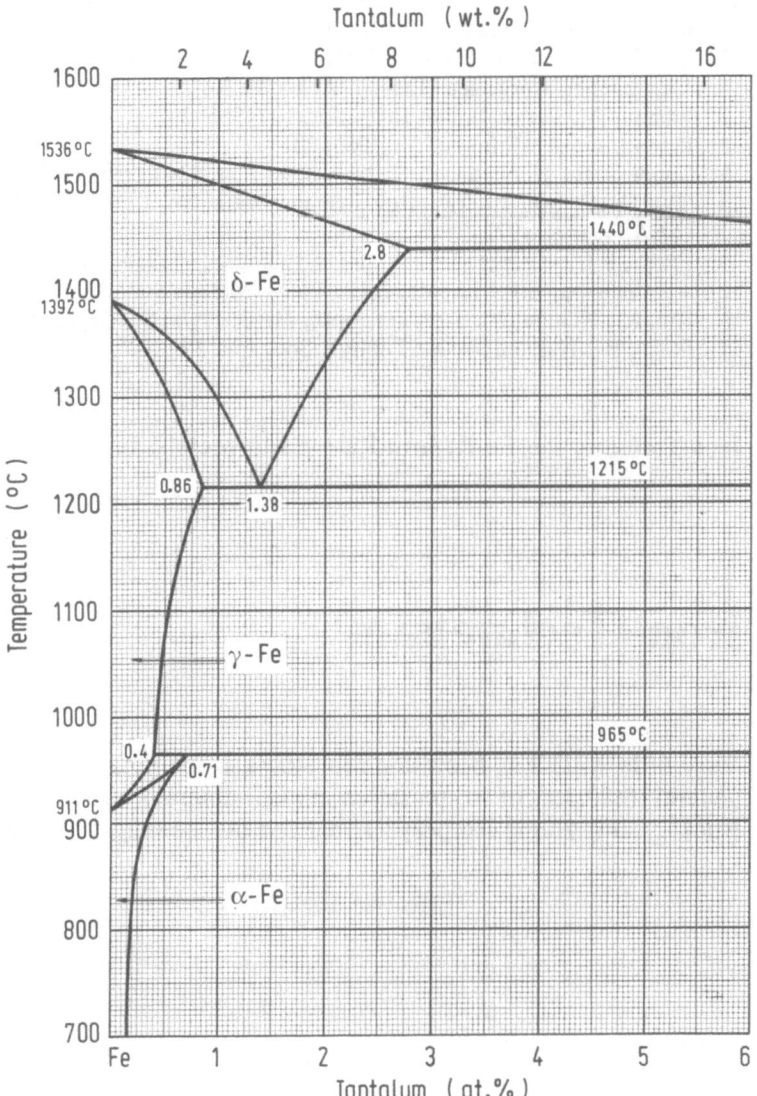

Fig. 83. Fe-Ta

energy and that of the neighbouring phases (see also Fe-Nb). The Fe-Ta system is similar to the Fe-Nb system. The structure of the σ phase determined in the latter, is closely related to μ, both are electron- rather than size factor-compounds, as has been shown by Frank and Kasper [11]. According to a suggestion by [7], the phase melts congruently at a temperature between 1,800 and 1,900 °C, whereas [3] propose a peritectic formation at a temperature of about 1,400 °C on the basis of metallographic evidence.

The position of the Ta-rich eutectic between FeTa and Ta solid solution is rather uncertain and needs further investigation. It was found to lie at about 60 at.% Ta [3] and 63 at.% Ta [7] respectively.

References

1 Sinha, K.A.; Hume-Rothery, W.: J. Iron Steel Inst. 205 (1967) 671
2 Fischer, W.A.; Lorenz, K.; Fabritius, H.; Schlegel, D.: Arch. Eisenhüttenwes. 41 (1970) 491

Fe–Ta

3 Burlakov, V.D.; Kogan, V.S.: Phys. Met. Metallogr. USSR 7 (1959) 67
4 Kuo, K.: Acta Metall. 1 (1953) 720
5 Elliott, R.P.; Rostocker, W.: Trans. ASM 50 (1958) 617
6 Unpublished work, see Hansen [A: Fe-Ta Ref. 7]
7 Raman, A.: Trans. Indian Inst. Met. 19 (1966) 202
8 Raman, A.: Z. Metallkd. 57 (1966) 301
9 Genders, R.; Harrison, R.: J. Iron Steel Inst. 134 (1936) 173
10 'Tantalum': At. Energy Rev., Spec. Issue No. 3, Part III Crystal Structures. IAEA, Vienna 1972
11 Frank, F.C.; Kasper, J.S.: Acta Crystallogr. 11 (1958) 184; 12 (1959) 483
12 Abrahamson, E.P.; Lopata, S.L.: Trans. Metall. Soc. AIME 236 (1966) 76
13 Kai, K.; Nakamichi, T.; Yamamoto, M.: J. Phys. Soc. Jpn. 29 (1970) 1094
14 Ikeda, K.; Nakamichi, T.: J. Phys. Soc. Jpn. 39 (1975) 963

Fe–Tb

Iron-Terbium, see Fe-R: Iron-Rare Earth Metals, Fig. 63

Fe–Tc   Iron–Technetium

Figs. 84, 85

Very little work has been done to establish the binary system. However, the general features of a tentative phase diagram are suggested in Fig. 84. Buckley and Hume-Rothery [1] carried out heating and cooling analyses on

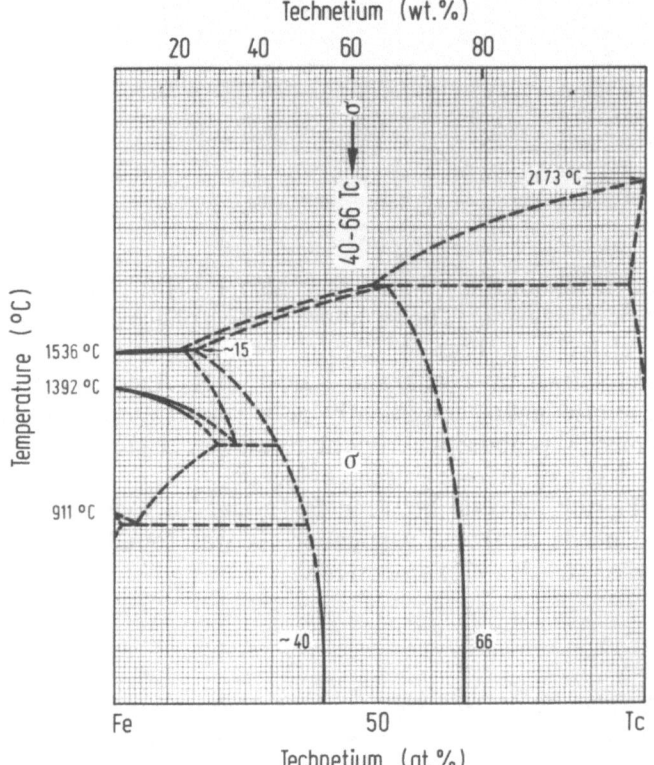

Fig. 84. Fe-Tc

146

alloys containing 1, 5, 10 and 15 at.% Tc using 99.95% pure (BISRA "H") iron and technetium supplied by the Oak Ridge National Laboratory. Details about its purity are not given. Great care was taken to ensure proper alloying as confirmed by autoradiographic analyses on sectioned and polished ingots. [1] have ascertained that Tc is extensively soluble in both γ- and δ(Fe) (see Fig. 84). The addition of Tc to Fe produces very little change in the melting point (Fe m. pt. 1,535 °C). The δ liquidus curve is very nearly horizontal (Fig. 85).

Darby and Lam [2] and Darby et al. [3] found only one intermediate phase in the system, a tetragonal σphase in the composition range ~40–66 at.% Tc. The parameters are as follows:

$$40 \text{ at.\% Tc} \quad a = 0.9010 \text{ nm, } c = 0.4713 \text{ [2]}$$
$$60 \text{ at.\% Tc} \quad a = 0.9130 \text{ nm, } c = 0.4788 \text{ [3]}$$

Their technetium had a purity of ca. 97.6%.

In a communication on melting points and composition of σ phases Raynor [4] concluded that the majority of σ phases are formed by peritectic reaction. Therefore the phase diagram adopted from a schematic model published by Moffatt [5] seems feasible.

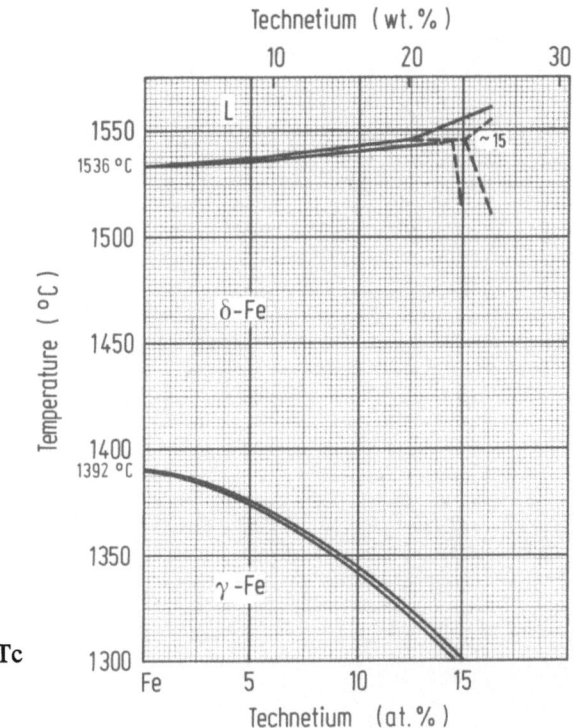

Fig. 85. Fe-Tc

References

1 Buckley, R.A.; Hume-Rothery, W.: J. Iron Steel Inst. London 201 (1963) 121
2 Darby, J.B.; Lam, D.J.: U.S. At. Energy Comm. ANL-6677 (1962) 255
3 Darby, J.B.; Lam, D.J.; Norton, L.J.; Downey, J.W.: J. Less Common Met. 4 (1962) 558.
4 Raynor, G.V.: J. Less Common Met. 29 (1972) 333
5 Moffatt, W.G.: The Handbook of Binary Phase Diagrams. General Electric Comp. 1976–1981

Fe–Te   Iron–Tellurium

Fig. 86

The phase diagram in Fig. 86 is based entirely on experimental results critically assessed in a systematic study of various investigations by Ipser et al. [1].

The system contains four 'compounds', $FeTe_{0.9}$ (β and β' tetragonal and rhomobohedral respectively), $FeTe_{1.2}$ (γ), δ and δ' a monoclinically distorted and a hexagonal NiAs phase respectively, and $FeTe_2$ (ε) with an orthorhombic structure.

Several authors have submitted results pertaining to the phase relationships in the system. Grønvold et al. [2] carried out accurate X-ray studies of the entire concentration range; Abrikosov et al. [3] used X-ray, thermal and

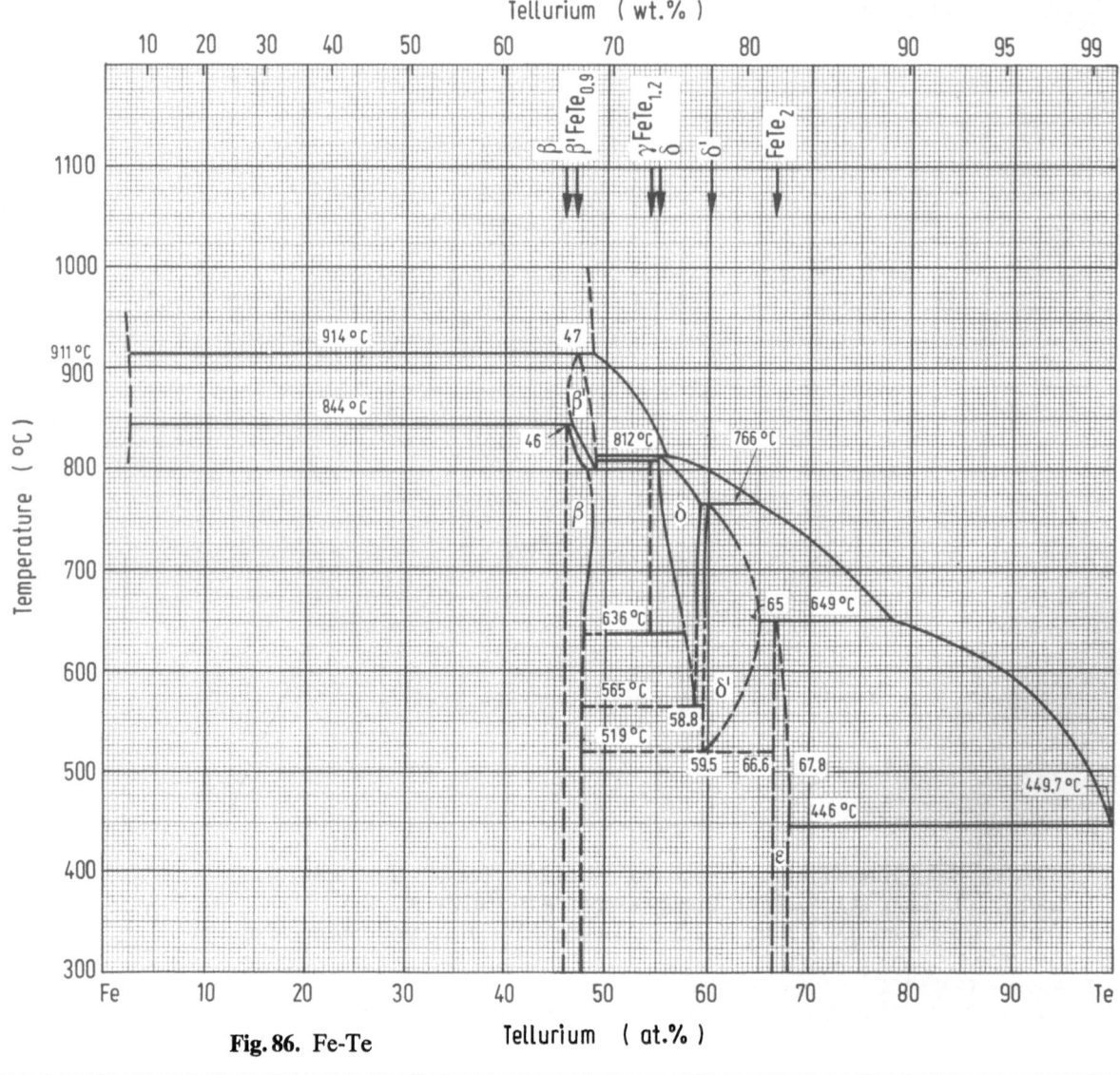

Fig. 86. Fe-Te

metallographic measurements on alloys covering the complete system; Chiba [4] investigated the system up to 50 at.% Te employing thermal, X-ray and magnetic techniques; Llewellyn and Smith [5] reported on the 62–69 at.% Te region. However, a comparison of these data revealed major discrepancies and a re-investigation of the system was undertaken by [1]. They used thermal, X-ray and isopiestic measurements of alloys prepared from 99.9% Fe and 99.99 and 99.999% Te. Samples were heated for ~ 15 hours at 900–1,000°C, annealed 1–3 weeks at 600–800°C and furnace cooled. Samples prepared for X-ray measurements were sealed under vacuum in quartz ampoules, heat-treated and quenched in ice water. Their results are represented in Fig. 86. The phases β, γ, δ and ε found by [2] and the high-temperature phase β' found by Røst and Webjørnsen have been confirmed. The additional high temperature phases observed by [5] and [3] were not discovered. $\beta(FeTe_{0.9})$ tetragonal is stable from room temperature to 844°C where it decomposes – probably by peritectoid reaction – into α-Fe and the high-temperature phase β'. However, it is possible that the reaction at 844°C is eutectoid with a congruent transformation point between β and β'; β shows a maximum range of homogeneity at 750°C of about 2.5 at.% Te; $\beta'(FeTe_{0.9})$ which is rhombohedral, is formed peritectically at 914°C. It decomposes eutectoidally at 800°C into β and $\gamma(FeTe_{1.2})$. The range of homogeneity extending from less than 46.1 to about 47.9 at.% Te at 870°C is based on isopiestic results. $\gamma(FeTe_{1.2})$ is stable between 809 and 636°C at a composition of 54.2 at.% and may exhibit a small range of homogeneity (between 54 and 55 at.% Te). The monoclinically distorted NiAs phase δ decomposes peritectically at 55.2 at.% Te and 812°C and is stable down to 565°C. It is separated from the hexagonal NiAs-type by a narrow two phase region. Both NiAs phases habe an appreciable range of homogeneity. Orthorhombic $\varepsilon(FeTe_{2.0})$ exists between room temperature and 649°C, is formed peritectically and has a probable range of homogeneity extending from 66.6 to 67.8 at.% Te at 500°C. $FeTe_{2.0}$ and Te form a degenerate eutectic at 446°C.

The solid solubility of Te in Fe has not been accurately determined but appears to be small. According to X-ray measurements, a sample of about 1.5 at.% Te equilibrated at 830°C showed the lattic parameter identical to that of pure $\alpha(Fe)$. [2] reported a low terminal solid solubility, which is in sharp contrast to results obtained by [3] who found a solubility of 18 at.% Te at 800°C.

References

1 Ipser, H.; Komarek, K.L.; Mikler, H.: Monatsh. Chem. 105 (1974) 1322
2 Gronvøld, F.; Haraldsen, H.; Vihovde, J.: Acta Chem. Scand. 8 (1954) 1927
3 Abrikosov, N.Kh.; Dyuldina, K.A.; Zhdanova, V.V.: Khal' kogenidy 2 (1970) 98; Chem. Abstr. 75 (1971) 91874a
4 Chiba, S.: J. Phys. Soc. Jpn. 10 (1955) 837
5 Llewellyn, J.P.; Smith, T.: Proc. Phys. Soc. (London) 74 (1959) 65
6 Røst, E.; Webjørnsen, S.: Acta Chem. Scand. A28 (1974) 361

Fe–Th   Iron–Thorium

Fig. 87

The phase relationships of the Fe-Th system have been established by Thomson [1, 2] who has studied the system over the entire range of compositions at temperatures between 800°C and the liquidus. Further clarification of the nature and crystal structure of the intermediate phases has been undertaken by Johnson et al. [3] ($Fe_{17}Th_2$), Smith and Hansen [4] and Matthias et al. [5] (Fe_3Th_7) and Buschow and van der Goot [6] [7–10] (Fe_7Th_2).

The phase diagram in Fig. 87 is based mainly on results obtained by [2]. The following five compounds have been identified:

$Fe_{17}Th_2$, rhombohedral, $a = 0.8565$ nm, $c = 1.2469$ nm [11];

Fe_5Th, $CaCu_5$-type, hexagonal with $a = 0.513$ nm, $c = 0.402$ nm [11];

Fe_7Th_2, crystalizing in two modifications, $\alpha(Fe_7Th_2)$, hexagonal Ce_2Ni_7-type, $a = 0.5193$ nm, $c = 2.4785$ nm;

$\beta(Fe_7Th_2)$, rhombohedral Gd_2Co_7-type, $a = 0.5195$ nm, $c = 3.719$ nm [11, 6];

Fe_3Th hexagonal or rhombohedral $PuNi_3$-type, $a = 0.522$ nm, $c = 2.496$ nm or $a = 0.518$ nm, $c = 2.52$ nm [4, 11];

Fe_3Th_7, hexagonal $a = 0.985–0.982$ nm, $c = 0.615–0.621$ nm [11].

Based on metallographic evidence, [2] concluded that the solubility of Fe in Th at 950°C is significantly lower than 2 at.%. Solid solubility data of Th in Fe have not been reported as yet.

The melting and transformation points of Th, $(1,750 \pm 10)$°C and 1,360°C respectively, have been taken from an evaluation of the properties of the Actinides by Oetting et al. [12].

The values for the allotropic temperature of Th reported in the literature vary considerably (1,330–1,405°C). According to [12], this wide range of values is due to the effect of impurities. The selected value has been taken from the extrapolated transition temperature of Th for zero carbon content [12].

Thomson [2] employed thermal analysis, metallographic and X-ray measurements of samples annealed for 14 days at temperatures up to 1,500°C and prepared from high purity iron and iodide thorium containing the following impurities: Fe: 20 ppm, N: 10 ppm, O: 190 ppm, C: 20 ppm.

The four intermetallic compounds (Fe_3Th_7, Fe_3Th, Fe_5Th and $Fe_{17}Th_2$) reported by Florio et al. [13] were confirmed. There is no evidence that any of the phases exist over a measurable range of compositions. The Fe-richest compound crystallizes directly from the melt at 1,462°C, whereas all other intermediate phases are formed by peritectic reaction. The compositions of the compounds and eutectics are considered to be accurate to $\pm 1\%$ and the accuracy of the melting temperatures was estimated to ± 12 K [2].

Buschow and van der Goot [6], using X-ray diffraction and metallographic examination, obtained evidence that the Fe-Th system actually contains one more compound, corresponding to the formula Fe_7Th_2 and existing in two modifications, with a crystal structure closely related to the other three Fe-rich phases. The two polymorphs are reported to co-exist in mixed amounts with no well defined temperature of transformation. They differ only in the sequence of layering along the unique crystallographic axes. The observations are based on samples prepared from 99.9% Th and 99.99% Fe which were arcmelted and vacuum-annealed for 3–4 weeks at temperatures ranging from 950–1,100°C. The simultaneous occurrence of the α and β forms resembles

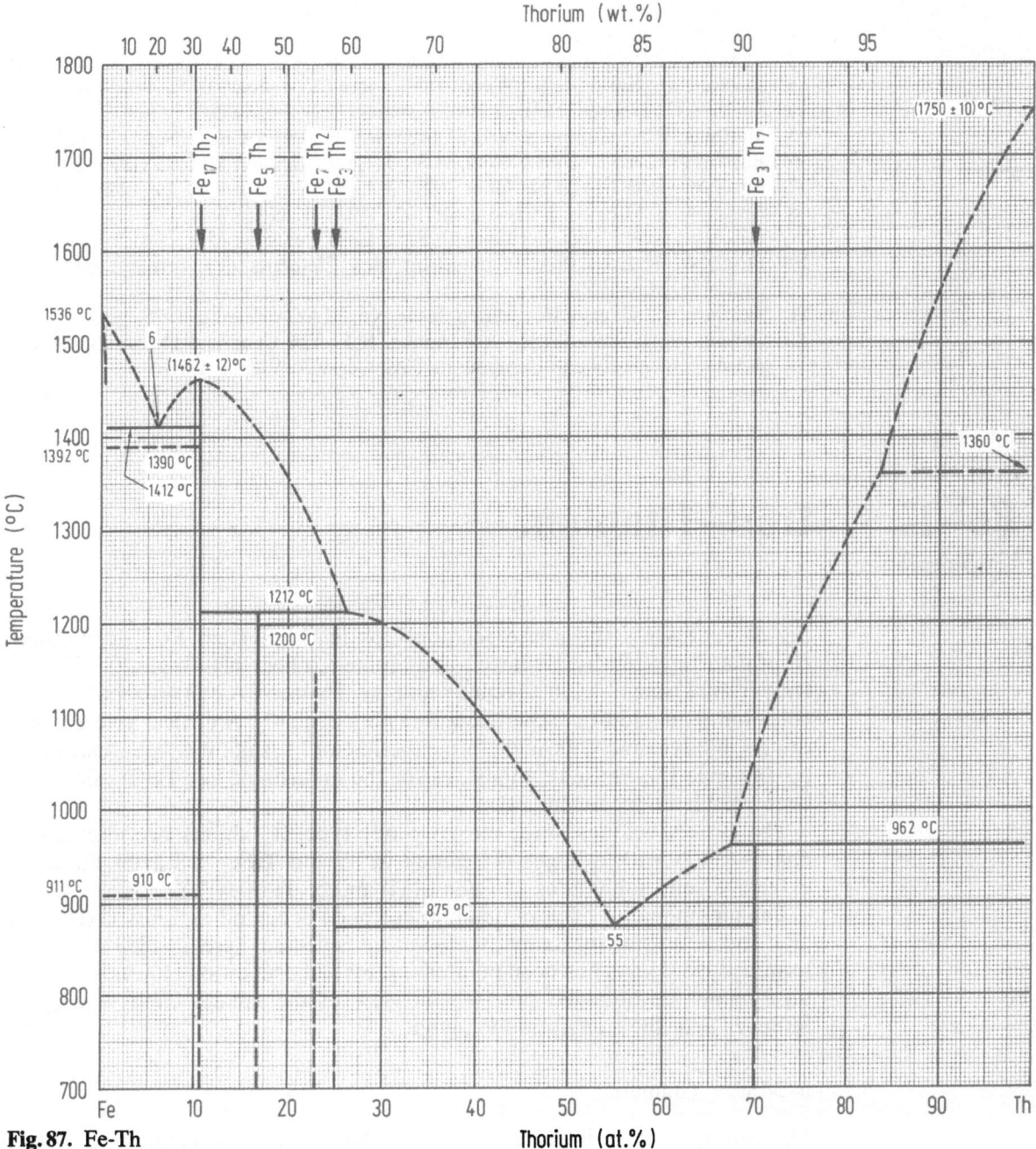

Fig. 87. Fe-Th

that found for Co_7Th_2 [8], Ni_7R_2 [9] and Co_7R_2 [10], where R is a medium-sized Rare Earth Metal (REM). The metallic radius of Th is approximately equal to that of the medium-sized REM and the phase relationships of Fe-Th show a certain resemblance to the medium-sized REM (see Fe-Tb, Fe-Dy, Fe-Ho, Fe-Er). However, there is one significant difference with respect to the formation of a compound of the type Fe_7R_2 if R represents the REM and M = Fe, Co, Ni, namely: Fe is unable to form the Fe_7R_2 phase (see Fe-R).

A Mössbauer study of the Fe-Th compounds has been carried out by [7]. Magnetic measurements revealed that the Th-richest phase is paramagnetic while the other four show ferromagnetism [11].

References

1   Thomson, J.R.: J. Nucl. Mater. 15 (1965) 88
2   Thomson, J.R.: J. Less Common Met. 10 (1966) 432
3   Johnson, Q.; Smith, G.S.; Wood, D.H.: Acta Crystallogr. B25 (1969) 464
4   Smith, J.F.; Hansen, D.H.: Acta Crystallogr. 19 (1965) 879
5   Matthias, B.T.; Compton, V.B.; Corenzwit, E.: Phys. Chem. Solids 19 (1961) 130
6   Buschow, K.H.J.; van der Goot, A.S.: J. Less Common Met. 23 (1971) 399
7   Buschow, K.H.J.; van der Goot, A.S.: J. Less Common Met. 20 (1969) 309
8   Buschow, K.H.J.: Acta Crystallogr. B26 (1970) 1389
9   Buschow, K.H.J.; van der Goot, A.S.: J. Less Common Met. 22 (1970) 419
10 Buschow, K.H.J.: Proc. Colloq. Int. des Elements des Terres Rares. Paris-Grenoble, May 1969
11 Thorium: At. Energy Rev., Spec. Issue No. 5 IAEA, Vienna 1975
12 Oetting, F.L.; Rand, M.H.; Ackermann, R.J.: The Chemical Thermodynamics of Actinide Elements and Compounds Part 1, IAEA, Vienna 1976

Fe–Ti Iron–Titanium

Figs. 88–90

Fig. 88 represents the phase relationships in the system Fe-Ti constructed from data reported by Hellawell and Hume-Rothery [1] (0–52 at.% Ti in the temperature range 1,250 °C-liquid) and by Hansen [A] (50–100 at.% Ti) incorporating in addition experimental results of [2–7]. A critical assessment of the thermodynamic data including experimental results has been carried out by Dinsdale [7]. Phase relationships for the metastable bcc Fe-Ti alloys have been calculated by Inden [24].

The system belongs to the closed γ-loop type. It exhibits two intermediary phases, FeTi with a slight deficiency in Ti and $TiFe_2$, a Laves phase, and extensive solid solubilities of Fe in β(Ti) and of Ti in α(Fe).

The melting point of pure titanium has been accepted as 1,668 °C (IPTS-68) [8]. The α/β transformation temperature selected here is (893 ± 7) °C determined by [9]. As has been pointed out by Kubaschewski [10], this temperature cannot be obtained precisely in view of the minute Gibbs energy differences near the transformation point.

0–50 at.% Titanium

The γ-loop was re-investigated by Fischer et al. [25] employing thermomagnetic measurements, microscopy and thermodynamic calculations to clarify the differences existing up to that time. The maximum solubility of Ti in γ(Fe) was found to be 0.8 at.% and the extent of the (α + γ) range at 1,150 °C about 0.6 at.%. Recently, Murray [26] re-assessed the (α + γ) phase boundaries by calculation. Her results are shown in Fig. 89 and agree very well with the more recent experimental observations.

The ranges 0–52 at.% Ti and 0–100% have been re-investigated, respectively, by [1] and [3] using thermal analysis [1] and thermal, metallographic and X-ray analyses [3] with high purity metals. [1] pointed out that alumina crucibles may be used for alloys containing up to 30 at.% Ti while samples richer in Ti require thoria or thoria-coated containers. The addition of Ti to Fe

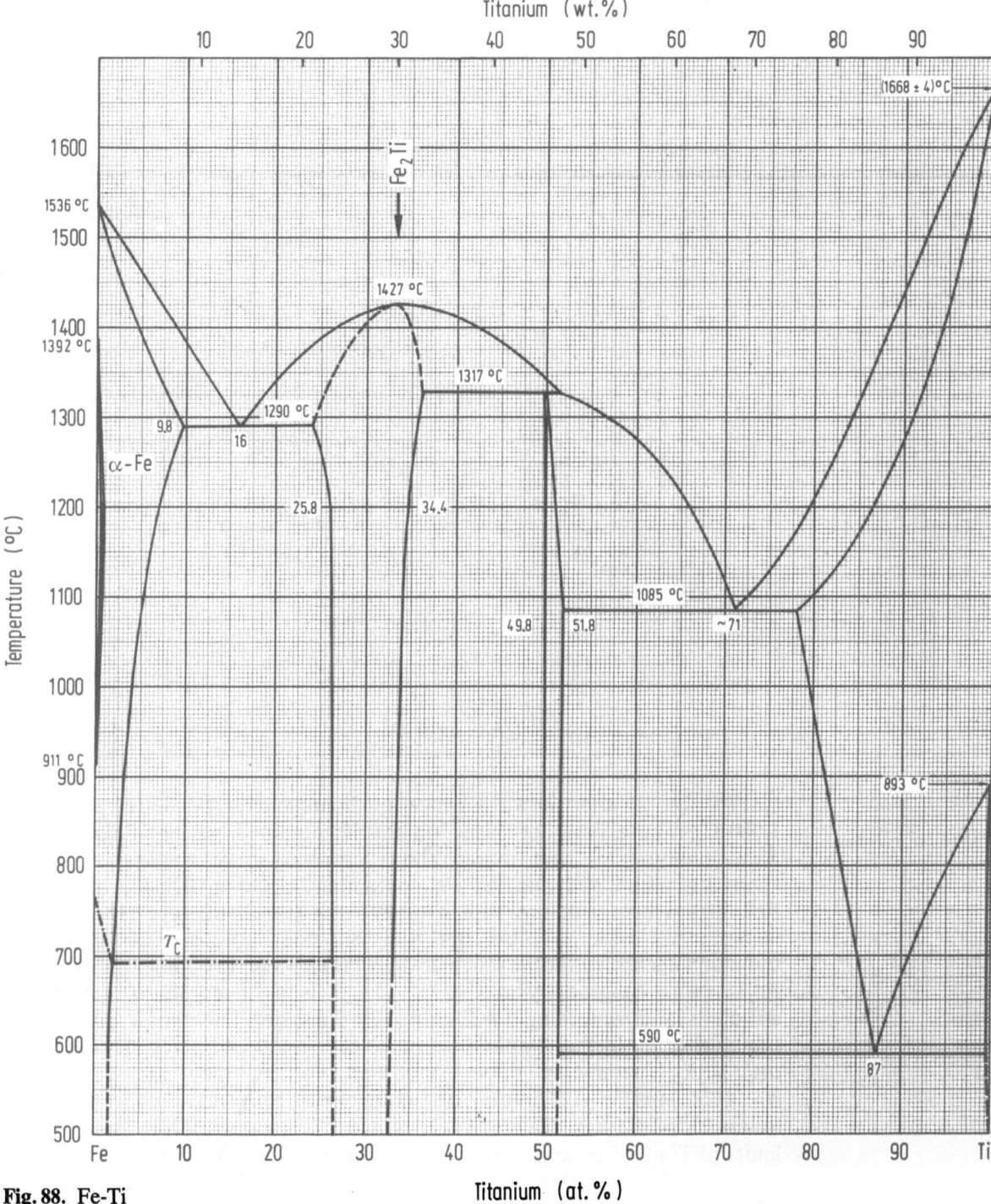

Fig. 88. Fe-Ti

causes a rapid fall of both the liquidus and solidus curves of δ(Fe) to a eutectic horizontal at 1,290°C corresponding to the reaction liq. (16 at.% Ti) ⇌ α, δ-Fe (9.8) + Fe₂Ti. From the eutectic point the liquidus rises to a rather flat maximum at 1,427°C, the melting point of Fe₂Ti. Both authors agree on the feature and on the existence of two compounds, but differ with regard to the melting of FeTi. This compound is formed by peritectic reaction at

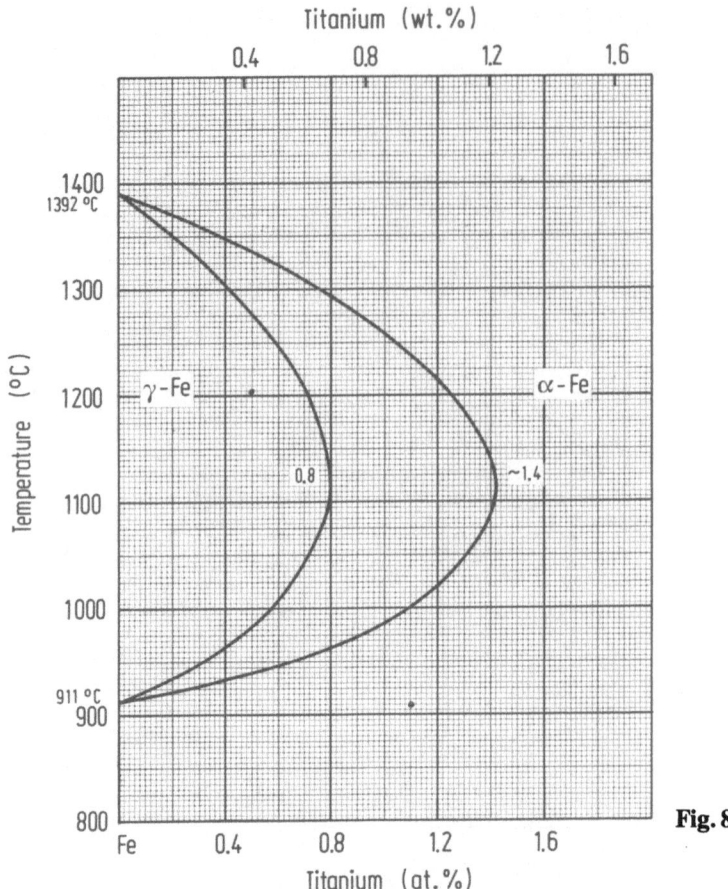

Fig. 89. Fe-Ti

1,317°C [1, 2] and shows a range of homogeneity (49.8–51.8 at.% Ti at 1,080°C). These data are based on results obtained by metallographic and X-ray analyses of heat-treated alloys prepared from magnesium reduced sponge titanium and purified Fe containing less than 0.004% oxygen [2]. FeTi has a bcc structure with $a = 0.2976$ nm (CsCl, B2 type [15, 16]). It is isomorphous with NiTi but unfortunately does not show its disordering transition necessary for the much desired "shape-memory alloys" (Källne [17]). Fe_2Ti melts congruently at 1,427°C [1]. It is a Laves phase, with $MgZn_2$ structure and according to Elliott [18] does not undergo a polymorphic transformation in the temperature range 600–1,400°C. Speich [19] confirmed the lattice parameter data established earlier: they are as follows: $a = 0.47857$ nm, $c = 0.7799$ nm. Its homogeneity range at 1,200°C was found to be 25.8–34.4 at.% Ti [2]. Fe_2Ti is antiferromagnetic [20]. The solubility limits of Ti in α-Fe, studied by lattice parameter measurements, are as follows: 600, 700, 800 and 900°C correspond to 1.88, 1.94, 2.52 and 3.08 at.% Ti [6] which is in fairly good agreement with [3, 19] for the lower temperatures.

50–100 at.% titanium range

The phase boundaries outlined in [A] are still valid. The liquidus of the Ti-rich region above 1,200°C has been taken from the data calculated by [7].

The iron solubility in α(Ti) has been re-investigated by two teams, Raub and Raub [4] and Matyka et al. [5]. The values obtained are similar (Fig. 90) but in contrast to [4], a solubility maximum was observed at 790°C by [5]. A eutectoid temperature close to 600°C was recorded by both teams. According to [5] this kind of solubility variation with temperature corresponds to a retrograde solubility which is sometimes observed in equilibrium with the liquidus.

Completely homogenous dilute solutions of Fe in α(Ti) are superconducting at about 0.42 K.

The metastable ω phase can be formed by quenching Ti-rich alloys. The structure is a complex bcc one [21]. The kinetics of transformation of ω have been studied by [22].

The thermodynamics of liquid binary Fe-Ti alloys have been investigated by mass-spectrometry by Wagner and St. Pierre [23], who found that the liquid system is symmetric and regular at 1,545°C and exhibits a marked negative deviation from the ideal solution behaviour.

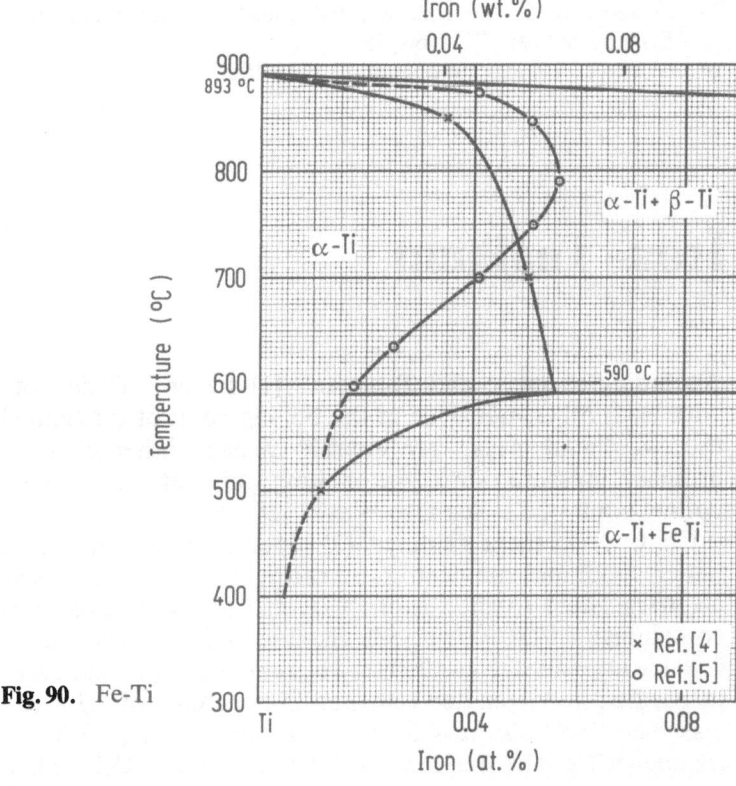

Fig. 90. Fe-Ti

References

1 Hellawell, A.; Hume-Rothery, W.: Philos. Trans. R. Soc., London A249 (1957) 417
2 Murakami, Y.; Kimura, H.; Nishimura, Y.: Trans. Nat. Res. Inst. Met. (Tokio) 1 (1959) 7
3 Kornilov, I.I.; Boriskina, N.G.: Dokl. Akad. Nauk SSSR 108 (1956) 1083
4 Raub, E.; Raub, Ch.J.; Röschel, E.: J. Less Common Met. 12 (1967) 36
5 Matyka, J.; Faudot, F.; Bigot, J.: Scr. Metall. 13 (1979) 645
6 Abrahamson, E.P.; Lopata, S.L.: Trans. AIME 236 (1966) 76
7 Dinsdale, A.T.: N.P.L. private communication (1980)
8 Kenisarin, M.M.; Chekhovskoi, V.Ya.; Berrezin, B.Ya.; Kats, S.A.: High Temp. High Press. 8 (1976) 367
9 Cezairliyan, A.; Müller, A.P.: 5. Res. Nat. Bur. Stand. 83 (1978) 127

Fe–Ti

10 Kubaschewski, O.: Phase Stability in Metals and Alloys. (Rudman, P.S.; Stringer, J.; Jaffee, R.I. Eds). Battelle Mem. Inst. Coll., Geneva 1967, p. 63
11 Wada, T.: Trans. Nat. Res. Inst. Met. (Tokio) 6 (1964) 43
12 Moll, S.H.; Ogilvie, R.E.: Trans. AIME 215 (1959) 613
13 Hillert, M.; Wada, T.; Wada, H.: Unpublished, quoted by [11]
14 Chapin, E.J.; Hayward, C.R.: Trans. ASM 38 (1947) 909
15 Philip, T.V.; Beck, P.A.: Trans. AIME 209 (1957) 1269
16 Dwight, A.E.: Trans. AIME 215 (1959) 283
17 Källne, E.: J. Phys. F. 4 (1974) 167
18 Elliott, R.P.: Armour Research Foundation, Chicago, Ill., Tech. Rep. 1, OSR Technical Note OSR-TN-247 (1954), 28
19 Speich, G.R.: Trans. AIME 224 (1962) 850
20 Nakamichi, T.: J. Phys. Soc. Jpn. 25 (1968) 1189
21 Yoshida, H.: Nippon Kinzoku Gakkaishi 20 (1956) 292
22 Gridnev, V.N.; Petrov, Yu.N.; Rafalovkiy, V.A.; Trefilov, V.I.: Akad. Nauk Ukr. SSSR, Sb. Nauchn. Rabot (1960) (11) 82
23 Wagner, S.; St. Pierre, G.R.: Metall. Trans. 5 (1974) 887
24 Inden, G.: CALPHAD 1 (1) (1977) 75
25 Fischer, W.A.; Lorenz, K.; Fabritius, H.; Hoffmann, A.; Kalwa, G.: Archiv Eisenhüttenwes. 37 (1966) 79
26 Murray, J.L.: Private communication 1981

Fe–Tl   Iron–Thallium

According to Isaac and Tammann [1] Fe and Tl do not measurably react with each other, not even at the boiling point of thallium (1,473°C). Whether the two liquid metals are soluble in each other cannot be determined at ordinary pressure, since the melting point of Fe is higher than the normal boiling point of Tl. See also Wever [2].

Contrary to these findings, [3] reported an interaction between Fe and Tl under conditions applicable to brazing. The materials used were low-carbon electrical steel (Armco iron) NZh MRTU 14-2-31-65, remelted in vacuum, containing: 0.025% C, 0.15% Si, 0.15% Mn, 0.1% Cu, 0.01% P, 0.01% S 0.004% O_2, 0.01% N_2, 0.0003% H_2, remainder Fe, and carbonyl iron, reduced in purified hydrogen and remelted in a vacuum of 1.3×10^{-4} mbar, containing less than 0.08% Si, less than 0.01% Mn, less than 0.001% each C, S and P, remainder Fe. The thallium used was grade TLOO RETU 87-59.

They found that at a temperature of 1,120°C a diffusion between Fe and Tl takes place. Tl penetrates the Fe grain boundaries and the microstructures of the weld showed that Tl dissolves a certain amount of Fe.

References

1 Isaac, E.; Tammann, G.: Z. Anorg. Chem. 55 (1907) 61
2 Wever, F.: Arch. Eisenhüttenwes. 2 (1928/29) 739
3 Grzhimal'skiy, L.L.; Petrunin, I.E.: Russ. Metall. (1968) 2/154

Fe–Tm   Iron–Thulium, see Fe-R: Iron-Rare Earth Metals, Fig. 67

A thermodynamic assessment of the system Fe-U has been published by Chiotti [1] who used the diagram proposed by Hultgren [D] with minor modifications of the melting and transition temperatures of uranium. An earlier evaluation of the Fe-U phase boundaries from thermochemical data was carried out by Kubaschewski [2]. In addition Kutaytsev [3] determined experimentally the entire system. The original publication not being available, the present compiler had to rely on a report by Lebedev et al. [4] which included the phase diagram mentioned.

The phase diagram in Fig. 91 agrees in its essential features with the one suggested by [D], i. e. it is based on the concordant results obtained by thermal, microscopic and X-ray investigations reported independently by two teams, Gordon and Kaufmann [5] and Grogan [6] and Clews [7]. The melting and transition points of uranium have been taken from an evaluation of the properties of the actinides by Oetting et al. [8] (IPTS-68) and are as follows: $(1,135 \pm 2)$ °C, (776 ± 1.6) °C and (669 ± 1.3) °C.

The system is characterized by two compounds, Fe_2U (a Laves phase) and FeU_6, two eutectics (at about 23 and 66 at.% U [5, 6] or 18 and 66 at.% U [3]) and limited terminal solubilities.

Gordon and Kaufmann [5] prepared the specimens from 99.9% pure uranium, containing as main impurities carbon and iron, and electrolyte iron with 0.012% C, 0.02% Ni, 0.009% Cu. Melting was carried out in beryllia or beryllia-lined crucibles using high frequency heating either in vacuo or under argon. Reaction between the metal and crucible was not observed at temperatures up to 1,800°C. As has been mentioned before, thermal analysis, X-ray and microscopic investigations of annealed alloys have been used to establish the complete phase diagram.

Grogan [6] used iron of 99.98% and uranium of high purity. Details about the nature of the impurities are not mentioned. The alloys were prepared by melting in a H. F. induction furnace in vacuo. Thermal and microscopic analyses, apparently on unannealed samples, have been used to study mainly the U-rich range of concentration (0–35 at.% Fe).

Fe-rich range

The addition of U to Fe lowers the liquidus temperature to 1,400°C, where the following peritectic reaction takes place: liq. (7.4 at.% U) + δ-Fe ⇆ γ-Fe. The only measurements pertaining to this temperature and concentration range have been reported by [5] who stated that the results are somewhat doubtful. The thermal points above the γ/δ(Fe) transformation temperature have therefore been plotted higher than the experimentally determined values, to bring the melting point of Fe up to the value of 1,535°C [5]. (See also [3].)

The compound Fe_2U and γ-Fe form a eutectic at 1,055°C [6] 1,080°C [5] or 1,080°C [3] (the latter being probably adopted from [5]). The α/γ(Fe) transformation has not been studied by [6]; according to thermal data obtained by [5] it occurs at 910°C.

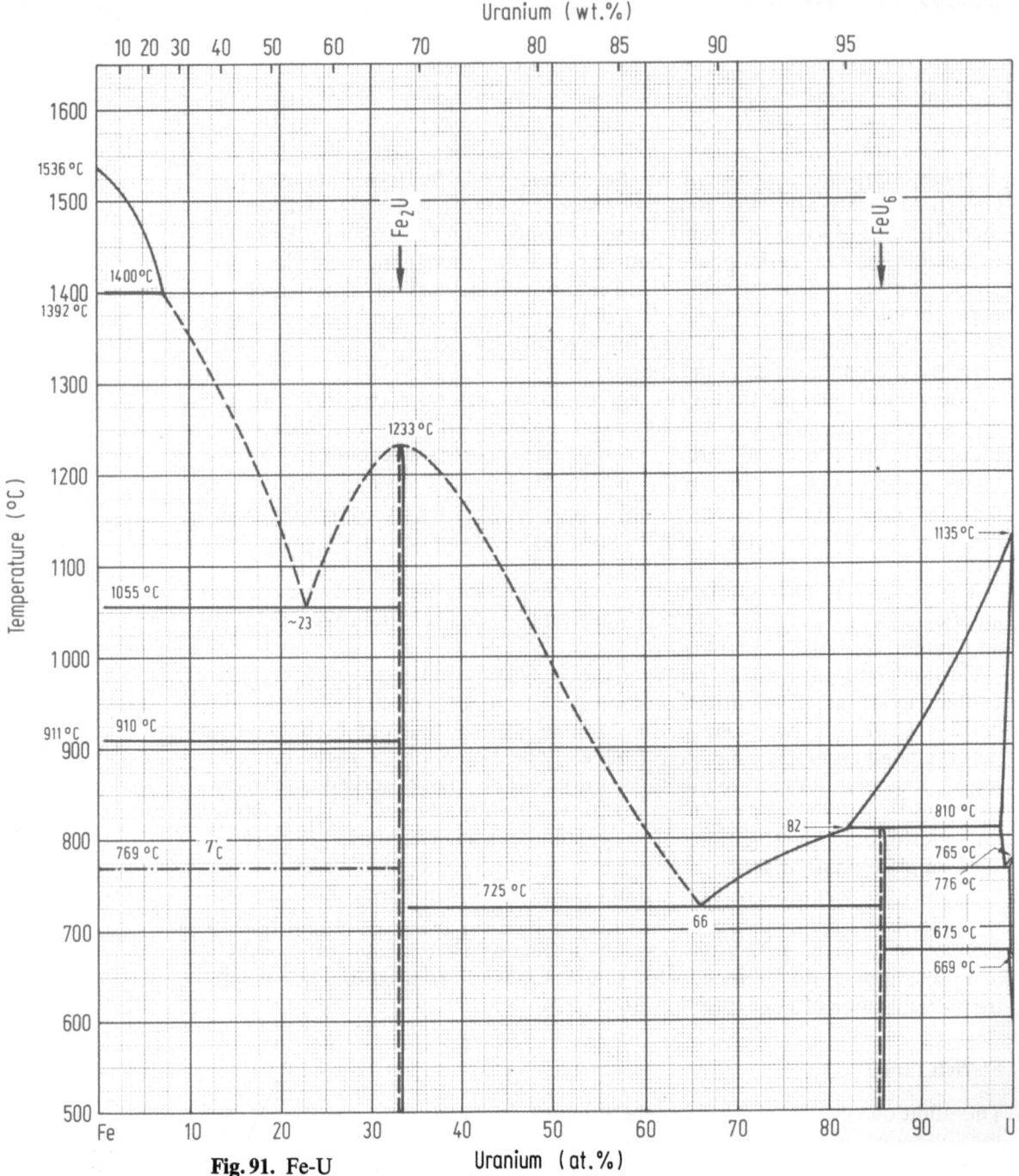

Fig. 91. Fe-U

Compounds

Fe₂U and FeU₆ show a very limited range of homogeneity. Fe₂U is a Laves phase with MgCu₂-type structure ($a = 0.7061$ nm) [9]; it melts congruently at 1,233 °C according to Kober and Nichov [10] who confirmed results observed by [6, 7]. Fe₂U is ferromagnetic below 195–172 K [D], which agrees very well with an estimated temperature of 193 K suggested by Campbell [11]. FeU₆ is formed peritectically at 810 °C and belongs to the tetragonal MnU₆-type

($a = 1.0289$ nm and $c = 0.5232$ nm) [12]. It is superconducting below $T_C = 3.86$ K [D]. The results for the temperature of the peritectic reaction quoted in the literature [3, 6, 10, 13, 14] vary between 805 and 815 °C. Observations by [10] based on re-measurements as well as solubility data [6, 13, 14] point to a temperature of formation of 810 °C.

U-rich range

Data on the U-rich region have been presented by several authors [4, 6, 13–16]. Figure 92, showing this area in detail, represents essentially results published by [6], supplemented by data of Straatmann and Neumann [13] and Russell [14]. [13] and [14] re-investigated the 0–3 at.% Fe range of concentration using optical and electron micrography [14], differential thermal [13] and electron microprobe analyses [13, 14] as well as electrical resistivity measurements [14] of alloys prepared from materials of at least 99.98% purity. The results submitted by both teams are in good agreement with those found by [6]. The solubility of Fe in solid uranium increases with temperature from about 0.018 at.% at 600 °C [6] to 1.36 at.% at 810 °C. On addition of Fe to γ(U) the solidus temperature increases with decreasing temperature, until a maximum is reached at 810 °C where the peritectic formation of FeU_6 occurs. The solubility data for γ(U) are due to [13] and are as follows: 0.21 0.72 and 1.15 at.% Fe corresponding to temperatures of 1,100, 1,000 and 900 °C. Values of [14] are approximately 25 K higher.

The maximum solubility of Fe in γ(U) was quoted as 1.36 at.% Fe (805 °C) [13] and ~1.3 at.% Fe at 815 °C [14, 6]. Grogan [6] suggested 1.5 at.% Fe at 805 °C. However, his findings, based on microscopical results only, allow a slightly different interpretation which would be in excellent agreement with [13] and [14]. (See Ref. [6, p. 573] "The Uranium-rich Area of the Uranium-Iron Diagram").

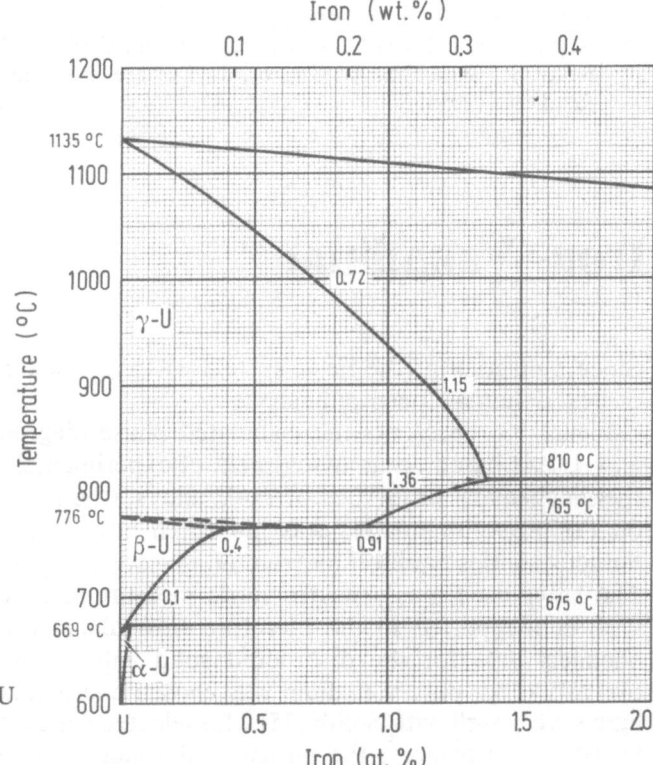

Fig. 92. Fe–U

The composition and temperature of formation of the U-rich eutectoid in Figs. 91, 92. 0.91 at.% Fe and 765 °C have been taken from results by [6] and are in good agreement with data reported by other teams, namely: 765 °C [16], 763 °C [12], 770 °C [3, 14] and 0.85 at.% Fe [16] and 0.82 at.% Fe [13].

The maximum solubility of Fe in β(U) has been found to be 0.38 at.% [13] 0.46 at.% [14] 0.42 at.% [6] or 0.43 at.% [16] at 765 °C. The course of the β-U and α-U solvus shown in Fig. 92 is based on data reported by [6, 13, 15]; 0.1 and 0.018 at.% Fe have been found for 700 and 600 °C respectively by [6]. Angermann [15] using electron microscopy studied the α-U solution field and observed a Fe solubility of 0.005–0.025 at.%.

References

1   Chiotti, P.; Akhachinskij, V.V.; Ansara, I.; Rand, M.H.: The Chemical Thermodynamics of Actinide Elements and Compounds, Part 5, IAEA, Vienna, 1981
2   Kubaschewski, O.: Thermodynamics of Nuclear Materials. IAEA, Vienna, 1962, p. 219
3   Kutaytsev, V.I.: Alloys of Thorium, Uranium, Plutonium. Gosatomizdat, 1962, p. 74
4   Lebedev, V.A.; Nazarov, N.V.; Pyatkov, V.I.; Nichkov, I.V.; Raspopin, S.P.: Izv. Akad. Nauk SSSR, Met. No. 2 (1973) 212
5   Gordon, P.; Kaufmann, A.R.: Trans. AIME 188 (1950) 182
6   Grogan, J.D.: J. Inst. Met. 77 (1950) 571
7   Clews, C.J.: J. Inst. Met. 77 (1950) 577
8   Oetting, F.L.; Rand, M.H.; Ackermann, R.J.: The Chemical Thermodynamics of Actinide Elements and Compounds, Part 1 IAEA, Vienna 1976
9   Katz, G.; Jacobs, A.J.: J. Nucl. Mater. 5 (1962) 338
10 Kober, V.I.; Nichkov, I.F.: Russ. Metall. (1974) 170
11 Campbell, G.M., Met. Trans. 8A (1977) 1493
12 Kehl, G.L.; Mendel, E.; Jaraiz, E.; Mueller, M.H.: Trans. ASM 51 (1959) 717
13 Straatmann, J.A.; Neumann, N.F.: U.S. At. Energy Comm., MCW-1487 (1964) 23
14 Russell, R.B.: U.S. At. Energy Comm., NMI-2813 (1964) 23
15 Angermann, C.L.: U.S. At. Energy Comm., DPST-64-219 (1964)
16 Bellot, J.; Blanchon, A.; Chazot, R.; Dosiere, P.; Henry, J.-M.; Colas, M.: Compt. Rend. 246 (1958) 3063

Fe–V   Iron–Vanadium

Figs. 93–95

Since a diagram of the Fe-V system was proposed by Hansen [A], a number of new data have been published.

Figure 93 represents the calculated phase diagram submitted by Hack et al. [1]. It is based on an assessment of experimental phase boundary data and thermochemical properties of Fe-V alloys, (the influence of magnetic effects on the phase boundaries was not taken into account) and incorporated the liquidus-solidus equilibrium from an investigation by Kubaschewski et al. [2]. The latter used mass spectrometric measurements on liquid alloys together with data reported in the literature to evaluate a complete set of thermochemical data for the disordered solid and liquid alloys. A melting point of 1,905 °C for pure vanadium was chosen and is taken over in Fig. 93, which agrees very well with Smith [15] who selected a melting point of (1,910 ± 10) °C based primarily upon the purity of the metal. The γ-loop, representing the

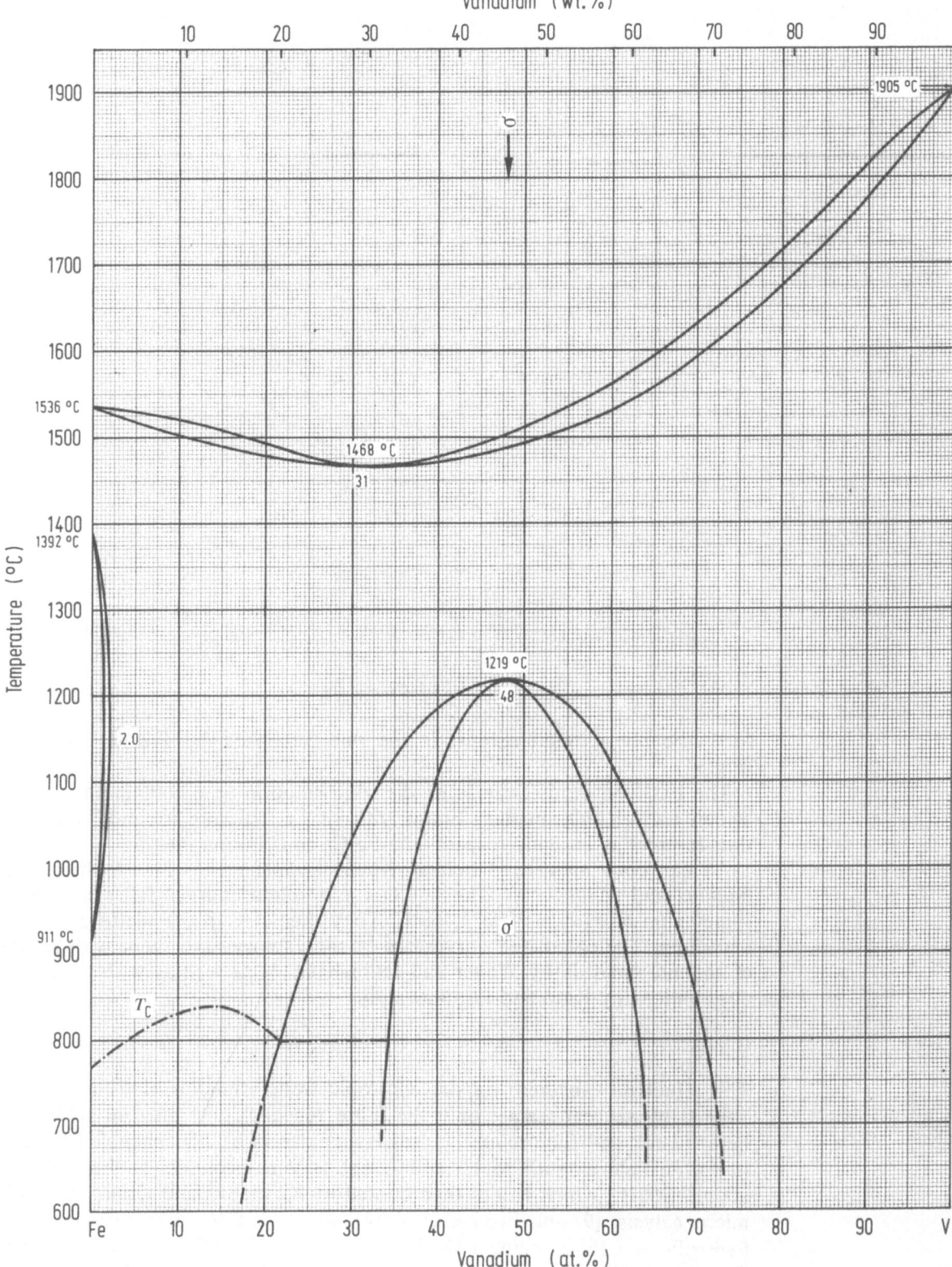

Fig. 93. Fe-V

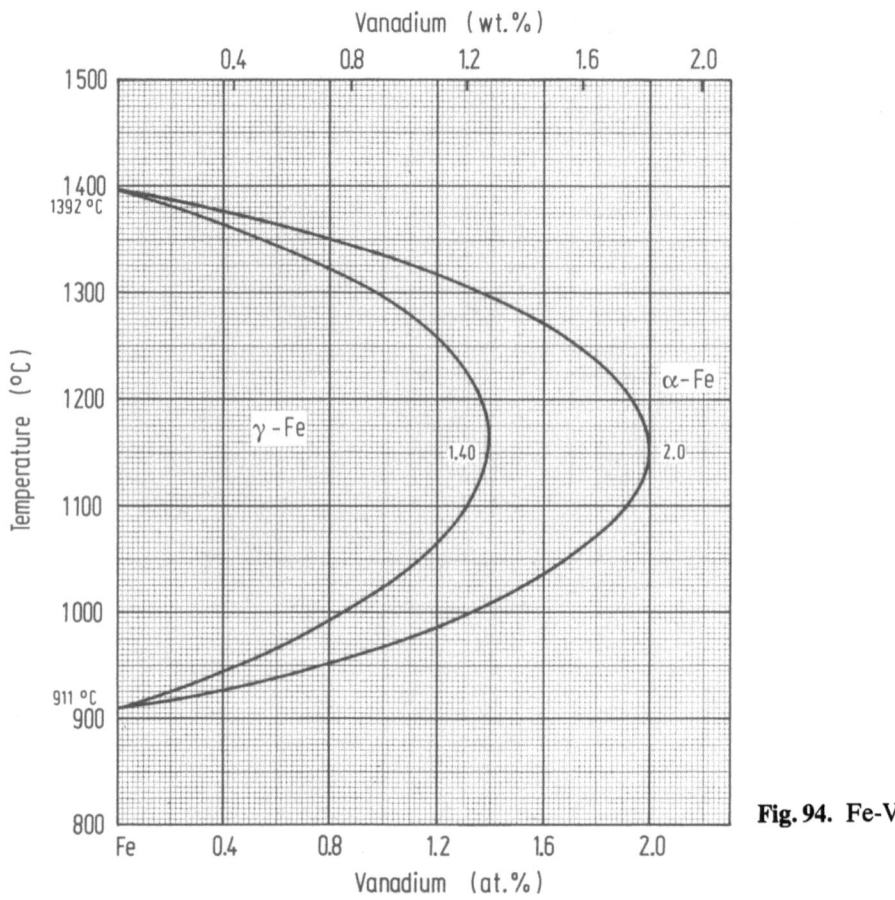

Vanadium (wt.%)

Fig. 94. Fe-V

(α + γ)Fe phase relationships, is the one published by Fischer et al. [4] (Fig. 94).

The Fe-V phase diagram is of the solid solution type containing a closed γ-loop and an intermediary phase with sigma structure around 50 at.% V, thus resembling in many respects the Fe-Cr constitution diagram.

As has been mentioned before, Kubaschewski et al. [2] established the liquidus and solidus gap. The liquidus and solidus curves fall smoothly to a minimum at 31 at.% V at 1,468°C. They then rise to the melting point of vanadium. The calculated curves are in satisfactory agreement with results of thermal analysis by Wever and Jellinghaus [5] as well as with the recent observations by Kato and Furukawa [6].

γ-loop

The effect of vanadium on the allotropic transformation of iron has been studied by several teams; only the more recent results may be mentioned here. Lisnik and Skvorchuk [7] and Hanneman et al. [8] concluded that the (α + γ)Fe field extends to 1.3 and 1.56 at.% V respectively at the vertex at about 1,150°C, which is less than quoted by Kovacova and Kralik (electron-microanalysis) [9] and Fischer et al. [4], who found the maximum region of (α + γ)Fe at 1,150°C being 1.38–1.8 at.% V and 1.40–2.0 at.% V respectively. Results of [4] have been chosen to construct the phase boundaries in Fig. 94. They used 16 alloys (impurities: C = 0.001–0.007, Si = <0.001–0.003, Mn = 0.0002–0.005, P = 0.002–0.005, S = 0.003–0.006, Al = <0.01, N = <0.001–

0.002 and O = 0.001–0.006%) and founded their results on susceptibility measurements as well as a calculation of the amount of ferrite and austenite of the two phase field present at each temperature.

Order/disorder transformation

The existence of an order/disorder transformation of the solution phase based on the equiatomic composition has been established by thermal analysis, microhardness and dilatometric measurements; however, conventional studies cannot provide accurate phase boundary limits. Therefore, Spencer and Putland [10] calculated these limits by using thermodynamic data obtained by adiabatic high-temperature calorimetry, samples being prepared from Fe containing about 0.029% oxygen and vanadium powder of 99.5% purity (oxygen and nitrogen: 0.43 and 0.036% respectively). Their description of the α/σ equilibrium (Fig. 93) is in satisfactory agreement with results of Chiriac and Inden [11] obtained by high temperature X-ray diffraction and X-ray analysis of quenched specimens after appropriate annealing.

The sigma phase is isotypic with σFeCr and extends over a wide range of compositions. It is asymmetric in shape, the concentration of the congruent formation of σ being 48 at.% V. The temperature of formation was found to be 1,219°C [1] which is in good agreement with Matveeva [12]. Although there are strong similarities between the Fe-Cr and Fe-V phase diagrams, observations differ in one respect, namely the closing of the σphase, at about 440°C established in the Fe-Cr system, does not occur in the Fe-V system [10].

Metastable order/disorder transformations

The bcc(α) phase can be retained by rapid cooling from the homogeneity range. It transforms to the CsCl-type ordered phase below 600°C [13]. The Fe-V ordered phase has a meta-stable critical transition temperature (between 850 and 880°C) at the equiatomic composition and decreases parabolically as

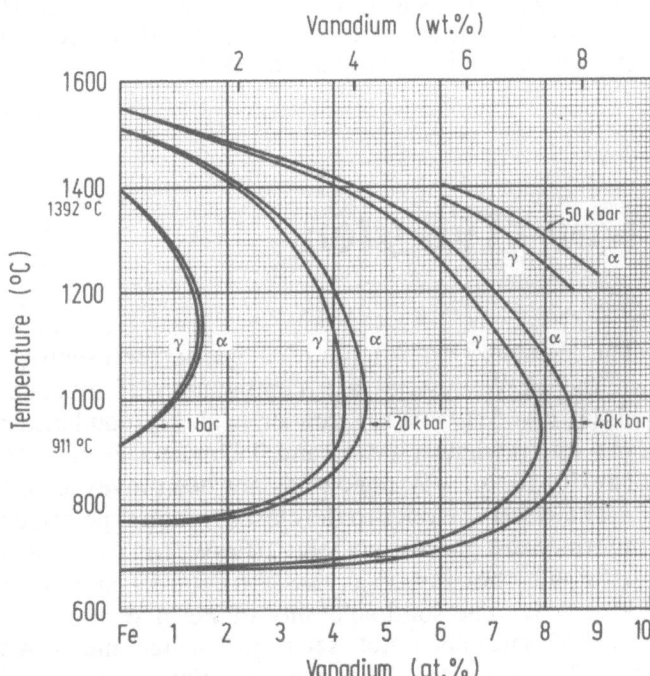

Fig. 95. Fe-V.
Pressure dependence of
the (α + γ) Fe region

the composition deviates from stoichiometry. The so-called "650°C anomaly" corresponds to the 550°C anomaly in the Fe-Co system and has been observed in the ordered Fe-V alloys during continuous heating of homogenized, quenched and annealed samples based on DTA and X-ray analysis by Seki et al. [14].

The effect of pressure on the stability of the $(\alpha + \gamma)$ phase (Fig. 95) and the σ phase transformation was studied by electron microbeam and X-ray analysis of binary diffusion couples diffused under simultaneous conditions of high pressures and temperature by Hanneman et al. [8]. The results show that the region of solubility of the σphase increases with pressure and the congruent transformation point of σ is moved to higher temperatures, which was already predicted by the application of the Clapeyron equation.

References

1 Hack, K.; Nüssler, H.D.; Spencer, P.J.; Inden, G.: CALPHAD VIII, Stockholm, May 1979, p. 244
2 Kubaschewski, O.; Probst, H.; Geiger, K.H.: Z. Phys. Chem. 104 (1977) 23
3 Kenisarin, M.M.; Chekhovskoi, V.Ya.; Berezin, B.Ya.; Kats, S.A.: High Temp. High Press. 8 (1976) 367
4 Fischer, W.A.; Lorenz, K.; Fabritius, H.; Schlegel, D.: Arch. Eisenhüttenwes. 41 (1970) 489
5 Wever, F.; Jellinghaus, W.: Mitt. Kaiser-Wilhelm-Inst. Eisenforschung Düsseldorf, 12 (1930) 317
6 Kato, E., Furukawa, T.: Tetsu To Hagane 61 (1975) 3050
7 Lisnik, A.G.; Skvorchuk, V.P.: Dopov. Akad. Nauk Ukr. RSR (1960) 1408
8 Hanneman, R.E.; Ogilvie, R.E.; Gatos, H.C.: Trans. Metall. Soc. AIME 233 (1965) 685
9 Kovacova, K.; Kralik, F.: Kovove Mater. 11 (1973) 93 (in slovak)
10 Spencer, P.J.; Putland, F.H.: J. Iron Steel Inst. 211 (1973) 911
11 Chiriac, M.; Inden, G.: To be published
12 Matveeva, P.M.: Akad. Nauk SSR (1955), cited by Vol, A.E.: Constitution and Properties of Binary Metallic Systems II, 9–20, Moscow, 1962
13 Bungardt, K.; Spyra, W.: Arch. Eisenhüttenwes. 30 (1959) 95
14 Seki, J.I.; Hagiwara, M.; Susuki, T.: J. Mater. Sci. (1979) 2404
15 Smith, J.: Private information 1982

Fe–W   Iron–Tungsten

Figs. 96, 97

The present (Fig. 96) phase diagram representing the phase relationships in the Fe-W system is a combination of evidence presented by Sinha et al. [1] (0–45 at.% W), Kirchner et al. [2] (equilibria between ferrite, austenite and intermediate phases) and Takayama et al. [11] (α(Fe) solubility limits). The 50–100 at.% W region has not been investigated fully.

The system belongs to the closed γ-loop type and entails two well established intermediate phases, namely: λ(Fe$_2$W, MgZn$_2$-type structure, a Laves phase), μ(Fe$_3$W$_2$, rhombohedral, D8$_5$ structure), and probably a σ phase around 50 at.% W. The solid solubility of Fe in W was found to be about 2.5 at.% at 1,640°C and does not seem to change much with temperature (Hansen [A: Refs. 11, 18]). The melting point of W, $(3,422 \pm 8)$ °C is a secondary

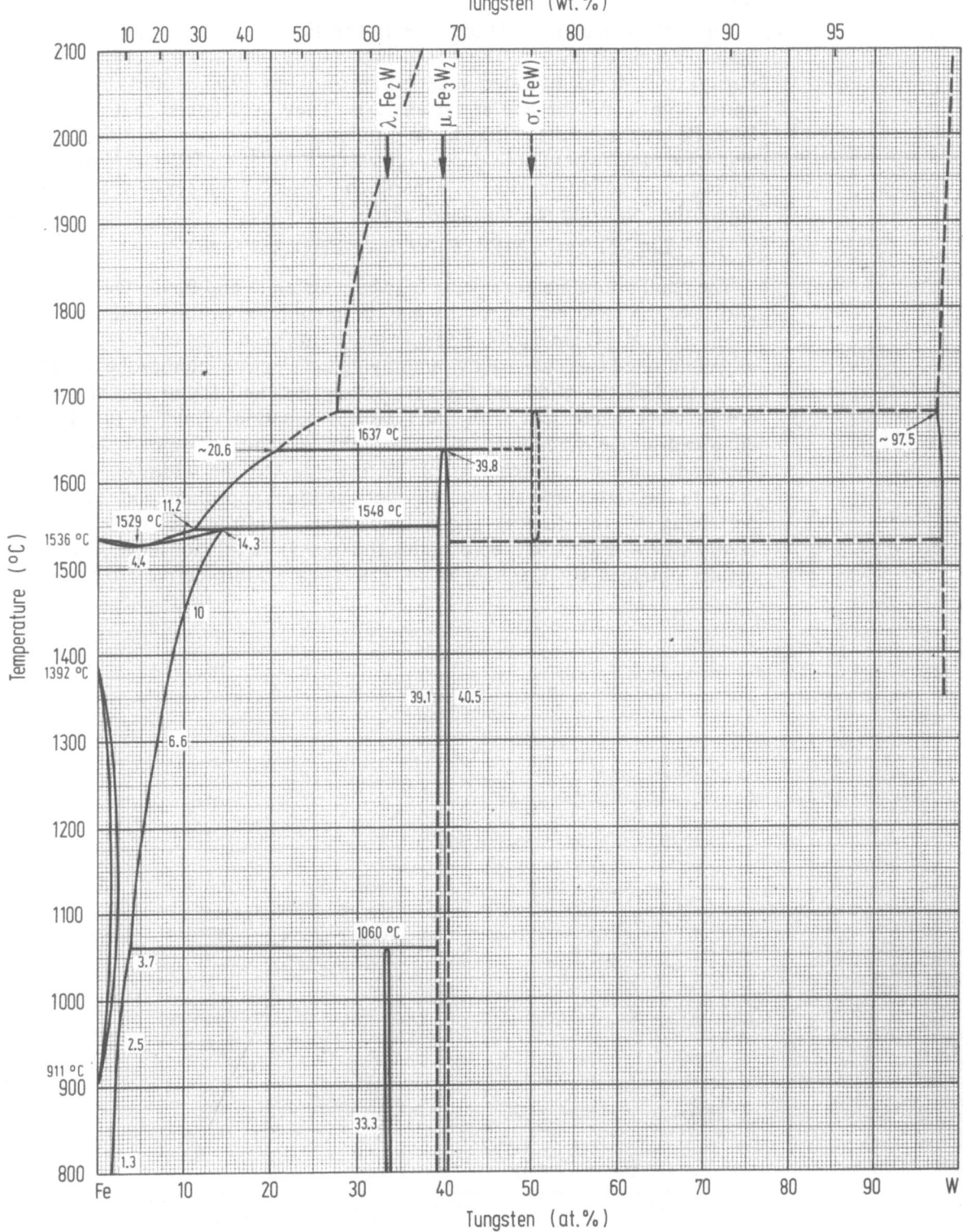

Fig. 96. Fe-W

reference point on the International Practical Temperature Scale (IPTS-68, revised 1975).

The liquidus/solidus relationships in the Fe-rich area up to 1,640°C have been determined by [1]. They used high purity alloys (Fe of 99.96–99.97% purity) and Murex tungsten sheet scrap of about 99.95% purity) arc-melted under argon, and thermal arrest analysis. They ascertained that addition of W to Fe lowers the δ(Fe) liquidus and solidus to a shallow minimum at 1,529°C and 4.4 at.% W where both curves intersect. Further addition of W raises the liquidus and solidus curves to a peritectic reaction temperature at 1,548°C, where the following reaction takes place:

$$\text{Liq. (11.2 at.\% W)} + Fe_3W_2 \leftrightarrows \alpha\text{-Fe(14.3 at.\% W)}.$$

The equilibrium measurements between austenite and ferrite are numerous and only the most recent results may be mentioned in this compilation.

Hillert et al. [4] have published measurements pertaining to the γ-loop based on equilibrated two phase alloys employing the electron microprobe technique. They studied the width of the two phase field and found that the results are in close agreement with van't Hoff's equation. Sinha et al. [1] using dilatometric measurements, estimated the α- and γ(Fe)phase boundaries thermodynamically. Fischer et al. [5] investigated the width of the γ-loop close to the vertex and applied the dilute solution approximation. The results pertaining to the (α + γ) phase boundaries confirm the applicability of Henry's Law. Figure 97, representing the (α + γ) Fe relationships, is based on results obtained by Kirchner et al. [2]. Their results derive from electron

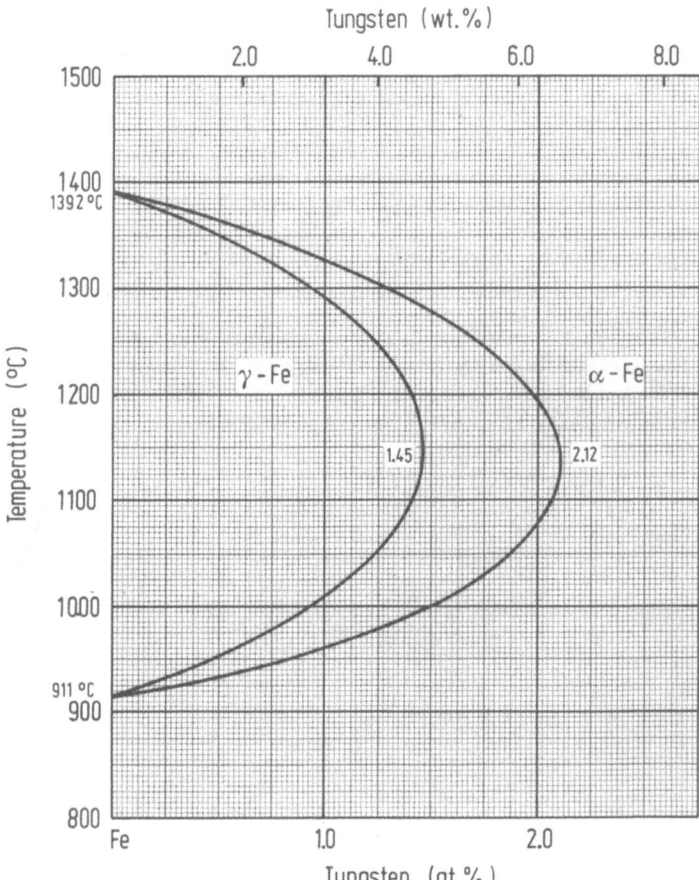

Fig. 97. Fe-W

microprobe analyses of specimens prepared from electrolytic iron and >99.97% W which had undergone carefully controlled equilibrium heat treatments and precise electron microprobe calibrations. The observations were analysed theoretically in terms of the regular solution model and the phase boundaries then calculated. The computed width of the ($\alpha + \gamma$) phase field agrees well with that determined experimentally. The width at the vertex (1,140°C) was found to be 1.45–2.12 at.% W which is around 0.3% higher than previous results but agrees well with measurements undertaken at the same time by Kovacova and Kralik [6], who used electron microanalysis and quoted the extent of the ($\alpha + \gamma$) field as being 1.44–2.0 at.% W at 1,150°C.

The solid solubility of W in α(Fe) in the temperature range 1,548–1,200°C was calculated on the basis of experimental results reported by [1] and [2]. The solubility limits in the temperature range 600–1,200°C were determined experimentally by Takayama et al. [11] (X-ray diffraction and electron microprobe analyses) on samples prepared from electrolyte iron with 99.95% and tungsten of 99.5% purity and annealed for 150–7,000 hours depending on temperature. They found that 4.33, 3.06, 2.51, 1.75, 1.33, 1.23 and 1.00 at.% W correspond to temperatures of 1,100, 1,000, 950, 850, 800, 750 and 700°C. These data confirm X-ray results for the range 600–900°C obtained by Abrahamson and Lopota [3]. The solubility does not appear to be affected by the magnetic transition [13].

μ(Fe$_3$W$_2$) is formed by peritectic reaction at about 1,637°C and 39.8 at.% W [2]. Between 1,300 and 1,600°C, the composition of the intermediary phase undergoes little change [1] and is restricted to 39.1–40.5 at.% W, which has been confirmed by [2]. The structure is isotypical with Fe$_7$W$_6$. That the ideal composition is not covered by its homogeneity range is not unusual with intermetallic compounds, the deviation depending on its Gibbs energy curve and that of the neighbouring phases. Specific heat measurements carried out by Proshina and Rezukhina [7] have shown that Fe$_3$W$_2$ undergoes no phase transformation in the temperature range 20–972°C.

Goldschmidt [8] claimed to have discovered evidence for the existence of a high temperature σ phase, but does not give further details. A similarity between the Fe-Mo and Fe-W phase diagram and estimations by Raynor [9] regarding the formation points and compositions of phases, lead to the assumption that a high temperature σ phase might exist. In accordance with these considerations, a tentative phase diagram has been constructed (Fig. 96).

For a calculated phase diagram, the reader may be referred to a publication by Kaufman and Nesor [10].

References

1 Sinha, A.K.; Buckley, R.A.; Hume-Rothery, W.: J. Iron Steel Inst. 205 (1967) 191, 1145
2 Kirchner, G.; Harvig, H.; Uhrenius, B.: Metall. Trans. 4 (1973) 1059
3 Abrahamson, E.P.; Lopata, S.L.: Trans. Metall. Soc. AIME 236 (1966) 76
4 Hillert, M.; Wada, T.; Wada, H.: J. Iron Steel Inst. 205 (1967) 539
5 Fischer, W.A.; Lorenz, K.; Fabritius, H.; Schlegel, D.: Arch. Eisenhüttenwes. 41 (1970) 489
6 Kovacova, K.; Kralik, F.: Kovove Mater. 11 (1973) 93
7 Proshina, Z.V.; Rezukhina, T.N.: Zh. Fiz. Khim. 36 (1962) 1749
8 Goldschmidt, H.J.: Research 8 (1951) 343
9 Raynor, G.V.: J. Less Common Met. 29 (1972) 333
10 Kaufman, L.; Nesor, H.: CALPHAD 2 (1) (1978) 62
11 Takayama, T.; Wey, M.Y.; Nishizawa, T.: Trans. Jpn. Inst. Met. 22 (1981) 315

Fe–Y   Iron–Yttrium

Fig. 98

The feature of the Fe-Y phase diagram shown in Fig. 98 is essentially the one suggested by Domagala et al. [1]. A summary and critical assessment of the literature and experimental results reported up to 1961 pertaining to this system has been carried out by Gschneidner [2]. More recent work has been concentrated on a re-investigation of the intermetallic compounds.

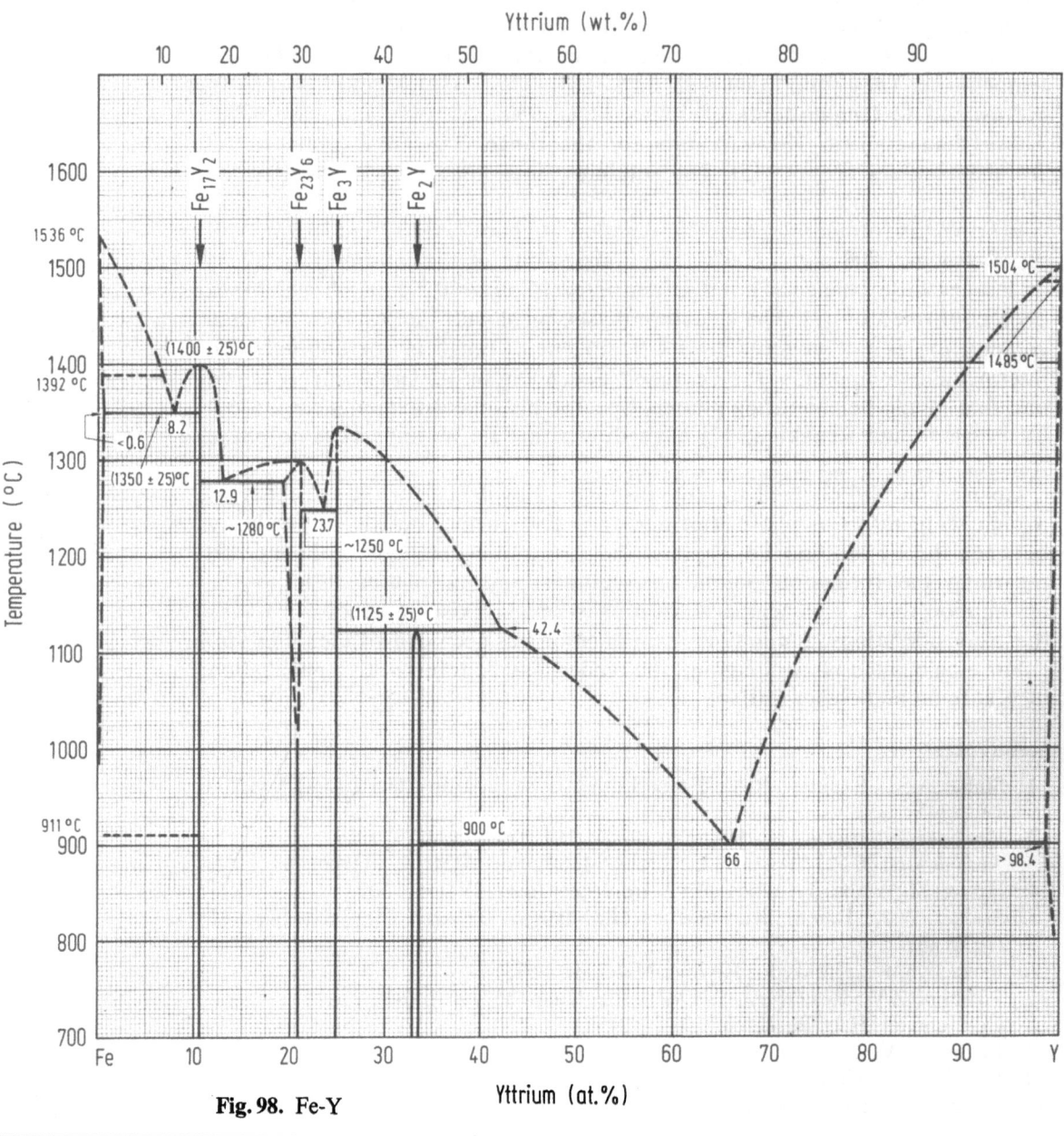

Fig. 98. Fe-Y

The existence of four compounds has been reliably established, namely: $Fe_{17}Y_2$ which appears in two modifications, a high temperature, hexagonal Th_2Zn_{17}-type and a low temperature, rhombohedral Th_2Zn_{17}-type; $Fe_{23}Y_6$, cubic with $Mn_{23}Th_6$-type structure; Fe_3Y isomorphous with $PuNi_3$ and the Laves phase Fe_2Y, cubic with $MgCu_2$-type structure.

The melting point of yttrium has been taken from a measurement by Ackermann and Rauh [3] ((1,504 ± 1.5) °C − IPTS-68), the transformation temperature ((1,485 ± 5) °C) from [2].

Two teams, Domagala et al. [1] and Farkas and Bauer [4], studied the phase relationships in the system. The former investigated the entire composition range carrying out thermal-, X-ray and microscopical analyses on very pure materials. The iron was stated to be 99.9% pure, and yttrium 99.0% pure, its major impurities being 0.5% Zr, 0.2% Ti and 0.12% O. [4] determined the Fe-rich region (about 70–100 wt.% Fe) using less pure component metals than [1], and their results were based on not homogenized specimens. In view of conflicting data and with the information mentioned, [2] decided that the diagram proposed by [1] is the more accurate one with the restriction that the experimental techniques employed do not justify showing the liquidus as a full line.

Buschow [5] confirmed [6] and re-determined the structure of $Fe_{17}Y_2$ which occurs in a high and low temperature modification. The lattice parameters were found to be $a = 0.8463$ nm, $c = 0.8282$ nm and $a = 0.8460$ nm, $c = 1.2410$ nm for the Th_2Ni_{17}- and Th_2Zn_{17}-type structure respectively. $Fe_{17}Y_2$ replaces the phase previously designated Fe_9Y [1, 7]. Yttrium is not an element belonging to the lanthanide-group elements, nevertheless size and valency of the ion allow it to behave like a trivalent Rare Earth Metal. From lattice constants observations, [5] conclude that Y has a metallic radius which lies in between that of terbium and dysprosium in transition metal compounds (i.e. Fe, Co, Ni).

Kripyakevich et al. [8] measured the lattice parameter of an Fe-Y compound containing about 20 at.% Y and found a Th_6Mn_{23}-type phase ($a = 1.2120$ nm). Kharchenko et al. [9] investigated the interactions of Y with Fe agree with these findings thus supporting the indexing of this phase as $Fe_{23}Y_6$ in favour of the previously described Fe_4Y [2, 7, 10]. The occurrence of a $CaCu_5$ structure, which has been reported by Wallace [11] and which has been discussed by [2] in his summary and could not be confirmed by Taylor and Poldy [12] seems to be doubtful in view of the general non-existence of FeR_5 compounds. The Fe-Y system differs substantially from the corresponding Y-Co and Y-Ni systems, in that a compound crystallizing with $CaCu_5$ structure has been definitely established for these systems.

The melting point of Fe_3Y reported by [1] as about 1,400°C seems rather high. The existence of this phase has been confirmed by van Vucht [13] and Buschow [14] who found a lattice parameter of $a = 0.5133$ nm, $c = 2.4600$ nm for the $PuNi_3$-type structure.

All investigators agree that Fe_2Y is a Laves phase with $MgCu_2$-type structure. The parameter reported agree very well, $a = 0.7355$ nm [15], 0.7356 nm [1], 0.7357 nm [16] except for that reported by Skrabek [17] who found $a = 0.729$ nm. Fe_2Y is formed by a peritectic reaction and shows a narrow homogeneity range [1, 16]. Its Curie temperature is 262°C [2].

The effect of Y on the α/γ(Fe) transformation is not known. Hellawell [18] reported that 1 at.% Y lowers the γ/δ(Fe) transformation by 3°C.

The terminal solid solubilities have not been measured. [2] conclude that for Y in Fe it is less than 1 wt.%.

References

1 Domagala, R.F.; Rausch, J.J.; Levinson, D.W.: Trans. ASM 53 (1961) 137
2 Gschneidner, K.A.: Rare Earth Alloys., Princeton: van Nostrand 1961
3 Ackermann, R.J.; Rauh, E.G.: High Temp. Sci. 4 (1972) 272
4 Farkas, M.S.; Bauer, A.A.: U.S. At. Energy Comm. BMI-1386, (1959) 20
5 Buschow, R.H.J.: J. Less Common Met. 11 (1966) 204
6 Zarechnyuk, O.S.; Kripyakevich, P.I.: Dopov. Akad. Nauk Ukr. RSR (12) (1964) 1593
7 Haefling, J.; Daane, A.H.: Unpublished data, quoted by Lundin, C.E.: In: Spedding, F.H.; Daane, A.H. (Eds.): The Rare Earths. New York: Wiley 1961
8 Kripyakevich, P.I. et al.: Poroshk. Metall. No. 11 (1965) 55
9 Kharchenko, O.I.; Bodak, O.I.; Gladyshevskiy, E.I.: Russ. Metall. 1976/170
10 Nassau, K.; Cherry, L.V.; Wallace, W.E.: Phys. Chem. Solids 16 (1960) 123
11 Wallace, W.E.: Private communication 1960 quoted by [2], see also [10]
12 Taylor, K.N.R.; Poldy, C.A.: J. Less Common Met. 27 (1972) 255
13 Van Vucht, J.H.N.: J. Less Common Met. 10 (1966) 146
14 Buschow, R.J.H.: Philips Res. Rep. 26 (1971) 49
15 Dwight, A.E.: Trans. ASM 53 (1961) 479
16 Beaudry, B.J.; Haefling, J.; Daane, A.H.: Acta Crystallogr. 13 (1960) 743
17 Skrabek, E.A.: University of Pittsburgh, 1962
18 Hellawell, A.: J. Less Common Met. 1 (1959) 110

Fe–Yb Iron–Ytterbium

Fig. 99

The structure and occurrence of stable Fe-Yb compounds have been investigated by Buschow [1] and Iandelli and Palenzona [2] without the intention of constructing a phase diagram. Figure 99 therefore represents only a tentative diagram, based on a schematic representation of [2]. [1] employed X-ray diffraction methods and [2] thermal, micrographical and X-ray analyses of alloys covering the complete range of concentration.

The system is characterized by restricted miscibility of liquid Yb and $Fe_{23}Yb_6$ and two compounds $Fe_{17}Yb_2$ (hexagonal with Th_2Ni_{17}-type structure) and $Fe_{23}Yb_6$ (cubic with Th_6Mn_{23}-type structure). Two thermal reactions have been observed at 1,310°C and 1,290°C. According to [2] the first one relates to the monotectic reaction whilst no explanation has been found so far for the second thermal reaction. The melting and transformation points of ytterbium, 815°C and 765°C, have been accepted from measurements by [2] carried out on purified ytterbium.

Owing to the high vapour pressure of Yb the preparation and thermal analysis of alloys proved to be very difficult. Buschow [1] using 99.9% pure starting materials sealed these into a molybdenum container and heated them subsequently at 1,200°C for 5 minutes. The samples were annealed, still inside the molybdenum container, and placed into evacuated quartz tubes at 700–800°C for three weeks.

[2] chose Mo- or Ta-containers which reacted with the samples only at higher temperatures. They were able to alloy the starting materials (99.999% Fe and de-hydrogenised Yb) but unable to carry out a thermal analysis at temperatures above 1,500°C. Thus only the Yb-rich region was investigated. The

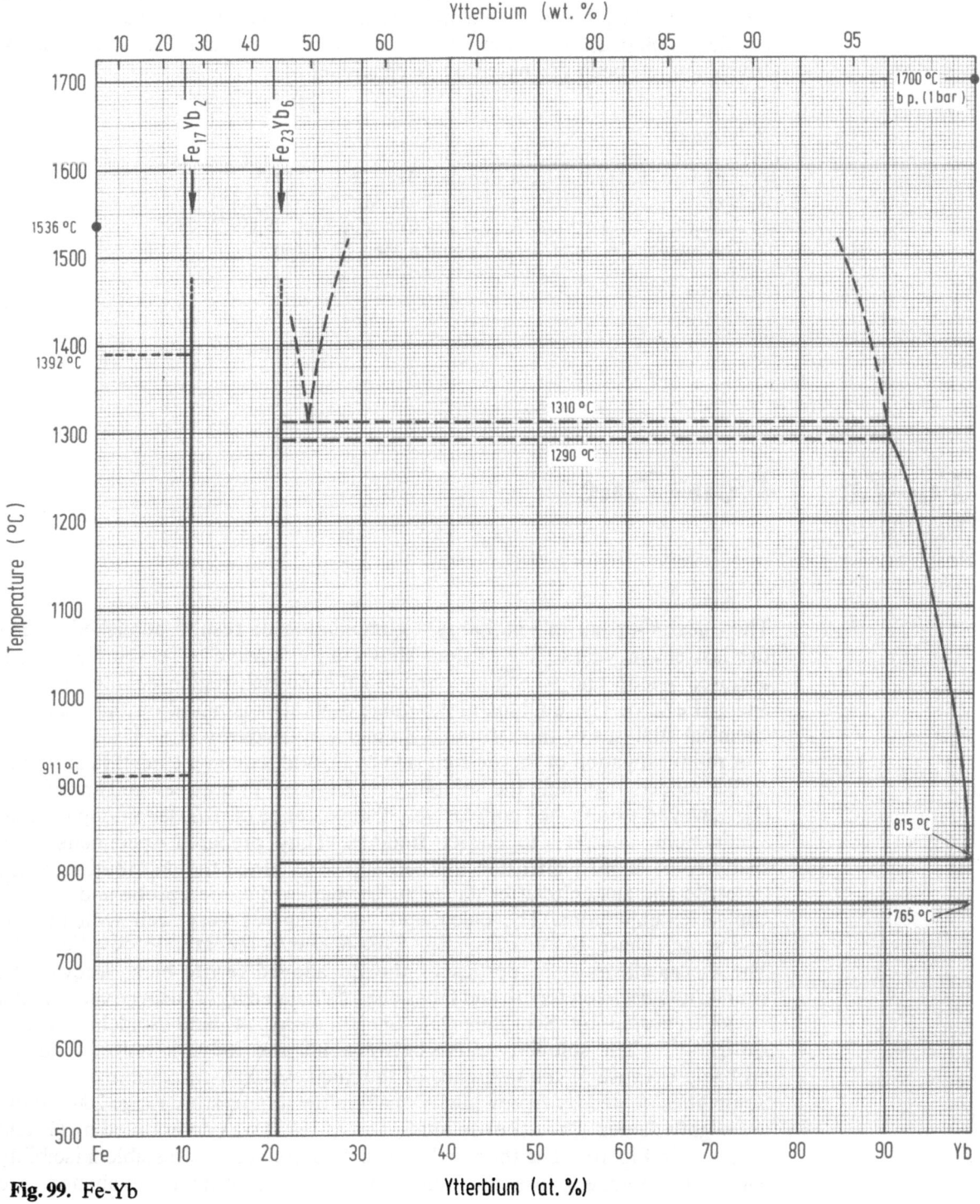

Fig. 99. Fe-Yb

results for the Fe-richer alloys are based on microscopical and X-ray analyses only.

Ytterbium and europium belong to the family of rare earth metals but differ in many ways from the other lanthanides owing to their different electronic structure. The melting-, boiling- and sublimation-points are much lower than those of the other REMs. Generally, Yb behaves in intermetallic compounds like a divalent alkaline earth metal. However, it is also able to form trivalent compounds. Conditions under which Yb (and Eu) may acquire trivalency has been discussed by Gschneidner [4].

References

1 Buschow, K.H.J.: J. Less Common Met. 26 (1972) 329
2 Iandelli, A.; Palenzona, A.: Rev. Chim. Miner. 13 (1976) 55
3 Iandelli, A.; Palenzona, A.: J. Less Common Met. 29 (1972) 293
4 Gschneidner, K.A.: J. Less Common Met. 17 (1969) 13

Fe–Zn   Iron–Zinc

Figs. 100, 101

The phase diagram in Fig. 100 is a combination of results published during the last seven years. The system shows an extensive α-Fe solubility area, a closed γ-loop and four compounds in the Zn-rich region denoted Γ, Γ_1, δ and ζ ($FeZn_{13}$). The liquidus in the range 0–92 at.% is based on the results of chemical analyses of the saturated Zn melts by Budurov et al. [1].

A re-investigation of the Fe-rich side by [1] and Kirchner et al. [2] confirmed earlier reports regarding the solubility of Zn in α(Fe) between 400 and 800°C. [2], however, established conclusively the existence of a closed γ-loop without a minimum, contrary to earlier findings. They prepared specimens by a specially developed diffusion method and determined the distribution of Zn in ferrite and austenite after annealing by means of a microprobe and microscopical examinations. The calculated γ-loop agrees very well with the experimental phase boundaries, e.g. 5.68–6.30 at.% Zn at 1,135°C (Fig. 101). The thermochemical properties of the Zn alloyed ferrite being known, the equilibrium between ferrite and Γ has been calculated by [2] using experimental values reported by Schramm [3], Speich et al. [4] and Stadelmaier and Bridgers [5]. The solubility curve between 400 and 780°C is not continuous, probably because of the influence of the magnetic transformation.
Solubilities quoted by Ageev [D(1974) 60] in the temperature range 200–500°C are slightly larger. The α(Fe) solidus (780–1,536°C) was determined by [1] and is shown in Fig. 100. The thermochemical data predict a metastable miscibility gap in the ferrite phase the boundary of which could be evaluated, see Nishizawa et al. [13].

Bastin et al. [6, 7] re-investigated the Zn-rich region and suggested a modified phase diagram which combined with recent results of Gellings et al. [8, 12] is incorporated in Fig. 100. Experimental difficulties in preparing Fe-Zn alloys (high vapour pressure of Zn) are probably responsible for the discrepancies encountered in the older literature. These difficulties can be largely avoided by employing isothermic diffusion. [6] re-investigating the Zn-rich

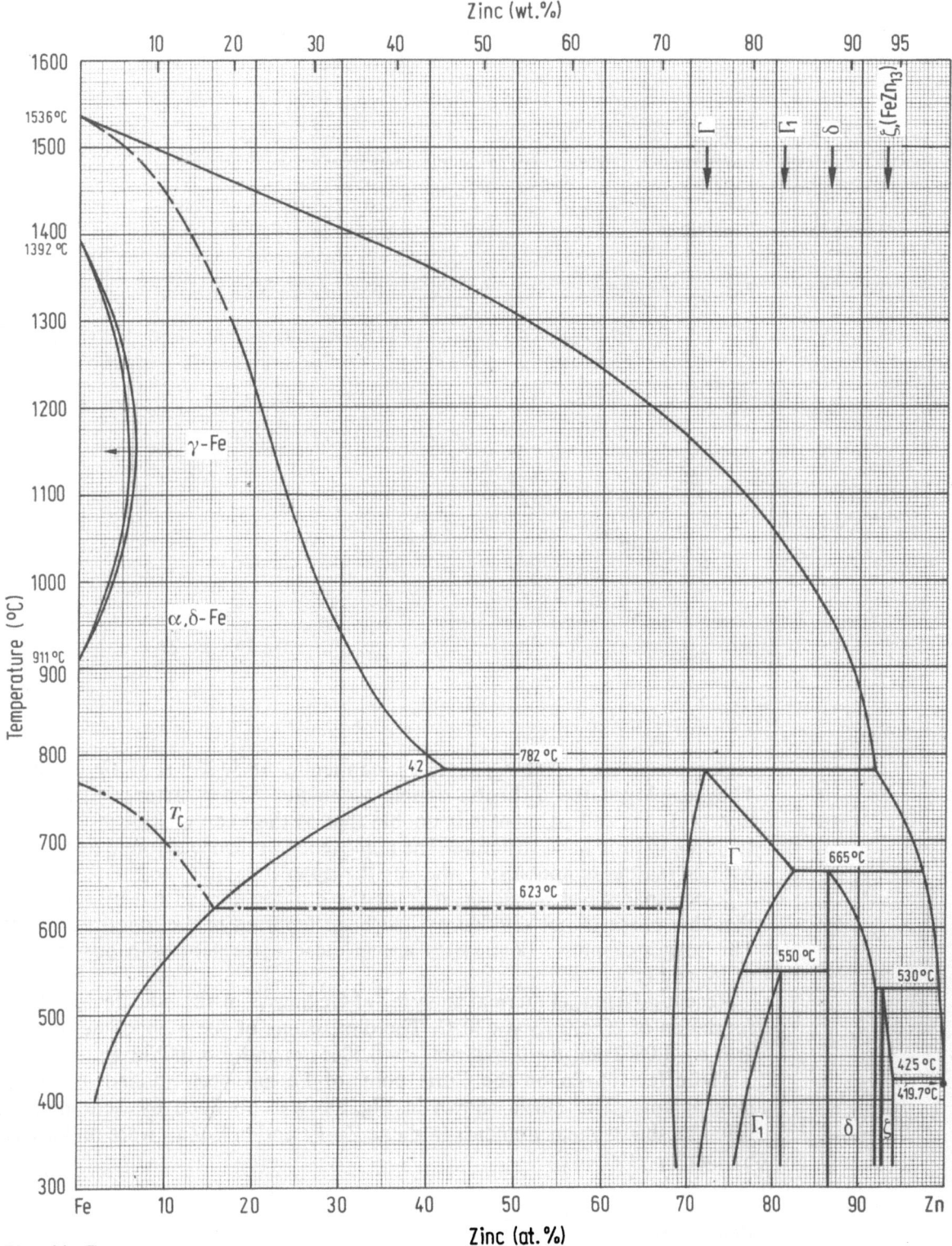

Fig. 100. Fe-Zn

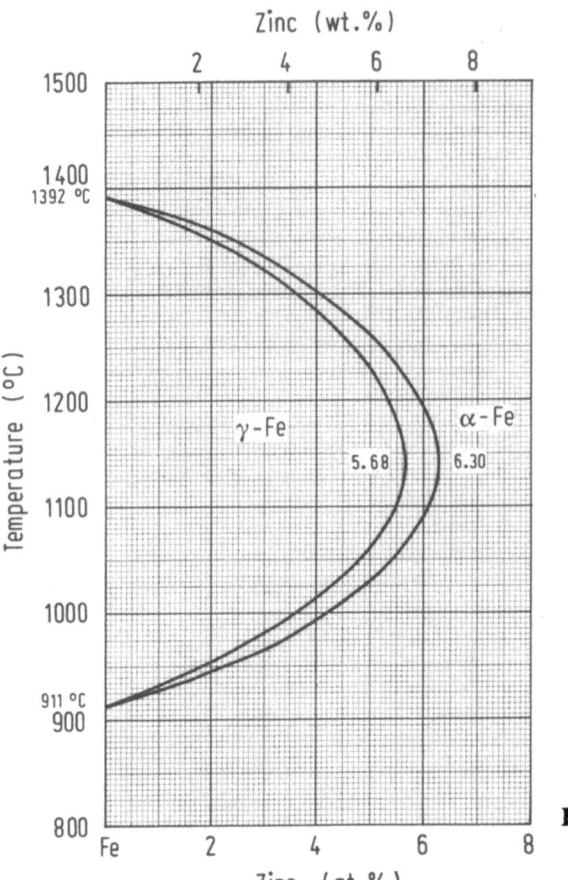

Fig. 101. Fe-Zn

region discovered a new phase, Γ_1, containing 18.5–23.5 at.% Fe at 380°C. They employed the diffusion couple technique. The structure is related to that of Γ. Its cell parameter is double that of the bcc unit cell of Γ namely 1.796 nm to form a fcc structure. The homogeneity range was found to decrease with increasing temperature, whereas the opposite applies to the Γ phase.

X-ray analysis of single crystals of the δ phase carried out by [7] proves that this phase is not divided as had been reported in the literature [9, 10]. The suggested phase boundaries are in close agreement with observations of [8] and [12].

Recent differential thermal analyses of Gellings et al. [8] indicate a peritectic formation of $\zeta(FeZn_{13})$ at 530°C. The dimensions of the monoclinic unit cell are: $a = 1.0862$ nm, $b = 0.7608$ nm, $c = 0.5061$ nm and $\beta = 100°32'$ [8].

The effect of Zn on the Curie point of Fe has been studied by several investigators. The curve shown in Fig. 100 was obtained by Schramm [3]. ^{57}Fe Mössbauer measurements in dilute Zn-Fe and in $FeZn_{13}$ were undertaken by Dunlop et al. [11].

References

1   Budurov, S.; Kovatchev, P.; Stojcev, N.; Kamenova, Z.: Z. Metallkd. 63 (1972) 348
2   Kirchner, G.; Harvig, H.; Moquist, K.R.; Hillert, M.: Arch. Eisenhüttenwes. 44 (1973) 227
3   Schramm, J.: Z. Metallkd. 30 (1938) 122, 131

4   Speich, G.R.; Zwell, L.; Wriedt, H.A.: Trans. Metall. Soc. AIME 230 (1964) 939
5   Stadelmaier, H.H.; Bridgers, R.K.: Metall 15 (1961) 761
6   Bastin, G.F.; van Loo, F.J.J.; Rieck, G.D.: Z. Metallkd. 65 (1974) 656
7   Bastin, G.F.; van Loo, F.J.J.; Rieck, G.D.: Z. Metallkd. 68 (1977) 359
8   Gellings, P.J.; de Bree, E.W.; Giermann, G.: Z. Metallkd. 70 (1979) 312, 315
9   Swetschnikov, W.N.; Gridnew, W.N.: Metallurgy 21 (1937) 35
10 Ghoniem, M.; Löhberg, K.: Metall 26 (1972) 1026
11 Dunlop, J.B.; Williams, J.M.; Longworth, G.: J. Phys. 8 (1978) 2115
12 Gellings, P.J.; Ierman, G.; Koster, D.; Kuit, J.: Z. Metallkd. 71 (1980) 70
13 Nishizawa, T.; Hasebe, M.; Ko, M.: Acta Metall. 27 (1979) 817

Iron–Zirconium

Fe–Zr

Figs. 102, 103

Numerous investigators have published data on the Fe-Zr system. Still, the phase relationships in the Zr-rich region require more experimental work.

Figure 102 represents a summary of the results, mainly obtained by Svechnikov et al. [1] (0–52 wt.% Zr), Fischer et al. [2] ($\alpha + \gamma$) Fe transformation, Rhines and Gould [3] and Tanner and Levinson [4] (0–100 at.% Zr).

The Fe-Zr system belongs to the contracted γ field type and displays four, probably five intermediate phases, namely: Fe_3Zr, fcc, isotypic with $Mn_{23}Th_6$; Fe_2Zr, fcc $MgCu_2$-type Laves phase; $FeZr_2$, bc tetragonal Al_2Cu-type; $(FeZr_3)$ and $FeZr_4$ (structures not determined). The melting point of Zr, 1,855°C, is a secondary reference point (IPTS-68 revised 1975), ((1,861 ± 4) °C [5]), the α/β transformation selected here is (863 ± 5) °C [D]. Thermal analyses by Vogel and Tonn [6] and Svechnikov and Spektor [7] ascertained that the addition of Zr to Fe lowers the δ(Fe) liquidus to a eutectic at 1,335°C [6, 2] or 1,325°C [7] where the following reaction takes place: Liq. ($\sim$ 8.8–10 at.% Zr) $\leftrightharpoons$ γ-Fe (0.2 at.% Zr) + Fe_3Zr. Further additions of Zr raise the liquidus temperature to the melting point of Fe_2Zr (1,675°C). Fischer et al. [2] investigated the 0–5.5 at.% Zr region in the temperature interval 900–1,400°C employing susceptibility measurements on high purity alloys (maximum impurity: 0.026 at.%) and also calculated the Gibbs energy changes. The temperature of the eutectoidal and the peritectoidal was found to be 1,355°C and 925°C respectively, which agrees well with results obtained by Svechnikov and Spektor [7], (1,350°C and 925°C respectively). The maximum solubility of Zr in γ(Fe) at 1,355°C and α(Fe) at 925°C has been determined as 0.2 and 0.2 at.% respectively [2]. The position of the eutectoidal point (δ-Fe $\leftrightharpoons$ γ-Fe + Liq.) is 0.4 at.% Zr which is at variance with results quoted by previous investigators [7, 6] who reported a concentration of 4.3 at.% Zr which seems questionable if thermodynamic considerations are applied [2] (Fig. 103).

The solubility of Zr in α(Fe) in the temperature range 600–900°C has been determined by Abrahamson and Lopata [16] (lattice parameter measurements). According to them, 0.044, 0.067, 0.082 and 0.160 at.% correspond to temperatures of 600, 700, 800 and 900°C.

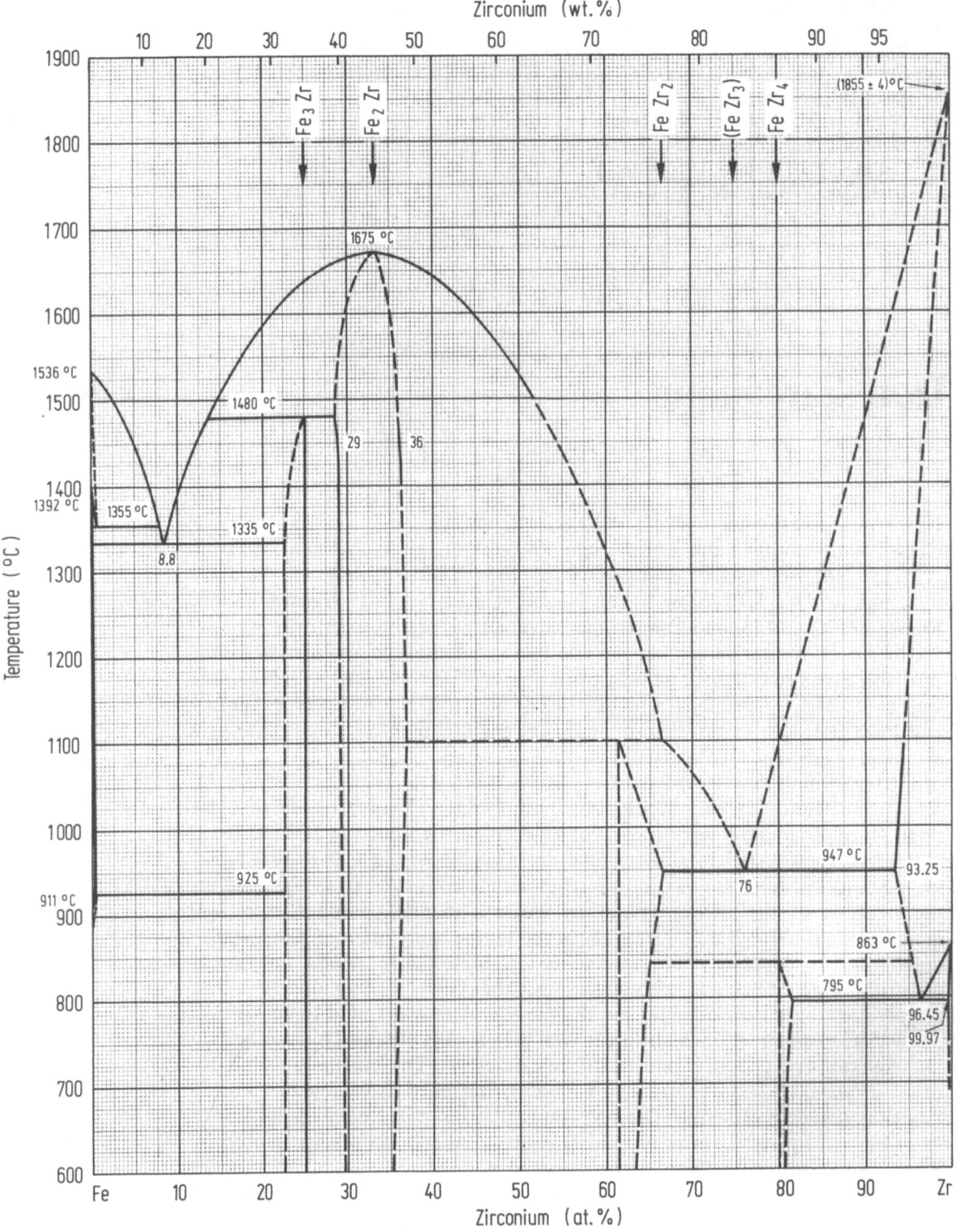

Fig. 102. Fe-Zr

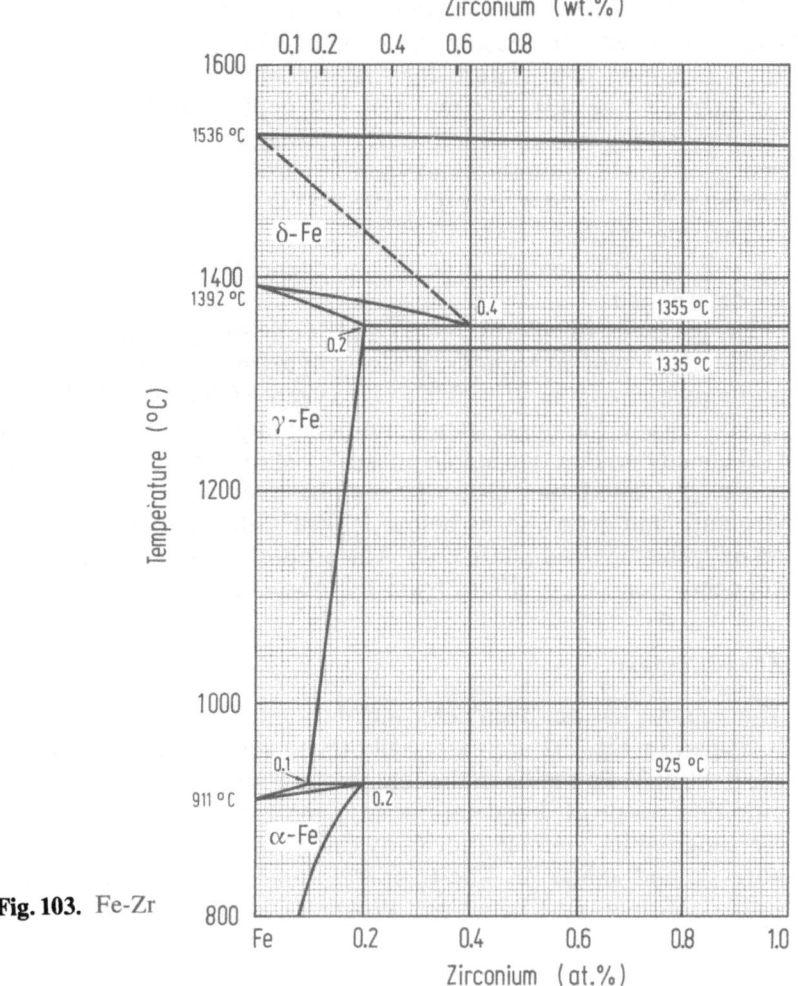

Fig. 103. Fe-Zr

As mentioned above the existence of four compounds (Fe_3Zr, Fe_2Zr, $FeZr_2$ and $FeZr_4$) has been reliably established. Metallographic, electron microprobe studies and X-ray investigations of Zr-rich alloys undertaken by Rhines and Gould [3], Levinson [8], Sweeney and Batt [9] indicate contrary to previous findings an additional stable Zr-rich phase, indexed $FeZr_3$, the structure of which has not been determined so far.

The phase, indexed Fe_3Zr, had been predicted by Wallbaum [10] and was confirmed by [1] who employed X-ray and metallography as well as differential thermal analysis of pre-annealed samples prepared from hydrogen-refined iron and iodide zirconium. Fe_3Zr is formed peritectically at 1,480°C and shows a range of homogeneity of about 3 at.% according to micro-analyses whereas X-ray data point to a much smaller region. Its structure was re-determined by Dripyakevich et al. [11] and found to be fcc, isotypic with $Mn_{23}Th_6$, rather than the complex $NiTi_2$ structure reported by [1]. Fe_3Zr is ferromagnetic below 275°C [1]. Fe_2Zr is a Laves phase with the $MgCu_2$-type structure and with an appreciable range of homogeneity extending from 29–36 at.% Zr at 1,450°C according to [1] who used metallographic and X-ray measurements for the determination of the phase boundaries (see Fe_3Zr). Information relating to the melting temperature varies somewhat; 1,675°C [1], 1,645°C [12] and 1,650°C [4] have been reported. Results by [1] have been incorporated in Fig. 102. Fe_2Zr is ferromagnetic with a Curie temperature decreasing from 475–310°C with increasing Zr content (29–36 at.% Zr) [1]. The magnetic transformation temperature for the stoichiometric composition is 364°C according to Craig [13]. Kuz'ma et al. [14] confirmed the existence of $FeZr_2$ and determined the crystal structure as bc tetragonal, isotypic with Al_2Cu. The presence of small amounts of oxygen apparently convert this structure to the complex fcc structure, isotypic with $NiTi_2$.

The intermediate phase $FeZr_4$ reported by [3] and [8] but not discovered by [14] at 800°C, entails a small homogeneity range. Its structure has yet to be determined.

On the basis of microscopical and X-ray analyses of 8 samples containing 4–33 at.% Zr and prepared from electrolyte iron (C = 0.02%, oxygen = 0.016%) and iodide zirconium (C = 0.021%, oxygen = 0.02%) in the temperature range 1,200–1,360°C (homogenization not being mentioned), Guseva and Malakhova [15] concluded that two stable phases exist in this area, Fe_2Zr and $Fe(Fe_{0.2}Zr_{0.8})$ = $Fe_{2.64}Zr$. The latter, also a Laves phase, exhibits a hexagonal, $MgZn_2$ type structure and decomposes below 1,200°C into $(\alpha + \eta)$. According to [15] η is an oxygen stabilized phase only. The proposed phase diagram for the $Fe-Fe_2Zr$ concentration range differs from the observations of other investigators.

Zr-rich region

The liquidus has not been determined as yet. The tentative phase relationships indicated in Fig. 102 are essentially those suggested by [3]. The equilibria for the 0–10 at.% Fe region in the temperature interval 700–1,050°C is the slightly modified result of a redetermination by Tanner and Levinson [4] who used metallographic, thermal and X-ray analyses of alloys prepared from 99.97% pure Fe and iodide Zr. The temperature of the Zr-rich eutectic reaction has been confirmed by Kuchar and Wozniakova [17].

References

1 Svechnikov, V.N.; Pan, V.M.; Spektor, A.T.: Russ. J. Inorg. Chem. 8 (1963) 110
2 Fischer, W.A.; Lorenz, K.; Fabritius, H.; Schlegel, D.: Arch. Eisenhüttenwes. 41 (1970) 489
3 Rhines, R.N.; Gould, R.W.: Adv. X-Ray Anal. 6 (1963) 62
4 Tanner, L.E.; Levinson, D.W.: Trans AIME 215 (1959) 1066
5 Ackermann, R.J.; Rauh, E.G.: High Temp. Sci. 4 (1972) 272
6 Vogel, R.; Tonn, W.: Arch. Eisenhüttenwes. 5 (1931–1932) 387
7 Svechnikov, V.N.; Spektor, A.T.: Dokl. Akad. Nauk SSSR 143 (1962) 613
8 Levinson, D.W.: Private communication, 1966. see Shunk [C: Ref. 6]
9 Sweeney, W.E.; Batt, A.P.: J. Nucl. Mater. 13 (1964) 84
10 Wallbaum, H.J.: Arch. Eisenhüttenwes. 14 (1941) 521
11 Kripyakevich, P.I.; Protasov, V.S.; Cherkashin, E.E.: Russ. J. Inorg. Chem. 10 (1965) 151
12 Elliott, R.P.; Rostoker, W.: Trans. ASM 50 (1958) 617
13 Craig, R.S.: U.S. At. Energy Comm. TID- 14769 (1962) 72
14 Kuz'ma, Yu.B.; Markiv, V.Ya.; Voroshilov, Yu.V.; Skolozska, R.V.: Izv. Akad. Nauk. SSSR, Neorg. Mater. 2 (1966) 259
15 Guseva, L.N.; Malakhova, T.O.: Akad. Nauk. Ukr. SSR, Metallofiz. abstracted in Ageev [E] 1973, p. 52
16 Abrahamson, E.P.; Lopata, S.L.: Trans. AIME 236 (1966) 76
17 Kuchar, L.; Wozniakova, B.: Hutn. Listy 5 (1976) 342

Appendix

Table 1. Physico-chemical properties of the elements

Symbol	Element	Atomic weight[a] Relative atomic mass	Modifi-cation	Crystal[b] System	Type	Melting and transformation points[c] K	°C	Refs.[d]	Density g/cm³ (20°C)
Ag	Silver	107.870	–	fcc	A1	MP[e] 1,235.08	961.93	1	10.47
Al	Aluminium	26.9815	–	fcc	A1	MP[f] 933.612	660.462	2	2.698
Am²⁴¹	Americium	241.13	α	dhcp	–	TP 923.0 ±50	650	6	–
			β	fcc	A1	TP 1,349 ± 5	1,076	6	–
			γ	bcc	A2	MP 1,449 ± 3	1,176	6	–
As	Arsenic	74.9216	α	rhombohed.	A7	MP[g] 1,081 ± 3	808	5	5.72
Au	Gold	196.967	–	fcc	A1	MP[e] 1,337.58	1,064.43	1	19.31
B	Boron	10.811	(β)	hexag.	Ag	MP 2,350	2,077	7	2.34
Ba	Barium	137.34	α	bcc	A2	TP 645	372	5	3.61
			β	hcp	A3	MP 1,002.5 ± 2	729.3	5	–
Be	Beryllium	9.0122	α	hcp	A3	TP 1,528.7 ± 5	1,255.5	5	1.839
			β	bcc	A2	MP 1,561.5 ± 5	1,288.3	5	–
Bi	Bismuth	208.980	–	rhombohed.	A7	MP[f] 544.592	271.442	2	9.88
C	Carbon	12.01	graphite	hexagonal	A9	MP[g] 4,020 ±20	3,747	19	2.22
			diamond	cubic	A4	– disintegrates to graphite below 2,000 °C			3.51
Ca	Calcium	40.08	α	fcc	A1	TP 720.0 ± 3	447	5	1.53
			β	bcc	A2	MP 1,112.7 ± 2	839.5	5	1:44
Cb	see Nb								
Cd	Cadmium	112.40	–	hexagonal	A3	MP[f] 594.258	321.108	2	8.646
Ce	Cerium	140.12	α	fcc	A1	TP 103	−170	8	8.24
			β	hcp	A3	TP 273	0	8	6.672
			γ	fcc	A1	TP 998	725	8	6.769
			δ	bcc	A2	MP 1,068	795	8	6.70
Co	Cobalt	58.9332	ε	hcp	A3	TP ~ 673–693	~ 400–420	10	8.93
			α,γ	fcc	A1	MP[f] 1,768	1,495	2	8.73
Cr	Chromium	51.996	–	bcc	A2	MP 2,130 ±20	1,857	5	7.22
Cs	Cesium	132.905	–	bcc	A2	MP 301.55 ± 0.05	28.40	5	1.963
Cu	Copper	63.54	–	fcc	A1	MP[f] 1,358.03	1,084.88	2	8.95
Dy	Dysprosium	162.50	α	hcp	A3	TP 1,663 ± 8	1,390	5,9	8.557
			β	bcc	A2	MP 1,685	1,412	5	–
Er	Erbium	167.26	α	hcp	A3	TP ~1,748	~1,475	9	9.060
			β	bcc	A2	MP 1,795	1,522	5	–
Eu	Europium	151.96	α	?	?	TP ?	?	–	–
			β	bcc	A2	MP 1,096 ± 5	823	10	5.244
Fe	Iron	55.847	α	bcc	A2	TP 1,184	911	2	7.866

Continued

ᵃ The atomic weights (relative atomic mass) are based on ¹²C = 12.000 g-atom as adopted by IUPAC, 1961

ᵇ The structural types accord to "Strukturberichte" (Leipzig) and Smithells, C.J.: Metals Reference Book, 3rd ed. London: Butterworths 1962

ᶜ The melting and transformation temperatures are based on the International Practical Temperature Scale, IPTS-68 (see 'Introduction' and Table 3)

ᵈ References see p. 182

ᵉ Primary fixed melting points

ᶠ Secondary reference points [F]

ᵍ Melting point under pressure

179

Symbol	Element	Atomic weight[a] Relative atomic mass	Modification	Crystal[b] System	Type		Melting and transformation points[c] K		°C	Refs.[d]	Density g/cm³ (20°C)
Fe	Iron	55.847	γ	fcc	A1	TP	1,665		1,392	2	8.33
			δ	bcc	A2	MP	1,809		1,536	2	7.86
Ga	Gallium	69.72	–	ortho-rhombic	A11	MP[f]	302.922		29.772	2	5.91
Gd	Gadolinium	157.25	α	hcp	A3	TP	1,533 ± 2		1,260	5,9	7.886
			β	bcc	A2	MP	1,586 ± 2		1,313	5,11	7.785
Ge	Germanium	72.59	–	diamond	A4	MP	1,210 ± 5		937	5,10	5.33
Hf	Hafnium	178.49	α	hcp	A3	TP	2,012 ±10		1,739	16	13.1
			β	bcc	A2	MP	2,495		2,222	4	–
Hg	Mercury	200.59	–	rhombohed.	A10	MP[f]	234.314		−38.836	2	13.546
Ho	Holmium	164.930	α	hcp	A3	TP	1,713		1,440	5	8.796
			β	bcc	A2	MP	1,745		1,472	5	
In	Indium	114.82	–	tetragonal	A6	MP[f]	429.784		156.634	2	7.34
Ir	Iridium	192.2	–	fcc	A1	MP[f]	2,720		2,447	2	22.35
K	Potassium	39.102	–	bcc	A2	MP	336.35		63.20	5	0.86
La	Lanthanum	138.91	α	dhcp	–	TP	550 ±40		277	5	6.163
			β	fcc	A1	TP	1,134 ± 5		861	5	6.190
			γ	bcc	A2	MP	1,194 ± 1		921	5	5.962
Li	Lithium	6.939	α	hcp	A3	TP	80 ±10		−193	5	0.534
			β	bcc	A2	MP	453.7 ± 0.5		180.5	5	–
Lu	Lutetium	174.97	α	hcp	A3	TP	~1,873 ±30		~1,600	9	9.846
			β	bcc	A2	MP	1,935		1,662	9,10	–
Mg	Magnesium	24.312	–	hcp	A3	MP	923.5		650.3	10	1.74
Mn	Manganese	54.938	α	cubic	A12	TP	980 ±20		707	5	7.526
			β	cubic	A13	TP	1,361 ±10		1,088	5	7.325
			γ	fcc	A1	TP	1,412 ± 5		1,139	5	7.277
			δ	bcc	A2	MP	1,519 ± 5		1,246	5	–
Mo	Molybdenum	95.94	–	bcc	A2	MP[f]	2,896 ± 3		2,623	2	10.20
Na	Sodium	22.9898	α	hcp	A3	TP	40 ± 5		−233	5	0.97
			β	bcc	A2	MP	371 ± 0.1		98	5	–
Nb	Niobium	92.906	–	bcc	A2	MP[f]	2,750		2.477	2	8.60
Nd	Neodymium	144.24	α	dhcp	–	TP	1,133 ± 5		860	5,10	7.005
			β	bcc	A2	MP	1,290.3 ± 5		1,017.1	5	6.803
Ni	Nickel	58.71	–	fcc	A1	MP[f]	1,728 ± 4		1,455	2	8.90
Os	Osmium	190.2	–	hcp	A3	MP	3,300 ±20		3,027	10	22.51
P	Phosphorus	30.975	β white	hexagonal	–	TP	195.4		−77.8	20	1.88
			α white	ortho-rhombic	–	MP	317.3		44.14	20	1.83
			"red"	cubic	–	MP	866 ± 5		593	19	–
Pb	Lead	207.19	–	fcc	A1	MP[f]	600.652		327.502	2	11.38
Pd	Palladium	106.4	–	fcc	A1	MP[f]	1,827		1,554	2	11.96
Pm	Promethium	(147.0)	α	dhcp	–	TP	~1,173 ±20		~ 900	9	(7.25)
			β	bcc	A2	MP	1,323 ± 5		1,050	5,9	(7.02)
Pr	Praseodymium	140.907	α	dhcp	–	TP	1,063 ± 3		790	10,12	6.768
			β	bcc	A2	MP	1,191 ± 4		918	12	6.647
Pt	Platinum	195.09	–	fcc	A1	MP[f]	2,042		1,769	2	21.44
Pu[239]	Plutonium	239.052	α	monoclinic	–	TP	395 ± 4		122	6	19.816
			β	bc mono-clinic	–	TP	480 ± 5		207	6	17.77
			γ	fc ortho-rhombic	–	TP	588 ± 3		315	6	17.14
			δ	fcc	A1	TP	730 ± 2		457	6	15.92
			δ'	bc tetrag.	–	TP	752 ± 4		479	6	16.0
			ε	bcc	A2	MP	913 ± 2		640	6	16.51

Symbol	Element	Atomic weight[a] (Relative atomic mass)	Modification	Crystal[b] System	Type	Melting and transformation points[c]	K	°C	Refs.[d]	Density g/cm³ (20°C)
Re	Rhenium	186.2	–	hcp	A3	MP	3,460 ±20	3,186	5	21.0
Rh	Rhodium	102.905	–	fcc	A1	MP[f]	2,236	1,963	2	12.40
Ru	Ruthenium	101.07	–	hcp	A3	MP	2,607 ±8	2,334	3	12.18
S	Sulphur	32.064	α (yellow)	ortho-rhombic	A16	TP	368.5	95.4	18	1.96
			β	monoclinic	–	MP	388.36	115.2	18	–
						BP[f]	717.824	444.674	2	–
Sb	Antimony	121.75	–	rhombohed.	A7	MP[f]	903.905	630.755	2	6.68
Sc	Scandium	44.956	α	hcp	A3	TP	1,609 ±3	1,336	5	2.985
			β	bcc	A2	MP	1,814 ±5	1,541	4	3.188
Se	Selenium	78.96	γ	hexagonal	A8	TP	482	209	5	4.82
			α	monoclinic	A_k					
			β	monoclinic	A_l	MP	494	221	5	–
Si	Silicon	28.086	–	diamond	A4	MP	1,685 ±2	1,412	13	2.33
Sm	Samarium	150.35	α	hcp	A3	TP	1,191	918	5	7.536
			β	bcc	A2	MP	1,346	1,073	5	7.406
Sn	Tin	118.69	α	diamond	A4	TP	286.2 ±0.5	13.0	5	5.76
			β	tetragonal	A5	MP[e]	505.118	231.968	1	7.30
Sr	Strontium	87.62	α	fcc	A1	TP	504	231	5	2.60
			β	hcp	A3	TP	896	623	5	2.61
			γ	bcc	A2	MP	1,042	769	5	2.52
Ta	Tantalum	180.984	–	bcc	A2	MP	3,293 ±10	3,020	5	16.68
Tb	Terbium	158.924	α	hcp	A3	TP	1,591 ±4	1,318	9,14	8.252
			β	bcc	A2	MP	1,633 ±6	1,360	5	–
Tc	Technetium	(98)	–	hcp	A3	MP	2,446 ±50	2,173	12	11.52
Te	Tellurium	127.60	–	hexagonal	A8	MP	722.7 ±0.2	449.55	5	6.24
Th^{232}	Thorium	232.038	α	fcc	A1	TP	1,633	1,360	6	11.50
			β	bcc	A2	MP	2,023 ±10	1,750	6	11.1
Ti	Titanium	47.90	α	hcp	A3	TP	1,166 ±7	893	17	4.44
			β	bcc	A2	MP	1,941 ±4	1,668	3	4.32
Tl	Thallium	204.37	α	hcp	A3	TP	507 ±2	234	5	11.85
			β	bcc	A2	MP	577 ±2	304	5	11.75
Tm	Thulium	168.934	α	hcp	A3	TP	~1,783 ±10	~1,510	9	9.318
			β	bcc	A2	MP	1,820	1,547	5	–
U^{238}	Uranium	238.03	α	ortho-rhombic	A20	TP	942 ±1.3	669	6	19.042
			β	compl. tetragonal	A_b	TP	1,049 ±1.6	776	6	–
			γ	bcc	A2	MP	1,408 ±2	1,135	6	18.89
V	Vanadium	50.942	–	bcc	A2	MP	2,183 ±10	1,910	21	6.17
W	Tungsten	183.85	–	bcc	A2	MP[f]	3,695 ±8	3,422	2	19.35
Y	Yttrium	88.905	α	hcp	A3	TP	1,758 ±6	1,485	5,10	4.472
			β	bcc	A2	MP	1,777 ±1.5	1,504	4	4.254
Yb	Ytterbium	173.04	α	fcc	A1	TP	1,038	765	15	6.958
			β	bcc	A2	MP	1,088	815	15	6.530
Zn	Zinc	65.37	–	hexagonal	A3	MP[e]	692.73	419.58	1	7.18
Zr	Zirconium	91.22	α	hcp	A3	TP	1,136 ±5	863	5	6.52
			β	bcc	A2	MP[f]	2,128	1,855	2	6.47

Appendix

References for Table 1

1 Henning, F.; Moser, H.: Temperaturmessung, 3. Aufl. Berlin, Heidelberg, New York: Springer 1977
2 Crovini, L.; Bedford, R.E.; Moser, A.: Extended List of Secondary Reference Points. Metrologia 13 (1977) 197
3 Kenisarin, M.M.; Chekhovskoi, V.Ya.; Berezin, B.Ya.; Kats, S.A.: High Temp. High Press. 8 (1976) 367
4 Ackermann, R.J.; Rauh, E.G.: High Temp. Sci. 4 (1972) 272
5 Hultgren, R.; Orr, R.L.; Anderson, P.D.; Kelley, K.K.: Selected Values of Thermodynamic Properties of Metals and Alloys. New York: Wiley 1963 and Supplements
6 Oetting, F.L.; Rand, M.J.; Ackermann, R.J.: The Chemical Thermodynamics of Actinide Elements and Compounds. IAEA, Vienna 1976
7 Chart, T.G.: Private communication
8 Gschneidner, K.A.; Verkade, M.E.: Selected Cerium-Phase Diagrams, Iowa State University, White Plains, New York 10604
9 Estimate (see Section "Rare Earth Metals")
10 Value selected by compiler
11 Beaudry, B.J.; Spedding, F.H.: Metall. Trans. 5 (1974) 1631
12 Moffatt, W.G.: The Handbook of Binary Phase Diagrams. General Electric Comp. 1976–1981
13 Chart, T.G.: High Temp. High Press. 2 (1970) 461
14 Dariel, M.P.; Holthuis, J.T.; Pickus, M.R.: J. Less Common Met. 45 (1976) 91
15 Iandelli, A.; Palenzona, A.: Rev. Chim. Miner. 13 (1976) 55
16 Cezairliyan, A.; McClure, J.L.: High Temp. High Press. 8 (1976) 461
17 Cezairliyan, A.; Müller, A.P.: J. Res. Nat. Bur. Stand. 83 (1978) 127
18 Mills, K.C.: Thermodynamic Data for Inorganic Sulphides, Selenides, Tellurides. London: Butterworths 1974
19 Glushko, V.P.; Medvedev, V.A.: Termicheskie konstanty veshchestv, Part 4, Section 2 (1971) Moscow, Akad. Nauk SSSR
20 Encyclopedia Britannica 17 (1964) 777
21 Smith, J.: Private information 1982

Table 2. Structural types of elements and compounds (According to "Strukturberichte" (Leipzig) and Smithells, C.J.: Metal Reference Book, 3rd ed. London: Butterworths 1962)

Name of type	Crystal system	No. of atoms per unit cell	Name of type	Crystal system	No. of atoms per unit cell
A1(Cu)	cubic	4	B37(TlSe)	tetragonal	16
A2(W)	cubic	2			
A3(Mg)	hexagonal	2	C1(CaF_2)	cubic	12
A4(diamond)	cubic	8	C2(FeS_2, pyrites)	cubic	12
A5(Sn, metallic)	tetragonal	4	C3(Cu_2O)	cubic	6
A6(In)	tetragonal	2	C4(TiO_2, rutile)	tetragonal	6
A7(As)	rhombohedral	2	C5(TiO_2, anatase)	tetragonal	12
A8(Se)	hexagonal	3	C6(CdI_2)	hexagonal	3
A9(graphite)	(α) hexagonal	4	C7(MoS_2)	hexagonal	6
	(β) rhombohedral	6	C8(SiO_2, quartz 575°C)	hexagonal	9
A10(Hg)	rhombohedral	1	(SiO_2 575°C)	orthorhombic	—
A11(Ga)	orthorhombic	8	C9(β–cristobalite)	cubic	24
A12(α–Mn)	cubic	58	C10(β–tridymite)	hexagonal	12
A13(β–Mn)	cubic	20	C11a(CaC_2)	tetragonal	6
A14(I_2)	orthorhombic	8	C11b($MoSi_2$)	tetragonal	6
A15(β–W or Cr_3Si)	cubic	8	C12($CaSi_2$)	rhombohedral	6
A16(S_8)	orthorhombic	128	C13(HgI_2)	tetragonal	6
A17(P_{black})	orthorhombic	8	C14($MgZn_2$)	hexagonal	12
A18(Cl_2)	tetragonal	16	C15($MgCu_2$)	cubic	24
A19(Po)	monoclinic	12	C16($CuAl_2$)	tetragonal	12
A20(α–U)	orthorhombic	4	C18(FeS_2, marcasite)	orthorhombic	6
			C19($CdCl_2$)	orthorhombic	3
A_a(Pa type)	tetragonal	2	C21(TiO_2, brookite)	orthorhombic	24
A_b(β–U type)	complex		C22(Fe_2P)	hexagonal	9
	tetragonal	30	C23($PbCl_2$)	orthorhombic	12
A_c(α–Np)	orthorhombic	8	C24($HgBr_2$)	orthorhombic	4
A_d(β–Np)	tetragonal	4	C27(β–CdJ_2)	hexagonal	3
A_g(B)	tetragonal	50	C28($HgCl_2$)	orthorhombic	12
A_k(α–Se)	monoclinic	32	C29(SrH_2)	orthorhombic	12
A_l(β–Se)	monoclinic	32	C32(AlB_2)	hexagonal	3
			C33(Bi_2Te_2S, tetradymite)	rhombohedral	5
B1(NaCl)	cubic	8	C34($AuTe_2$, calaverite)	monoclinic	6
B2(CsCl)	cubic	2	C36($MgNi_2$)	hexagonal	24
B3(ZnS, zincblende)	cubic	8	C37(Co_2Si)	orthorhombic	12
B4(ZnS, wurtzite)	hexagonal	4	C38(Cu_2Sb)	tetragonal	6
B8(α) NiAs	hexagonal	4	C40($CrSi_2$)	hexagonal	9
(β) Ni_2In	hexagonal	6	C42(SiS_2)	orthorhombic	12
B9(HgS, cinnabar)	hexagonal	6	C43(ZrO_2)	monoclinic	12
B10(LiOH, PbO)	tetragonal	4	C44(GeS_2)	orthorhombic	72
B11(γ–TiCu)	tetragonal	4	C46($AuTe_2$, krennerite)	orthorhombic	24
B12(BN)	hexagonal	2	C49($ZrSi_2$)	orthorhombic	12
B13(NiS, millerite)	rhombohedral	6	C52(TeO_2)	orthorhombic	24
B16(GeS)	orthorhombic	8	C54($TiSi_2$)	orthorhombic	24
B17(PtS, cooperite)	tetragonal	4			
B18(CuS, covellite)	hexagonal	12	C_a(Mg_2Ni type)	hexagonal	18
B19(AuCd)	orthorhombic	4	C_b(Mg_2Cu type)	orthorhombic	48
B20(FeSi)	cubic	8	C_c($ThSi_2$ type)	tetragonal	12
B26(CuO)	monoclinic	8			
B27(FeB)	orthorhombic	8	$D0_2$($CoAs_3$)	cubic	32
B29(SnS)	orthorhombic	8	$D0_3$(BiF_3 or Li_3Bi)	cubic	16
B31(MnP)	orthorhombic	8	$D0_9$(ReO_3 or Cu_3N)	cubic	4
B32(NaTl)	cubic	16	$D0_{11}$(Fe_3C)	orthorhombic	16
B34(PdS)	tetragonal	16	$D0_{17}$(BaS_3)	orthorhombic	16
B35(CoSn)	hexagonal	6			

Continued

Name of type	Crystal system	No. of atoms per unit cell	Name of type	Crystal system	No. of atoms per unit cell
$DO_{18}(Na_3As)$	hexagonal	8	$D7_1(Al_4C_3)$	rhombohedral	7
$DO_{19}(Mg_3Cd$ or $Ni_3Sn)$	hexagonal	8	$D7_2(Co_3S_4)$	cubic	56
$DO_{20}(NiAl_3)$	orthorhombic	16	$D7_3(Th_3P_4)$	cubic	28
$DO_{21}(Cu_3P)$	hexagonal	24	$D8_1(Fe_3Zn_{10})$	cubic	52
$DO_{22}(TiAl_3)$	tetragonal	8	$D8_2(Cu_5Zn_8)$	cubic	52
$DO_{23}(ZrAl_3)$	tetragonal	16	$D8_3(Cu_9Al_4)$	cubic	52
$DO_{24}(TiNi_3)$	hexagonal	16	$D8_4(Cr_{23}C_6)$	cubic	116
$D1_3(BaAl_4)$	tetragonal	10	$D8_5(W_6Fe_7)$	rhombohedral	13
$D2_1(CaB_6)$	cubic	7	$D8_6(Cu_{15}Si_4)$	cubic	76
$D2_3(NaZn_{13})$	cubic	112	$D8_7(V_2O_5)$	orthorhombic	14
$D2_9(Fe_8N)$	tetragonal	18	$D8_8(Mn_5Si_3)$	hexagonal	16
$D5_1(\alpha-Al_2O_3,$ corundum)	rhombohedral	10	$D8_9(Co_9S_8)$	cubic	68
$D5_2(La_2O_3)$	hexagonal	5	$D8_{10}(Cr_5Al_8)$	rhombohedral	26
$D5_3(Mn_2O_3)$	cubic	80	$D8_{11}(Co_2Al_5)$	hexagonal	28
$D5_4$ or $D6_1(Sb_2O_3,$ senarmonite)	cubic	80	$D8_a(Th_6Mn_{23})$	cubic	116
$D5_8(Sb_2S_3)$	orthorhombic	20	$D8_b(\sigma\ type)$	tetragonal	30
$D5_9(Zn_3P_2)$	tetragonal	40			
$D5_{10}(Cr_3C_2)$	orthorhombic	20	$L1_0(CuAu\ type)$	tetragonal	4
$D5_{11}(Sb_2O_3,$ valentinite)	orthorhombic	20	$L1_1(PtCu\ type)$	rhombohedral	32
$D5_{12}(\beta-Bi_2O_3)$	tetragonal	40	$L1_2(Cu_3Au\ type)$	cubic	4
$D5_{13}(Ni_2Al_3)$	hexagonal	5	$L2_2(Tl_7Sb_2)$	cubic	54
			$L'1(Fe_4N)$	cubic	5
			$L'2$ (Martensite type)	tetragonal	$2Fe+(up\ to)\,0.12C$

Table 3. Numerical differences between the International Practical Temperature Scale of 1968 and that of 1948. Values of the numerical differences between IPTS-1968 and IPTS-1948, over the temperature range 90–10,000 K, are reported as $\Delta = (T_{68} - T_{48})$. The values in this table stem from the tabulation of Douglas[a]

T_{68}/K	Δ/mK	T_{68}/K	Δ/mK	T_{68}/K	Δ/mK	T_{68}/K	Δ/mK	T_{68}/K	Δ/K	T_{68}/K	Δ/K
90	+ 8	170	20	400	10	820	98	1,600	1.87	3,800	7.6
92	11	175	24	410	15	840	113	1,650	1.96	3,900	8.0
94	13	180	27	420	19	860	134	1,700	2.05	4,000	8.3
96	13	185	30	430	24	880	160	1,750	2.15	4,100	8.7
98	12	190	32	440	28	900	194	1,800	2.24	4,200	9.1
100	11	195	33	450	32	920	245	1,850	2.34	4,300	9.4
102	9	200	34	460	37	940	300	1,900	2.44	4,400	9.8
104	6	210	33	470	41	960	354	1,950	2.54	4,500	10.2
106	4	220	30	480	45	980	409	2,000	2.65	4,600	10.6
108	1	230	25	490	49	1,000	464	2,100	2.86	4,700	11.0
110	− 1	240	20	500	53	1,020	519	2,200	3.08	4,800	11.4
112	− 4	250	14	520	60	1,040	575	2,300	3.31	4,900	11.8
114	− 6	260	7	540	66	1,060	631	2,400	3.55	5,000	12.3
116	− 8	270	2	560	70	1,080	687	2,500	3.79	5,500	14
118	−10	280	− 3	580	74	1,100	743	2,600	4.0	6,000	17
120	−11	290	− 7	600	76	1,150	886	2,700	4.3	6,500	19
124	−13	300	− 9	620	77	1,200	1,029	2,800	4.6	7,000	22
128	−14	310	−10	640	77	1,250	1,173	2,900	4.8	7,500	25
132	−13	320	−10	660	76	1,300	1,319	3,000	5.1	8,000	27
136	−11	330	−10	680	76	1,337.58	1,430	3,100	5.4	8,500	30
140	− 9	340	− 9	700	75			3,200	5.7	9,000	33
145	− 5	350	− 7	720	74	1,350	1,450	3,300	6.0	9,500	37
150	0	360	− 4	740	75	1,400	1,530	3,400	6.3	10,000	40
155	+ 5	370	− 1	760	77	1,450	1,610	3,500	6.6		
160	10	380	+ 2	780	81	1,500	1,700	3,600	7.0		
165	15	390	6	800	88	1,550	1,780	3,700	7.3		

[a] Douglas, T.B.: Conversion of existing calorimetrically determined thermodynamic properties to the basis of the International Practical Temperature Scale of 1968. J. Res. Nat. Bur. Stand. 73A (1969) 451